Optics and Lasers

Matt Young

Optics and Lasers

Including Fibers and Optical Waveguides

Fourth Revised Edition

With 188 Figures

Springer-Verlag

Berlin Heidelberg New York
London Paris Tokyo
Hong Kong Barcelona
Budapest

Matt Young, Ph. D.

3145 Fremont, Boulder, CO 80304, USA

The first three editions appeared as
Springer Series in Optical Sciences, Vol. 5

Second Corrected Printing 1993

ISBN 3-540-55010-0 4. Aufl. Springer-Verlag Berlin Heidelberg New York
ISBN 0-387-55010-0 4th Ed. Springer-Verlag New York Berlin Heidelberg

ISBN 3-540-16127-9 3. Auflage Springer-Verlag Berlin Heidelberg New York
ISBN 0-387-16127-9 3rd Edition Springer-Verlag New York Heidelberg Berlin

Library of Congress Cataloging-in-Publication Data. Young, Matt, 1941 – Optics and lasers: including fibers and optical waveguides / Matt Young. – 4th rev. ed. p. cm. Includes bibliographical references and index. ISBN 3-540-55010-0 (Berlin). – ISBN 0-387-55010-0 (New York) 1. Optics. 2. Lasers. 3. Fiber optics. 4. Optical waveguides. I. Title. QC355.2.Y68 1992   621.36–dc20   91-42910

The use of general descriptive names, registered names, trademarks, etc. in this publication does not imply, even in the absence of a specific statement, that such names are exempt from the relevant protective laws and regulations and therefore free for general use.

Typesetting: Macmillan India Ltd., India

54/3140/SPS – 5432 – Printed on acid-free paper

For my father,
Professor Arthur K. Young,
from whom I am
still learning the art
of clear thinking

Preface

In this fourth edition of *Optics and Lasers*, I have added sections on scanning confocal microscopy, video microscopy, digital image processing, edge response, optical-fiber connectors, and liquid crystals. In addition, I have rewritten many paragraphs to improve their clarity or precision and, further, corrected minor errors of punctuation and taken care of other similarly small details.

The book now includes over 100 problems which, I hope, will make it more useful in the classroom. As before, some of the problems derive an especially important or useful result; these I have integrated within the body of the book. In such cases, I always state the result and, often, give it an equation number and a citation in the index. Teachers who adopt the book may obtain solutions to the problems by writing me and asking for them on letterhead stationery.

Optics has been changing greatly for over 30 years: since the invention of the laser. Partly because of the applied or engineering nature of much of modern optics, there has been a need for a practical text that surveys the entire field. Such a book should not be a classical-optics text but, rather, should be strong on principles, applications, and instrumentation; on lasers, holography, and coherent light; and on optical-fiber waveguides and integrated optics. On the other hand, it should concern itself relatively little with such admittedly interesting topics as the formation of the rainbow or the precise determination of the speed of light.

My purpose, therefore, has been to write an up-to-date textbook that surveys applied or engineering optics, including lasers, optical processing, optical waveguides, and other areas that might be called modern optics. I have attempted to treat each topic in enough depth to give it considerable practical value, while keeping it as free from mathematical detail as possible. Because I have surveyed applied optics in a very general way (including much more than I would attempt to incorporate into a single, one-semester college course), this book should also be a useful handbook for the practicing physicist or engineer who works from time to time with optics. Any of the material is appropriate to an introductory undergraduate course in optics; the work as a whole will be useful to the graduate student or applied physicist with scant background in optics.

The book originated in class notes for several one-semester courses that I offered in the Electrical-Engineering Curriculum at Rensselaer Polytechnic Institute and in the Physics Department of the University of Waterloo (Canada), before I joined the Electromagnetic Technology Division of the National Institute of Standards and Technology. Most of the courses were at the second- and fourth-year level, but I have drawn much additional material from graduate courses I have offered in lasers and related areas. I have also used the book as a textbook for courses in the Electrical and Computer Engineering Department of the University of Colorado and in the Electronics Department of the Weizmann Institute of Science. To make the book useful to as large an audience as possible, I have included short reviews of such subjects as complex-exponential notation, superposition of waves, and atomic energy levels. The book is a private venture, written in my basement, so to speak, and has no connection whatsoever with the National Institute of Standards and Technology.

Nearly all the references are to books or reviews and are chosen to allow the reader to explore any topic in greater detail. The problems are designed to help increase the reader's understanding and, sometimes, to derive a useful result. Certain portions of the text are largely descriptive; there I have used comparatively few problems.

It is my very great pleasure to acknowledge the invaluable assistance of the first editor of this book, David MacAdam, whose guidance and comments have led to a clearer, more readable, and more complete work. My former officemate at Rensselaer Polytechnic Institute, William Jennings, read the early versions with great care, offered excellent suggestions, and occasionally made me rewrite the same passage several times with very salutary results. Helmut Lotsch of Springer-Verlag has ably supervised the production of the book and adhered only to the highest standards.

I also acknowledge my debt to my former professors and fellow students at the Institute of Optics of the University of Rochester. My closest advisers there were Michael Hercher and Albert Gold; I also have warm memories of Philip Baumeister, Parker Givens, and others. My first optics course was Rudolf Kingslake's introductory optical-engineering course, and I still occasionally refer to his duplicated course notes.

I have been working, on and off, with optical-fiber communications since about 1972; the number of people I have learned from is, as a practical matter, nondenumerable. However, I want to single out for acknowledgement my former colleagues and co-editors of the Optical Waveguide Communications Glossary and, in particular, Robert Gallawa and Gordon Day of NIST in Boulder. Neither of these able scientists ever lets me get away with anything, and Bob Gallawa has offered many pithy comments on the chapters on optical waveguides. I am equally grateful to Ernest Kim for his critical reading of the entire third edition. Kevin Malone and Steven Mechels of NIST read most of the material that is new to this edition and suggested many worthwhile improvements. Roberto Forneris and Yara Forneris of the University of Sao Paulo, Brazil, and Burton Brody of Bard College pointed out a number of errors.

Finally, I thank Theodor Tamir, the editor of the second edition, for dozens of helpful suggestions and also acknowledge my very good fortune to have been a Visiting Scientist at the Weizmann Institute of Science. A course I taught there gave me the impetus to organize, edit, and supplement my problems and led to their inclusion in this volume.

Boulder, Colorado, August 1991   M. Young

Contents

And God said: "Let there be light." And there was light.
And God saw the light, that it was good;
and God divided the light from the darkness.

THE TORAH

And the light is sweet, and a pleasant
thing it is for the eyes to behold the sun.

KOHELETH (ECCLESIASTES)

But soft! what light through yonder window breaks?
It is the east, and Juliet is the sun!

SHAKESPEARE

Light breaks where no sun shines.

DYLAN THOMAS

Sadness flies on the wings of morning and
out of the heart of darkness comes light.

GIRAUDOUX

Let us bathe in this crystalline light!

POE

... On a river of crystal light,
Into a sea of dew.

EUGENE FIELD

I see a black light.

VICTOR HUGO (last words)

Do not go gentle into that good night ...
Rage, rage against the dying of the light.

DYLAN THOMAS

1. Introduction

This is an applied optics book. It is written for physics or engineering students who will incorporate optical instruments into practical devices or who will use optical components in their laboratories or their experiments. My aim is to present as complete a picture of modern applied optics as possible, while going into as much depth as possible, yet using a minimum of advanced mathematics.

In much of the book, we will consider a beam of light as a collection of rays. When it is necessary to understand interference and diffraction, we will, in effect, add a wave motion to the rays. Less often, we will use the particle nature of light and, in effect, consider the rays as if they were streams of particles. If you like, you can call this the *triplicity* of light – rays, waves, particles. We will use the wave and particle natures of light without justification and without philosophical foundation: that is, as heuristic devices that enable us to understand certain kinds of phenomena in as much depth as we require for designing and understanding optical instruments and systems.

Deeper understanding of the wave and particle natures of light is presented in courses in quantum electrodynamics. Here, let me just say, without apology, that sometimes it is convenient to consider light as a wave motion and sometimes as a stream of particles, depending on the kind of experiment we are performing. Still, there is something mysterious about performing an experiment, like the double-slit experiment (Chap. 5), in which the light propagates and exhibits interference precisely as if it were a wave, and yet detecting the interference pattern with a quantum detector (Chap. 4), which interacts with the light as if it were a series of particles. The most common explanation, that particles in the subatomic world behave in a way that we do not find intuitive, is not very satisfying and gets us back where we began: we must, to some extent, consider the light as a wave when it propagates, but as a particle when it is absorbed by matter. When the wave motion is not important, as in many simple lens instruments, we ignore it and use a formalism based on rays.

The book begins with two chapters based on geometrical, or ray, optics. In Chap. 2, I treat as much ray optics as I find necessary for a complete understanding of the optical instruments introduced in Chap. 3. In particular, Chap. 2 derives the *lens equation*, which allows calculation of object and image positions, and shows geometrical constructions for tracing rays through lens systems.

The instruments in Chap. 3 are described almost entirely by the *paraxial approximation*. This is the approximation that all rays are infinitesimally close to the axis and results in essentially perfect imaging. I use this approximation on the supposition that *aberration* theory has little practical interest to the non-specialist. This chapter includes, however, a short warning that lenses must be used the way they are designed: microscope objectives for nearby objects and photographic objectives for distant objects, for example.

More specifically, Chap. 3 begins by treating the human eye as if it were an optical instrument, largely stripped of most of its physiological or psychological aspects. The chapter goes on to describe the basic camera, including the important aspects of the photographic emulsion. Detailed treatments of the telescope, the microscope, and the relatively new *scanning confocal microscope* follow. The chapter concludes by anticipating a result from wave optics and using it to calculate the theoretical resolution limits of the microscope and the telescope, as well as to derive practical upper limits for their magnifying powers.

Chapter 4, "Light Sources and Detectors", begins with *radiometry* and *photometry*, which concern the propagation and measurement of optical power, as from a source to a screen or a detector. Radiometry concerns the measurement of radiant power in general; photometry, by contrast, implies visual or *luminous* power, or power that is referred to the human eye as a detector. This chapter explains this sometimes confusing topic by sticking to a consistent set of units and making no formal distinction between radiant power, which is what a detector sees, and luminous power, which is what the human eye sees. The section on radiometry and photometry concludes with an explanation of image *luminance* (loosely, brightness) and explains why brightness cannot be increased with lenses.

The remainder of Chap. 4 surveys light sources: blackbodies, continuous sources, and line sources. The chapter concludes with a section on detectors for visible and near-infrared spectra, and shows, for example, how to calculate the lowest power that can be discerned by a specific detector.

I have deliberately omitted electromagnetic theory from Chap. 5, "Wave Optics". This chapter develops the elements of interference and diffraction, mostly in preparation for their application to the next chapter. Here, we discuss interference that is brought about by two or more reflections from partially reflecting surfaces and interference that is brought about by geometrically dividing a beam into segments, as in the double-slit experiment. These are treated by the mathematically simple *far-field* or *Fraunhofer*-diffraction theory, a formalism that is appropriate for a majority of applications, including lens optics. Chapter 5 includes, however, enough *near-field* diffraction theory to allow an understanding of the role of the *Fresnel zone plate* in holography.

Coherence relates to the ability of a beam of light to form an interference pattern. Incoherent light, for example, is wholly incapable of forming an interference pattern. The coherence properties of a beam of light influence the kind of image that will be formed; even the resolution is different in coherent light than in incoherent light. The discussion of coherence and of imaging and resolution in coherent light in Chap. 5 is unusual, perhaps unique, for a book of this level.

Chapter 6, "Interferometry and Related Areas", includes *diffraction gratings* and *interferometers*. These are instruments that can be used to *disperse* light so that its constituent wavelengths can be distinguished; other interferometers are used for measuring distance, for example, or for testing optics. The chapter concludes with a description of *multilayer mirrors*, which may be used for very efficient reflectors, and *interference filters*, which may be used to transmit a very narrow band of wavelengths.

Chapter 7, "Holography and Image Processing", begins by describing holography almost entirely in terms of Fresnel zone plates and diffraction gratings. It uses simple arguments to derive, for example, the maximum field of view of a hologram as a function of the resolving power of the recording medium, and the minimum angle between the *reference* and the *object* beam. Chapter 7 goes on to describe *Fourier-transform optics* and includes the *Abbe theory* of the microscope and ways such as *phase-contrast microscopy* and *spatial filtering* for manipulating an image. Phase-contrast microscopy, for example, can be used to make visible the image of an object that consists entirely of structures that are transparent but have different index of refraction from their surrounding. The chapter continues with sections on transfer functions and concludes with sections on scanning microscopy, video microscopy, and digital image processing. The treatment of scanning confocal microscopy includes what I think is a new heuristic derivation of the impulse response of this instrument.

Chapter 8, "Lasers", is intended to introduce the terms and concepts related to lasers and *optical resonators*. It begins by discussing the dynamics of exciting a laser material to operate as a continuous laser; a *Q-switched* or *giant-pulse* laser, which emits pulses several nanoseconds in duration and with peak powers of hundreds of megawatts; and a *mode-locked* laser, which can generate pulses shorter than 1 ps. This chapter goes on to describe optical resonators; laser *modes*, or allowed radiation patterns; and the propagation of *Gaussian beams*, which is fundamentally different from propagation of the uniform beams encountered in ordinary, incoherent optics. The chapter concludes with a discussion of the most important solid, liquid, and gas lasers, and with a discussion of laser safety.

Chapter 9, "Electromagnetic and Polarization Effects", begins by explaining that light is a *transverse wave* and showing some of the consequences of this fact. For example, *Brewster's angle* is the angle of incidence at which one *polarization* displays little or no reflection. The chapter continues by describing reflection from a dielectric interface. *Total internal reflection* is treated in some detail, because of the connection with optical waveguides, and the phase change on reflection is presented in a new way that is designed to show clearly how that phase change varies from 0 to π within the region of total reflection. Chapter 9 concludes with discussions of polarization optics, nonlinear optics, and electro-optics, magneto-optics, and acousto-optics.

Among the major advances in optics are the maturing of optical communications and the development of integrated optics. Chapter 10, "Optical Wave-guides", develops optical waveguide theory primarily on the basis of ray optics and interference in planar waveguides. When necessary, however, I quote more precise results or results for waveguides with circular cross sections. Prism and grating couplers are included in this chapter for tutorial reasons; they could as logically have found their way into the chapter on integrated optics. Chapter 10 discusses modes in waveguides, single-mode waveguides, graded-index fibers, leaky rays, and the kind of *launch conditions* used for many measurements. It concludes with a discussion of losses in splices or connectors between both multimode and single-mode fibers.

Chapter 11, "Optical-Fiber Measurements", deals with attenuation, electrical bandwidth, index profile and core diameter, optical time-domain reflectometry, numerical aperture, and the techniques for measuring them. For index-profile measurements, I have chosen to dwell particularly on near-field scanning and the refracted-ray method, because these are becoming widespread and are recommended procedures of the Telecommunications Industry Association.

Chapter 12, "Integrated Optics", makes a slightly artificial distinction between *optical integrated circuits* and *planar optical devices*. Optical integrated circuits perform functions analogous to electronic and microwave circuits, whereas planar optical devices are planar versions of optical devices like lenses, diffraction gratings, and optical processors. The section on optical integrated circuits is largely descriptive, since I have anticipated the necessary physics in previous chapters. It includes descriptions of *channel* or *strip waveguides*, or waveguides that confine the light in two dimensions; and branches, couplers, and modulators for manipulating light within a circuit and coupling the light into or out of the circuit.

The section on planar devices uses a planar *spectrum analyzer* as a starting point from which to discuss some of the components, especially planar lenses. The chapter next describes a variety of uses for gratings in planar optics: as lenses, as couplers to other waveguides, and as couplers out of the plane of the device, as between integrated (electronic) circuits. The chapter concludes by describing *surface-emitting lasers*, which are distinguished from the more conventional edge-emitting lasers and can be made into arrays or connected efficiently to optical fibers.

2. Ray Optics

2.1 Reflection and Refraction

In this chapter, we treat light beams as *rays* that propagate along lines, except at interfaces between dissimilar materials, where the rays may be bent or *refracted*. This approach, which had been assumed to be completely accurate before the discovery of the wave nature of light, leads to a great many useful results regarding lens optics and optical instruments.

2.1.1 Refraction

When a light ray strikes a smooth interface between two transparent media at an angle, it is refracted. Each medium may be characterized by an *index of refraction n*, which is a useful parameter for describing the sharpness of the refraction at the interface. The index of refraction of air (more precisely, of free space) is arbitrarily taken to be 1. n is most conveniently regarded as a parameter whose value is determined by experiment. We know now that the physical significance of n is that it is the ratio of the velocity of light *in vacuo* to that in the medium.

Suppose that the ray in Fig. 2.1 is incident on the interface at point O. It is refracted in such a way that

$$n \sin i = n' \sin i' \, , \tag{2.1}$$

no matter what the inclination of the incident ray to the surface. n is the index of

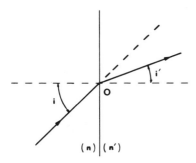

Fig. 2.1. Refraction at an interface

refraction of the first medium, n' that of the second. The *angle of incidence i* is the angle between the incident ray and the normal to the surface; the *angle of refraction i'* is the angle between the refracted ray and the normal. Equation (2.1) is known as the *law of refraction*, or *Snell's law*.

2.1.2 Index of Refraction

Most common optical materials are transparent in the visible region of the spectrum, whose wavelength ranges from 400 to 700 nm. They exhibit strong *absorption* at shorter wavelengths, usually 200 nm and below.

The index of refraction of a given material is not independent of wavelength, but generally increases slightly with decreasing wavelength. (Near the *absorption edge* at 200 nm, the index of glass increases sharply.) This phenomenon is known as *dispersion*; dispersion curves of several common glasses are given in Fig. 2.2. Dispersion can be used to display a spectrum with a prism; it also gives rise to unwanted variations of lens properties with wavelength.

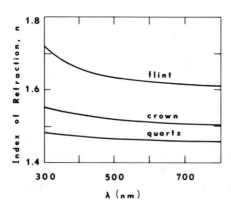

Fig. 2.2. Index of refraction of several materials as a function of wavelength

Table 2.1. Index of refraction of various optical materials

Material	Index of refraction n_D	Material	Index of refraction n_D
air	1.0003	sodium chloride	1.54
water	1.33	light flint glass	1.57
methanol	1.33	carbon disulfide	1.62
ethanol	1.36	medium flint glass	1.63
magnesium fluoride	1.38	dense flint glass	1.66
fused quartz	1.46	sapphire	1.77
Pyrex glass	1.47	extra-dense flint glass	1.73
benzene	1.50	heaviest flint glass	1.89
xylene	1.50	zinc sulfide (thin film)	2.3
crown glass	1.52	titanium dioxide (thin film)	2.4–2.9
Canada balsam (cement)	1.53		

Optical glasses are generally specified both by index n (see Table 2.1) and by a quantity known as *dispersion* v,

$$v = \frac{n_F - n_C}{n_D - 1} .$$
(2.2)

The subscripts F, D, and C refer to the indices at certain short, middle, and long wavelengths (blue, yellow, red).

2.1.3 Reflection

Certain highly polished metal surfaces and other interfaces may reflect all or nearly all of the light falling on the surface. In addition, ordinary, transparent glasses reflect a few percent of the incident light and transmit the rest.

Figure 2.3 depicts a reflecting surface. The angle of incidence is i and the *angle of reflection* i'. Experiment shows that the angles of incidence and reflection are equal, except in a very few peculiar cases. We shall later adopt the convention that i is positive as shown; that is, if the acute angle opens counterclockwise from the normal to the ray, i is positive. The sign of i' is clearly opposite to that of i. We therefore write the *law of reflection* as

$$i' = -i .$$
(2.3)

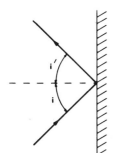

Fig. 2.3. Reflection at an interface

2.1.4 Total Internal Reflection

Here we consider a ray that strikes an interface from the high-index side, say, from glass to air (not air to glass). This is known as *internal refraction*. The law of refraction shows that the incident ray is in this case bent away from the normal when it crosses the interface (Fig. 2.4). Thus, there will be some angle of incidence for which the refracted ray will travel just parallel to the interface. In this case, $i' = 90°$, so the law of refraction becomes

$$n \sin i_c = n' \sin 90° ,$$
(2.4)

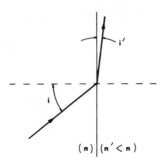

Fig. 2.4. Refraction near the critical angle

(n)|$(n' < n)$

where i_c is known as the *critical angle*. Since $\sin 90° = 1$,

$$\sin i_c = (n'/n) . \tag{2.5}$$

If i exceeds i_c, then $n \sin i > n'$, and the law of refraction demands that $\sin i$ exceed 1. Because this is impossible, we conclude that there can be no refracted ray in such cases. The light cannot simply vanish, so we are not surprised that it must be wholly reflected; this is indeed the case. The phenomenon is known as *total internal reflection*; it occurs whenever

$$i > \sin^{-1}(n'/n) . \tag{2.6}$$

The reflected light obeys the law of reflection.

For a typical glass-air interface, $n = 1.5$; the critical angle is about 42°. Glass *prisms* that exhibit total reflection are therefore commonly used as mirrors with angles of incidence of about 45°.

2.1.5 Reflecting Prisms

There are many different types of reflecting prism. The most common are prisms whose cross sections are right isosceles triangles. Figure 2.5 shows such a prism being used in place of a plane mirror. One advantage of a prism over a metal-coated mirror is that its reflectance is nearly 100% if the surfaces normal to the light are antireflection coated (Sect. 6.4). Further, the prism's properties do not change as the prism ages, whereas metallic mirrors are subject to oxidation and are relatively easy to scratch. A glass prism is sufficiently durable that it can withstand all but the most intense laser beams.

In image-forming systems, these prisms must be used in parallel light beams to avoid introducing defects into the optical image.

Figure 2.5 also shows the same prism being used to reflect a beam back into the direction from which it originated. Prisms used in this manner are often called *Porro prisms* or *roof prisms*. It is left as a problem to show that an incoming ray is always reflected parallel to itself, provided only that the incident ray lie in a plane perpendicular to the face of the prism (Problem 2.1).

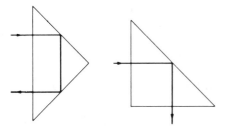

Fig. 2.5. Reflecting prisms

A *cube corner* or *retroreflector* is a prism with three edges that intersect at right angles to one another, as the edges of a cube intersect at right angles at the corners. Such a prism is a sort of generalization of the roof prism and reflects any ray parallel to itself, independent of its orientation. An observer looking at a cube-corner reflector sees only the pupil of his eye at the center of the reflector.

2.2 Imaging

2.2.1 Spherical Surfaces

Because a simple lens consists of a piece of glass with, in general, two spherical surfaces, we will find it necessary to examine some of the properties of a single, spherical refracting surface. If the reader will pardon an almost unconscionable pun, we will for brevity call such a surface, shown in Fig. 2.6, a "len". Two of these form a lens. To avoid confusion, we will always place "len" in quotes.

We are interested in the imaging property of the "len". We consider a bright point A and define the axis along the line AC, where C is the center of the spherical surface. We examine a particular ray AP that strikes the "len" at P. We shall be interested in the point A' where this ray intersects the axis.

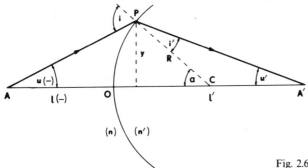

Fig. 2.6. Spherical refracting surface

Before proceeding any further, we must adopt a *sign convention*. The choice of convention is, of course, arbitrary, but once we choose a convention, we shall have to stick with it. The convention we adopt appears, at first, quite complicated. We choose it at least in part because it is universally applicable; with it we will not need to derive a special convention for spherical mirrors.

To begin, imagine a set of Cartesian coordinate axes centered at O. Distances are measured from O. Distances measured from O to the right are positive; those measured from O to the left are negative. Thus, for example, OA' and OC are positive, whereas OA is negative. Similarly, distances measured above the axis are positive; those below are negative. This is our first sign convention.

We now adopt a convention for the signs of angles such as angle OAP or $OA'P$. We determine their signs by trigonometry. For example, the tangent of angle OAP is approximately

$$\tan OAP \cong y/OA , \tag{2.7}$$

where y is the distance indicated between P and the axis. Our previous convention shows that y is positive, and OA, negative. Thus, $\tan OAP$ is negative and so is OAP itself. Similarly, $OA'P$ and OCP are positive.

This is our second sign convention. An equivalent statement is that angle $OA'P$ (for example) is positive if it opens clockwise from the axis, or negative otherwise. It is probably simplest, however, merely to remember that angle OAP is negative as drawn in Fig. 2.6.

Finally, we deal with angles of incidence and refraction, such as angle CPA'. It is most convenient to define CPA' to be positive as shown in Fig. 2.6. This convention has already been stated formally in connection with Fig. 2.3. The angle of incidence or refraction is positive if it opens counterclockwise from the normal (which is, in this case, the radius of the spherical surface).

Unfortunately, when the last convention is expressed in this way, the statement differs from that which refers to angles (such as OAP) formed by a ray crossing the axis. It is best to learn the sign convention by remembering the signs of all of the important angles in Fig. 2.6. Only angle OAP is negative.

Let us now assign symbols to the more important quantities in Fig. 2.6. The point A' is located a distance l' to the right of O, and the ray intersects the axis at A' with angle u'. The quantities u and l are defined analogously. The radius R through the point P makes angle α with the axis. The angles of incidence and refraction are i and i'.

Parameters in *image space* are indicated by primed characters; those in *object space* are indicated by unprimed characters. This is another convention. The object and the image may be on opposite sides of the lens or, as we shall see (Fig. 2.10), they may lie on the same side. Thus, the convention does not imply, for example, that distances to the left of the lens are unprimed and distances to the right are primed. In fact an object or an image may lie on either side of the lens; therefore, primed or unprimed quantities may be either positive or negative.

We must be careful of the signs of l and u, both of which are negative according to our sign convention. This is indicated in Fig. 2.6 with parenthetical minus signs. We shall later find it necessary, after a derivation based on geometry alone, to go through our formulas and *change the signs* of all quantities that are algebraically negative. This is so because our sign convention is not that used in ordinary geometry. To make our formulas both algebraically and numerically correct, we must introduce our sign convention, which we do as indicated, by changing signs appropriately.

2.2.2 Object-Image Relationship

We now attempt to find a relationship between the quantities l and l' for a given geometry. First, we relate angles u and i to angle α. The three angles in triangle PAC are u, α, and $\pi - i$. Because the sum of these angles must be π, we have

$$u + \alpha + (\pi - i) = \pi, \quad \text{or} \tag{2.8}$$

$$i = \alpha + u . \tag{2.9a}$$

Similarly,

$$i' = \alpha - u' . \tag{2.9b}$$

At this point, it is convenient to make the *paraxial approximation*, namely, the approximation that the ray AP remains sufficiently close to the axis that angles u, u', i, and i' are so small that their sines or tangents can be replaced by their arguments; that is,

$$\sin \theta = \tan \theta = \theta , \tag{2.10}$$

where θ is measured in radians.

It is difficult to draw rays that nearly coincide with the axis, so we redraw Fig. 2.6 by expanding the vertical axis a great amount, while leaving the horizontal axis intact. The result is shown in Fig. 2.7. The vertical axis has been

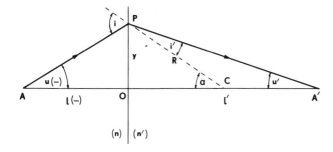

Fig. 2.7. Spherical refracting surface in paraxial approximation

stretched so much that the surface looks like a plane. In addition, because only one axis has been expanded, all angles are greatly distorted and can be discussed only in terms of their tangents. Thus, for example,

$$u = y/l \quad \text{and} \tag{2.11a}$$

$$u' = y/l' \tag{2.11b}$$

in paraxial approximation. Also, large angles are distorted. Although the radius is normal to the surface, it does not look normal in the paraxial approximation.

To return to the problem at hand, the law of refraction is

$$ni = n'i' \tag{2.12}$$

in paraxial approximation; from (2.9) and (2.12) we write

$$n(\alpha + u) = n'(\alpha - u') . \tag{2.13}$$

Because $OC = R$, we write α as

$$\alpha = y/R . \tag{2.14}$$

Equation (2.13) therefore becomes

$$n\left(\frac{y}{R} + \frac{y}{l}\right) = n'\left(\frac{y}{R} - \frac{y}{l'}\right) . \tag{2.15}$$

A factor of y is common to every term and therefore cancels. We rewrite (2.15)

$$\frac{n'}{l'} + \frac{n}{l} = \frac{n' - n}{R} . \tag{2.16}$$

At this point, we have made no mention of the sign convention. We derived the preceding equation on the basis of geometry alone. According to our sign convention, all of the terms in the equation are positive, except l, which is negative. To make the equation algebraically correct, we must, therefore, *change the sign* of the term containing l. This change alters the equation to

$$\frac{n'}{l'} - \frac{n}{l} = \frac{n' - n}{R} , \tag{2.17}$$

which we refer to as the "len" equation.

There is no dependence on y in the "len" equation. Thus, in paraxial approximation, every ray leaving A (and striking the surface) crosses the axis at A'. We therefore refer to A' as the *image* of A. A and A' are called *conjugate points*, and the *object distance* l and *image distance* l' are called *conjugate distances*.

Had we not made the paraxial approximation, the y dependence of the image point would not have vanished. Rays that struck the lens at large values of y would not cross the axis precisely at A'. The dependence on y is relatively small, so we would still refer to A' as the image point. We say that the image suffers from *aberrations* if all of the geometrical rays do not cross the axis within a specified distance of A'.

2.2.3 Use of the Sign convention

A word of warning with regard to the *signs* in algebraic expressions: Because of the sign convention adopted here, derivations based solely on geometry will not necessarily result in the correct sign for a given term. There are two ways to correct this defect. The first, to carry a minus sign before the symbol of each negative quantity, is too cumbersome and confusing for general use. Thus, we adopt the second, which is to go through the final formula and *change the sign* of each negative quantity. This procedure has already been adopted in connection with the "len" equation and is necessary, as noted, to make the formula algebraically correct. It is important, though, not to change the signs until the final step, lest some signs be altered twice.

2.2.4 Lens Equation

A *thin lens* consists merely of two successive spherical refracting surfaces with a very small separation between them. Figure 2.8 shows a thin lens in air. The index of the lens is n. The two refracting surfaces have radii R_1 and R_2, both of which are drawn positive.

We can derive an equation that relates the object distance l and the image distance l' by considering the behavior of the two surfaces separately. The first surface alone would project an image of point A to a point A'_1. If A'_1 is located at a distance l'_1 to the right of the first surface, the "len" equation shows that, in paraxial approximation,

$$\frac{n}{l'_1} - \frac{1}{l} = \frac{n-1}{R_1} ,$$ (2.18)

because n is the index of the glass (second medium) and 1, the index of the air.

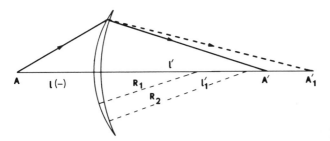

Fig. 2.8. Thin lens

The ray does not ever reach A_1', because it is intercepted by the second surface. Hitting the second surface, however, the ray behaves as if an object were located at A_1'. The object distance is l_1', if we neglect the thickness of the lens. In applying the "len" equation to the second surface, we must realize that the ray travels across the interface from glass to air. Thus, n is now the index of the first medium and 1, that of the second. The final image point A' is also the image projected by the lens as a whole. If we call the corresponding image distance l', then the "len" equation yields

$$\frac{1}{l'} - \frac{n}{l_1'} = \frac{1-n}{R_2} \tag{2.19}$$

for the second surface.

If we add the last two equations algebraically to eliminate l_1', we find that

$$\frac{1}{l'} - \frac{1}{l} = (n-1)\left(\frac{1}{R_1} - \frac{1}{R_2}\right), \tag{2.20}$$

which is known as the *lens-maker's formula*. The lens-maker's formula was derived from the "len" equation by algebra alone. There are no signs to change because that step was included in the derivation of the "len" equation.

We may define a quantity f' whose reciprocal is equal to the right side of the lens-maker's formula,

$$\frac{1}{f'} = (n-1)\left(\frac{1}{R_1} - \frac{1}{R_2}\right). \tag{2.21}$$

The lens-maker's formula may then be written as

$$\frac{1}{l'} - \frac{1}{l} = \frac{1}{f'}, \tag{2.22}$$

where f' is the *focal length* of the lens. We call this equation the *lens equation*.

We may see the significance of f' in the following way. If the object is infinitely distant from the lens, then $l = -\infty$ (Fig. 2.9). The lens equation then shows that the image distance is equal to f'. If the object is located along the axis of the lens, the image also falls on the axis. We call the image point in this case the *secondary focal point* F'. Any ray that travels parallel to the axis is directed by the lens through F', an observation that we will later find particularly useful.

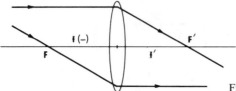

Fig. 2.9. Primary and secondary focal lengths

We define the *primary focal point F* in a similar way. The primary focal length f is the object distance for which $l' = \infty$. Thus, the lens equation shows that

$$f' = -f; \tag{2.23}$$

the primary and secondary focal lengths have equal magnitudes. Any ray that passes through F from the left will be directed by the lens parallel to the axis. When a ray enters the optical system from the right, use f instead of f' in (2.22). That is, use the negative of f'.

Finally, in the general case, a lens may have different media on opposite sides. In this case, the lens equation is

$$\frac{n'}{l'} - \frac{n}{l} = \frac{n'}{f'} = -\frac{n}{f}, \tag{2.24}$$

where n and n' are the indices in the first and second media, respectively. The primary and secondary focal lengths are not equal, but are related by

$$f'/f = -n'/n . \tag{2.25}$$

Problem. Show that the combined focal length of two thin lenses in contact is

$$1/f_e' = 1/f_1' + 1/f_2' . \tag{2.26}$$

Begin with a finite object distance l and show that the combination obeys the lens equation with the proper *effective focal length* f_e'.

More generally, it is possible to show that the effective focal length of two thin lenses separated by a distance d is given by

$$1/f' = 1/f_1' + 1/f_2' - d/f_1' f_2' . \tag{2.27}$$

Problem. **Huygens Eyepiece.** It is possible to make an *achromatic eyepiece* of two thin lenses even if the lenses are made of the same glass, with the same index of refraction and dispersion. That is, to first approximation, the power of the eyepiece will not vary with wavelength.

Beginning with the formula for the power of two thin lenses separated by a distance d, show that the power of the eyepiece is roughly independent of index of refraction when the separation between the lenses is

$$d = \tfrac{1}{2}(f_1' + f_2') . \tag{2.28}$$

Because the index of refraction depends on wavelength, this is equivalent to showing that the power is independent of a small change in index of refraction when d is given by (2.28). An eyepiece made according to this formula is called a *Huygens eyepiece*.

2.2.5 Classification of Lenses and Images

This is very largely self explanatory and is illustrated in Fig. 2.10. *A positive lens is a lens that will cause a bundle of parallel rays to converge to a point.* Its secondary focal point lies to the right of the lens, and f' is therefore positive. It may be regarded as a lens that is capable of projecting an image of a relatively distant object on a screen. An image that can be projected on a screen is called a *real image*. In general, a positive lens projects a real, *inverted* image of any object located to the left of its primary focal point F. When an object is located at F, the image is projectd to ∞. The lens is not strong enough to project an image when the object is inside F. In that case, an erect image appears to lie behind the lens and is known as a *virtual image*.

A positive lens need not have two *convex* surfaces (like those in Fig. 2.10). It may have the *meniscus* shape of Fig. 2.8. If the lens is thickest in the middle, the lens-maker's formula (2.20) will show it to be a positive lens.

A negative lens, shown also in Fig. 2.10, has its secondary focal point located to its left. Its secondary focal length f' is negative, and it cannot project a real image of a *real object*. Rather, it displays an erect, virtual image of such an object. In only one instance can a negative lens display a real image. This is the case when a positive lens projects a real image that is intercepted by a negative lens located to the left of the image plane. Because the rays are cut off by the negative lens, the real image never appears, but behaves as a *virtual object* projected by the negative lens as shown in the figure.

Like a positive lens, a negative lens need not be *concave* on both surfaces, but may be a meniscus. If the lens is thinnest in the center, f' will prove to be negative and the lens, also negative.

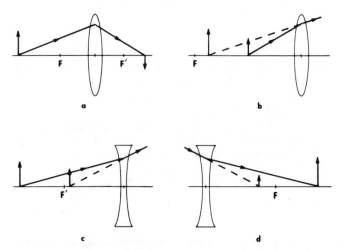

Fig. 2.10. (a) Positive lens; real, inverted image. (b) Positive lens; virtual, erect image. (c) Negative lens; virtual erect image. (d) Negative lens; real, erect image

2.2.6 Spherical Mirrors

Our formalism allows mirror optics to be developed as a special case of lens optics. We notice first that the law of reflection $i' = -i$ can also be written

$$(-1)\sin i' = 1 \sin i , \tag{2.29}$$

which is precisely analogous to the law of refraction, with $n' = -1$. We may therefore regard a mirror as a single refracting surface, across which the index changes from $+1$ to -1. It is left as a problem to apply the "len" equation to this case. We find that the focal length of a mirror is

$$f' = R/2 ,$$ (2.30)

where R is the radius of curvature. In addition, the focal points F and F' coincide. The formula that relates the conjugates for a curved-mirror system is

$$(1/l') + (1/l) = 2/R .$$ (2.31)

Mirrors are usually classified as *concave* and *convex*. Figure 2.11 shows that a concave mirror usually projects a real, inverted image, whereas a convex mirror forms an erect, virtual image of a real object.

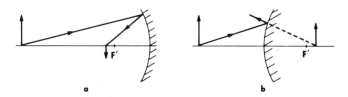

Fig. 2.11. (a) Concave mirror; real, inverted image. (b) Convex mirror; virtual, erect image

2.2.7 Thick Lenses

To this point, we have discussed only single-element lenses and neglected their thickness. The thin-lens approximation is not always applicable, but fortunately the formalism requires relatively little modification to accommodate *thick lenses* with many elements.

To begin, consider the fat lens in Fig. 2.12. We limit ourselves to the paraxial approximation and, accordingly, work with the tangents to the actual surfaces

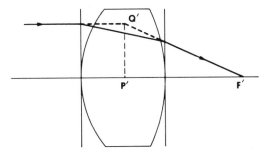

Fig. 2.12. Thick lens; construction of secondary principal plane

of the lens. Knowing the radii of the surfaces, we trace the path of a ray travelling originally parallel to the lens axis. The ray is refracted downward (in this case) at each surface and crosses the axis at F'. If we knew nothing about the lens but examined the incident and emerging rays, we might state that the refraction of the incident ray appeared to take place at Q'. The locus of such points Q' that correspond to rays incident at different heights is known as the *equivalent refracting surface*. In paraxial approximation, the equivalent refracting surface is a plane known as the *secondary principal plane*. The secondary principal plane intersects the axis at the *secondary principal point P'*.

We could also trace a ray that originates from the primary focal point F and construct the *primary principal plane* and the *primary principal point P* in precisely the same way. In general P and P' do not coincide; either or both may well lie outside the lens itself.

Precisely the same kind of construction can be performed on a complicated multi-element lens. For our convenience, we replace the lens by its focal points and principal planes, as shown in Fig. 2.13. We shall call these four points the *cardinal points* of the lens.

The principal planes were generated by examining, respectively, rays incident and emerging parallel to the axis. It is possible to show further that arbitrary rays can also be constructed with the aid of the principal planes. Figure 2.14 illustrates the construction of an image point. All rays behave as if they intersect the primary principal plane, jump the gap between the planes with no change of height, and are directed at the secondary principal plane toward the proper image point. The region between P and P' is, in a sense, dead space.

Formulas such as the lens equation may be applied to thick lenses, provided that *object and image distances are measured from P and P'*, respectively. The lens

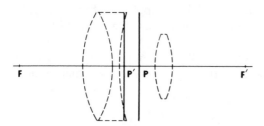

Fig. 2.13. Cardinal points of an optical system

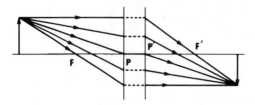

Fig. 2.14. Formation of the image

focal lengths are thus PF and $P'F'$. The distance from the rear surface of the lens to F', for example, is almost never equal to the focal length. This distance is known as the *back focal length* or *working distance*. As the latter name implies, it may be extremely important in the mechanical design of optical instruments.

The focal length of a thick, single-element lens whose index of refraction is n is given implicitly by the equation

$$\frac{1}{f'} = (n-1)\left(\frac{1}{R_1} - \frac{1}{R_2}\right) + \frac{d(n-1)^2}{nR_1R_2} , \tag{2.32}$$

where d is the thickness of the lens.

2.2.8 Image Construction

For making certain computations, an image is most easily constructed by tracing two or three particular rays. For example, to locate the image in Fig. 2.14, it is necessary to trace only two of the many rays that originate from the arrowhead. Their intersection locates the image of the arrowhead in paraxial approximation.

For our construction, we choose, first, the ray that leaves the arrowhead parallel to the axis. It is directed through F'. The second ray is the one that passes through F. It is directed parallel to the axis. Where the rays intersect, we draw the arrowhead and construct the image, as shown.

Besides these two rays, it is often useful to trace the ray directed at P. This ray emerges from P'. To trace the path of this ray, first consider a thin lens in the paraxial approximation. Consider a ray directed with angle w toward the center of the lens. Because the lens is infinitesimally thin, the ray is equivalently directed toward the vertex of the first surface, that is, toward the intersection of the first surface with the axis. Near the axis, however, the surface appears to be a plane perpendicular to the axis. The ray is refracted with angle of refraction r. It passes through the lens with no change of height, because the lens is infinitesimally thin. The second surface also appears to be a plane perpendicular to the axis, and the angle of emergence w' from the lens is, by a second application of Snell's law (2.1), equal to the angle of incidence w, or

$$w' = w . \tag{2.33}$$

In other words, near its vertex, a thin lens in paraxial approximation looks like a thin parallel plate, so any ray directed toward the center of the lens is undeviated by the lens.

A lens may be immersed in fluids that have different indices of refraction; typically, this means air on one side and water or oil on the other. The argument that led to (2.33) shows that, in the general case,

$$n'w' = nw , \tag{2.34}$$

where n is the index of refraction on the left side of the lens, and n' is the index on the right side.

Equation (2.34) may be applied to a thick lens simply by replacing the center of the thin lens with the two principal planes. That is, in Fig. 2.15, one ray is directed toward the primary principal plane P and makes angle w with the axis. The angle of the ray that emerges from the secondary principal plane P' is the angle w' given implicitly by (2.34). This is an important result for the ray-tracing applications that follow.

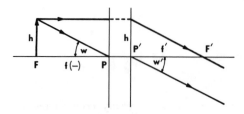

Fig. 2.15. Ray directed toward the primary principal point

A ray directed at P emerges from P' unchanged in direction, provided that the lens is wholly immersed in air or some other fluid. This property defines the *nodal points* of a lens. If the lens is immersed in air the nodal points coincide with the principal points; otherwise, they do not.

We may use the property of the nodal points to locate the principal planes of a thick lens: If a ray were to be directed precisely at P, it would emerge from P' unchanged in direction, even if we were to swivel the lens about an axis that passed through P, that is, if we were to scan through a range of values of w by rotating the lens rather than the incident beam. This is so because, according to (2.33) $w' = w$, irrespective of the value of w.

To locate the nodal points, swivel the lens around an axis perpendicular to the axis of the lens. Then, translate the lens parallel to its own axis until the direction of the emergent ray remains unaffected by the rotation. The axis of rotation now passes through the primary principal point. An optical bench that performs these operations – translating and swiveling the lens – is called a *nodal slide*, because it locates the principal points by using their nodal-point property. The secondary principal plane may be located by reversing the direction of the lens in the nodal slide.

2.2.9 Magnification

We construct an image, using the two rays shown in Fig. 2.16. We require the fact that $w' = w$, as we have just shown. From simple geometry, we find that

$$w = h/l \quad \text{and} \tag{2.35a}$$

$$w = w' = h'/l' , \tag{2.35b}$$

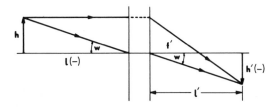

Fig. 2.16. Magnification

where h is the object height and h', the image height. Equating w' and w, as indicated, we find

$$h'/h = l'/l .$$

(2.36)

We define *magnification* as the ratio of image size to object size,

$$m = h'/h .$$

(2.37)

In terms of l and l', therefore,

$$m = l'/l .$$

(2.38)

The definition of m includes the signs of l and l'. Although this is a small point, it makes m negative when the image is inverted, as in Fig. 2.16, and positive when the image is erect.

In addition to magnification m, there is a quantity known as *longitudinal magnification* μ. The concept is illustrated in Fig. 2.17, where a small object with length Δl lies along the optical axis. Its image is shown as a real image and has length $\Delta l'$. The ratio of these lengths is μ. The easiest way to calculate μ is to begin with the lens equation in air,

$$(1/l') - (1/l) = 1/f' .$$

(2.39)

We differentiate both sides with respect to l and find that

$$-\frac{1}{l'^2} \frac{\Delta l'}{\Delta l} + \frac{1}{l^2} = 0 ,$$

(2.40)

because f' is a constant.

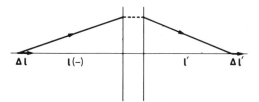

Fig. 2.17. Longitudinal magnification

Defining μ by

$$\mu = \Delta l'/\Delta l , \qquad (2.41)$$

we find immediately that

$$\mu = (l'/l)^2 \quad \text{or} \qquad (2.42)$$

$$\mu = m^2 . \qquad (2.43)$$

The longitudinal magnification in air is the square of the magnification.

Problem. Photographers use *effective F number* $(f'/D)(1 - m)$, where m is negative for a real, inverted image, to help determine the correct exposure for close-up photography when m is of the order of -1. Help establish the validity of this parameter by showing that $l' = f'(1 - m)$. Incidentally, on the other side of the lens, $l = f(1 - 1/m)$. [In photography, the relation is usually written $(f'/D)(1 + m)$, where m is here the *magnitude* of the magnification.]

2.2.10 Newton's Form of the Lens Equation

This is an alternative form of the lens equation and is most useful when either conjugate point is close to a focal point.

We define the distance x between F and the object point, as in Fig. 2.18. Because the opposite angles formed at F are equal,

$$h/x = h'/f . \qquad (2.44a)$$

Similarly, we define x' and find that

$$h'/x' = h/f' . \qquad (2.44b)$$

Extracting a factor of h'/h from each equation, we find that

$$xx' = ff' . \qquad (2.45)$$

Because f and x are both negative, it is not necessary to alter any signs in this equation. For a lens in air, $f' = -f$, and the equation becomes

$$xx' = -f'^2 , \qquad (2.46)$$

which is *Newton's form* of the lens equation.

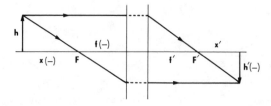

Fig. 2.18. Newton's form of the lens equation

2.2.11 Lagrange Invariant

To derive a quantity known as the *Lagrange invariant*, we trace the two rays shown in Fig. 2.19. We consider the general case of a lens with different media on either side. It is straightforward to generalize our earlier theorem (2.34) and show that the relation

$$nw = n'w' \tag{2.47}$$

pertains to any ray directed at the primary principal point P. Because $w = h/l$ and $w' = h'/l'$, we find that

$$n(h/l) = n'(h'/l') \ . \tag{2.48}$$

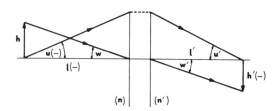

Fig. 2.19. Lagrange invariant

We trace a second ray that originates on the axis (at the tail of the arrow). If the ray intersects the principal plane at height y, then l and l' are related to u and u' by $u = y/l$ and $u' = y/l'$. Thus, we find that

$$hnu = h'n'u' \ . \tag{2.49}$$

That is, the quantity (hnu) remains constant as we trace a given ray through an optical system.

hnu is known as the *Lagrange invariant*. When we avoid the paraxial approximation, we find that a well corrected optical system should obey the sine condition, that

$$hn\sin u = h'n'\sin u' \ . \tag{2.50}$$

$hn\sin u$ becomes hnu in paraxial approximation. u is usually taken as the greatest angle that will allow the ray to enter the optical system.

2.2.12 Aberrations

The aberrations of simple, single-element lenses can be quite severe when the lens is comparatively large (with respect to image or object distance) or when the object is located far from the lens axis. A detailed discussion of aberrations is not appropriate here; let it suffice to say that when a simple lens is incapable of performing a certain task, it will be necessary to employ a lens, such as a camera lens, whose aberrations have been largely corrected. For especially demanding functions, special lenses may have to be designed and built.

As a rule of thumb, a simple lens may be adequate for general-purpose optics provided that its diameter is less than, say, one tenth of the object or image distance (whichever is smaller) and that the object is relatively close to the axis. With these constraints, performance may be close to the theoretical limit discussed in Chap. 3. When the conjugates are unequal, the best shape for a simple lens is very nearly plano-convex, the plane side facing the short conjugate; when the conjugates are roughly equal, the lens should be double convex.

A telescope objective that consists of two elements (usually cemented together) will give somewhat better performance than a simple lens, especially in white light. The main advantage of using such a lens is that it is partially corrected for *chromatic aberration*, which results from the variation of index of refraction with wavelength. Many telescope objectives are nearly plane on one side; this is the side that should face the short conjugate.

Photographic objectives are usually designed to have one conjugate at infinity, the other near the focal point. A good photographic objective may project an adequate image over a 20° to 25° angular field. If it is necessary to use a photographic objective with the object near the focal point and the image a great distance away, care should be taken to ensure that the lens is oriented with the short conjugate on the side of the lens that normally faces the camera. (With certain modern, high-aperture lenses, there may be problems getting a flat image plane in this configuration.)

Ordinary camera lenses are designed specifically for distant objects; they do not perform well at magnifications near 1. Close-up (macro) lenses, copying lenses, or enlarging lenses are preferable. Similarly, collimating lenses are designed specifically for rays parallel to the axis and should not be used to project a high-quality image over a wide field.

Highly specialized lenses, such as those used in aerial photography, can project nearly perfect images over the entirety of a fairly large image plane.

The average user of optics need not have any knowledge of aberration theory; nevertheless, he is well advised to bear in mind the purpose for which a given lens was originally intended.

Problems

2.1 *Constant-Deviation System.* Two plane mirrors are fixed at an angle α ($< 180°$) to one another. A ray is incident on one of the mirrors at angle a. We define the *deviation angle* δ as the acute angle between the incident and emergent rays.

(a) Show that δ is independent of a. Two such mirrors are an example of a *constant-deviation system*, so called because the deviation is independent of a.

(b) Show further that a glass reflecting prism may be used as a constant-deviation system only if the prism angle is 90° and then only in the paraxial approximation.

(c) Under what conditions will an incident ray fail to emerge from a prism after the second bounce?

2.2   Show that, in the paraxial approximation, a slab of glass moves an image away from a lens by the distance $d(1 - 1/n)$, where d is the thickness of the slab and n its index of refraction.

2.3   Show by geometrical construction, not by the "len" equation, that the object-image relationship for a curved mirror is given by (2.31).

2.4.   *Thick Mirror.* Show that the effective focal length f_e' of a lens and mirror in close contact is given by

$$1/f_e' = (2/f') + (2/R) \,,$$

where f' is the focal length of the lens and R is the radius of curvature of the mirror.

2.5   *Telephoto Lens.* A *telephoto lens* consists of a positive lens, before whose focal point is located a negative lens. Sketch such a lens and show that its effective focal length is greater than the distance between the positive lens and the focal plane of the telephoto lens. What is the value of such a lens design? (The principle of the telephoto lens is used in astronomy to increase the focal length of a telescope objective lens. The negative lens is called a *Barlow lens.*)

A *reverse-telephoto lens* consists of a positive lens preceded by a negative lens. Sketch this lens and locate its secondary principal plane. What is the value of this design? (In photography, wide-angle lenses often use reverse-telephoto designs.)

2.6   A thin lens projects an image of a tilted object plane. Show that the image will be in focus at all points, provided that the extensions of the object plane, the principal plane of the lens, and the image plane meet along a line. (This arrangement does not guarantee rectilinearity of the image, however.) Hint: All that is needed here is a qualitative or geometrical argument.

2.7   (a) A thin lens has focal length f' in air. Its index of refraction is n. What is its focal length when it is immersed in a medium whose index of refraction is n'?

(b) Can you in general determine the focal length of a thin lens that has air on one side and a high-index fluid on the other?

(c) Trace the first, third, and fifth rays in Fig. 2.14 through a lens that has air on the left side a fluid with index n' on the right. Show that all three rays meet at one point.

2.8 A beam of parallel rays is said to be *collimated*. We wish to collimate a beam of light by locating a source at the primary focal point F of a lens that has a 50-mm focal length. To do so, we focus the source on the wall of the laboratory. Unfortunately, the wall is only 3 m from the lens. By what distance will the lens have to be translated toward the source to achieve true collimation?

This is a practical method for collimating a beam in cases where the requirements are only modestly severe.

2.9 A transparent sphere has a uniform index of refraction n. The image of a distant object lies on the surface of the sphere. What is the index of refraction of the sphere?

2.10 A perfect lens projects a cone of rays toward point O a distance l across an air-glass interface (Fig. 2.20). If the numerical aperture is high, a marginal ray crosses the axis at point O'', distance $L' > l'$ from the interface rather than at the paraxial image point O'. The radius δ of the circle of confusion in the paraxial focal plane is called the *transverse spherical aberration*. Show that the transverse spherical aberration is given by $\delta = (n \tan U - n' \tan U')(l'/n')$, where the quantities are defined by Fig. 2.20 and lowercase letters mean paraxial approximation. Suppose that $\lambda = 0.55\ \mu m$, $\sin U = 0.65$, $l' = 170\ \mu m$, $n = 1$, and $n' = 1.5$. Show that $\delta \cong 16\ \mu m$. Compare with the diffraction-limited resolution limit $0.61\ \lambda/NA$ (Sect. 3.8).

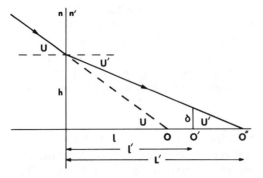

Fig. 2.20. Spherical aberration of a converging beam at a planar interface

2.11 A concave mirror in a dye laser has a radius of curvature of 10 cm. In this laser, the dye jet is located at the center of curvature of the mirror. The mirror is not highly reflective of light from an argon laser, so we decide to pump the dye by focusing the argon laser beam directly through the mirror. Assume that the mirror is flat on the nonreflecting surface. What focal length lens will we need to project a collimated beam through the mirror so that it will focus at the center of curvature of the mirror?

3. Optical Instruments

In this part, we discuss optical instruments that rely, so to speak, on geometrical optics; they are primarily lens devices. Our purpose is not merely to describe the instruments, but to present enough material to permit intelligent use of them.

3.1 The Eye (as an Optical Instrument)

For our purposes, the optical system of the human eye (Fig. 3.1) is the spherical, transparent *cornea* and behind it the *lens* or *crystalline lens*. The interior of the eye is a liquid with index of refraction 1.33, equal to that of water. The lens has a slightly higher index, 1.4; the eye focuses itself by varying the power of the lens with involuntary muscular contractions.

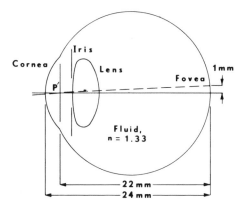

Fig. 3.1. Optical system of the human eye

The principal points of the eye are nearly coincident and are located about 2 mm behind the cornea and 22 mm from the photosensitive image surface, the *retina*. In addition, the eye has a variable aperture, the *iris*, in front of the lens. The iris controls, also by involuntary muscular contractions, the amount of light falling on the retina.

In physiological optics, we normally speak of the *power* of a lens, not its focal length. Power is defined as

$$P = n'/f' = n/f \, ;$$

the units of P are *diopters* (D), when f' is measured in meters. Power is useful because the powers of thin lenses in contact (or nearly in contact) add algebraically [see (2.26) and (2.27)].

The power of the cornea of a normal eye is about 43 D; that of the lens, 17 D. The total power of the eye is thus 60 D. The eye focuses on nearby objects by changing the power of the crystalline lens with a muscular contraction. This is known as *accommodation*. Even the normal eye loses its ability to accommodate as it ages, and people over 40 or so often require corrective lenses for reading. The phenomenon is a gradual one, but is generally not noticed until suddenly "my arms got too short!"

The nearest the eye can focus comfortably depends on its maximum accommodation. We refer to this distance as the shortest distance of distinct vision, or the *near point* d_v. It is customary to define the near point as 25 cm (or 10 in), which corresponds to a standard 40-year-old eye; in reality, the near point may be much less – a few centimeters for a small child.

Besides losing its ability to accommodate, a condition known as *presbyopia*, the eye may have an optical system whose power is abnormal in relation to the size of the eyeball. If the power of the eye is too great or the eyeball too large, the eye cannot focus on distant objects; the condition is called *myopia* or *near-sightedness*. Conversely, if the optical system of the eye is not sufficiently strong, the eye is said to have *hyperopia* or *far-sightedness*. Presbyopia is sometimes confused with far-sightedness, but far-sighted people cannot focus accurately at ∞, whereas *presbyopes*, or people who suffer from presbyopia, have a relatively distant near point, even when their eyes focus correctly at a great distance.

A myopic eye is corrected with a negative lens; a hyperopic eye, with a positive lens. A presbyopic eye is also corrected with a positive lens, but this lens is intended for close vision; many presbyopes need correction only for reading and close work.

Someone who is both myopic and presbyopic may wear *bifocals* or even *trifocals*. These are lenses that are divided into two (or three) regions of different power; usually, the upper part of the lens corrects the eye for distance vision, whereas the lower part is slightly less negative and corrects for near vision. Likewise, someone who is both hyperopic and presbyopic may wear bifocals or trifocals. Finally, an *astigmatic* eye is one whose refracting surfaces are slightly ellipsoidal, rather than spherical; astigmatism is corrected with a combination of cylindrical and spherical lenses.

Problem. Show that the lens equation for the eye may be written

$$1/l = -(P_c + P_a) \,, \tag{3.1}$$

where P_a is the power of accommodation, or the amount by which the lens of the eye has increased its power, and P_c is the refractive error, or the amount by which the eye is too strong or weak. (If a person is nearsighted by $+4$ D, then that is the value of P_c.)

Contractions of the iris help the eye *adapt* to different light levels. An opening, such as the iris, that limits the amount of light entering a system is known as an *aperture stop*; the iris is the aperture stop of the eye. The diameter of the iris varies from perhaps 2 mm in bright light to 8 mm in darkness. For many purposes, its diameter may be taken as 5 mm.

The retina consists of small, light-sensitive detectors called *rods* and *cones*. The cones are responsible for color vision and operate at high light levels. The rods take over at low luminances and cannot distinguish among colors. (This is one reason that things look uniformly colorless in dim light.) The eye switches from rod vision to cone vision by a mechanism called *neural inhibition*, which in effect switches off the rod vision; in this way, the eye achieves a dynamic range of 1 million to 1.

The region of greatest visual acuity is called the *macula lutea*, or *macula*. It is located just off the optic axis of the optical system. Near the center of the macula is a region called the *fovea centralis*, or *fovea*. The fovea contains no rods, but rather a high density of cones, and offers the very highest visual acuity. When the eye *fixates* on a certain spot, it is focusing that spot onto the fovea. Visual acuity decreases rapidly with distance from the fovea, as you can easily demonstrate by fixating on the center of one line on this page and trying to read the whole line.

The rods, on the whole, are located outside the fovea, where the number of cones is small. There are few rods in the fovea itself; it is therefore difficult to see clearly at night.

The cones are closely packed in the fovea; each cone subtends slightly less than 0.15 mrad (0.5′) at the principal points. The eye can distinguish two points only if their images are separated by one cone; otherwise, the two points will appear as only one. Therefore, the eye can at best distinguish points that subtend about 0.3 mrad (1′) at the principal points. For an object located at a distance d_v, this corresponds to a *resolution limit* of about 0.1 mm. Visual acuity falls off rapidly outside the fovea.

The resolution limit estimated in this way is remarkably close to the theoretical resolution limit calculated by diffraction theory for a 2 mm pupil; this is about the diameter of the pupil in bright sunlight. If the cones were packed much less closely, resolution would suffer. On the other hand, if the cones were much more closely packed, diffraction would cause the light from a single object point to fall on several cones, rather than one. The result would be a dim image, because relatively little light would fall on a given cone. Evolution has thus been very efficient at packing cones into the fovea at nearly the optimum density. When the pupil of the eye is larger than 2 mm, the resolution limit becomes larger owing to aberrations; the angular resolution limit of a 6-mm pupil is about 0.8 mrad (3′).

The visible portion of the spectrum is usually said to include wavelengths between 400 and 700 nm, although the eye is slightly responsive outside this region. Cone vision, known as *photopic* or *bright-adapted* vision, is most sensitive to green light whose wavelength is about 550 nm. Relative emission from the sun is also greatest at about this wavelength. Figure 3.2 shows how the cones' relative sensitivity V_λ decreases gradually and approaches 0 at the ends of the visible spectrum.

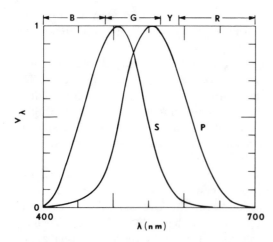

Fig. 3.2. Relative sensitivity of the human eye. *P*, photopic or bright-adapted. *S*, scotopic or dark-adapted

At twilight, there is little direct sunlight, and we see only because of light scattered by the atmosphere. Such light is bluish, and the rods are accordingly more blue sensitive than the cones. This is called *scotopic* or *dark-adapted* vision; the wavelength dependence of scotopic vision is shown also in Fig. 3.2. Both curves are normalized to 1; in fact, the rods operate only at far lower intensity than the cones.

3.2 Basic Camera

Like the eye, the camera consists of a lens, an aperture stop, and a light-sensitive screen, in this case, *film*. The camera shown in Fig. 3.3 uses a landscape lens located behind the aperture stop. Modern cameras nearly always have much more complex lenses with the aperture stop located between two elements. The

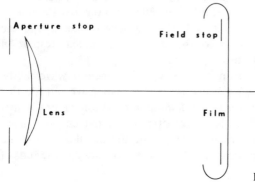

Fig. 3.3. Optical system of basic camera

aperture stop itself is an *iris diaphragm* whose diameter may be varied to control the exposure of the film.

A camera photographing an extended scene will not record everything before it. The film occupies a fixed area; objects that are imaged outside that area are not recorded. This area is defined by a rectangular opening that serves in part to hold the film in place and to keep it flat. Because such an opening limits the *field of view* to those objects that fall within a certain angle (as seen from the aperture stop), it is known as a *field stop*.

3.2.1 Photographic Emulsion

The film is made up of a thin, light-sensitive *emulsion* on a rigid *base*, or *support*, of glass or some more flexible material (Fig. 3.4). The film usually has an *antihalation dye* behind it, to reduce the blurring effects of scattered light. The dye is bleached or washed off when the film is processed.

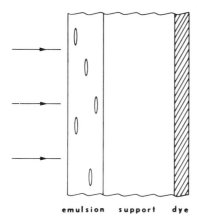

emulsion   support   dye

Fig. 3.4. Cross section of typical film or plate

The term "emulsion" is actually a misnomer, because the emulsion is a suspension of small *silver-halide grains* in gelatin. The grains are light sensitive in that they are *rendered developable* by exposure to light. Grains that are exposed become developable, whereas grains that are not exposed remain relatively undevelopable. An image formed in the emulsion, but not yet developed, is known as a *latent image*. It cannot be detected except by developing the film.

Development takes place when the grains are immersed in *developer*, a chemical solution that reduces the silver-halide grains to *metallic silver*. Exposed grains are reduced much more quickly than unexposed grains, and development is stopped long before many of the unexposed grains are reduced. The latent image is thus turned into a visible, *silver image*. The image is *grainy*, because in general a grain is reduced entirely to silver or it is not reduced at all, and except in the most strongly exposed areas there are gaps between neighboring developed grains.

To make the image permanent, the undeveloped grains are chemically dissolved away. The film is then no longer light sensitive. The process is known as *fixing*, and the chemical bath, *fixer* or *hypo*.

The image is thus made up of metallic silver and is nearly opaque where the exposure was greatest and transparent where the exposure was least. Most modern films record continuous-tone objects as continuous shades of gray. The image, however, is a *negative*, because bright parts of the object are recorded as black and dim parts as white. We most often obtain a positive by *printing* or *enlarging*, which is basically photographing the negative with an emulsion on a paper backing. (Color slides are made by a *direct positive* process which removes the developed silver and, prior to fixing, renders the remaining unexposed grains developable by exposure to light or by chemical means.)

Untreated silver-halide grains themselves are sensitive only to the blue and ultraviolet parts of the spectrum as shown in Fig. 3.5. Film can be made to respond to the entire visible and near-infrared spectra by coating the grains with *sensitizing dyes*. *Orthochromatic* films are made green sensitive; they can be examined during processing with special red lights called *safe lights*. Such films are often used in copy work, where response to different colors is unimportant and sometimes undesirable. Most films designed for pictorial use are *panchromatic* and respond to the entire visible spectrum. Other films can be made infrared sensitive, though they retain their sensitivity to the short wavelengths as well.

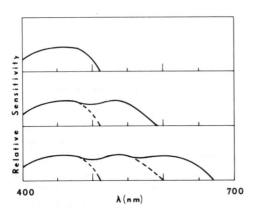

Fig. 3.5. Relative spectral sensitivities of typical photographic emulsions. *Top*, raw silver halide. *Center*, orthochromatic. *Bottom*, panchromatic

3.2.2 Sensitometry

The response of film to exposure by light is usually expressed in logarithmic units. This is convenient for several reasons. First, the mass of silver per unit area in the developed image is related to the logarithm of the transmittance of the film. Second, a characteristic curve plotted in logarithmic units has a long linear portion. Finally, the response of the human eye is approximately logarithmic in bright light.

To plot the characteristic curve of the film, we define two quantities, *optical density D* and *exposure $\mathcal{E}$*. Exposure is the quantity of light that falls on a certain area of film. Photographers generally write exposure as

$$\mathcal{E} = Et ,\tag{3.2}$$

the product of *irradiance* or *illuminance E* (sometimes called intensity in the past) and *exposure time t*. This definition is based on the *reciprocity law*, which states that, over a wide range of irradiance and exposure time, a given exposure produces a given response. Extremely long or short exposure times produce lower response than the reciprocity law suggests. This is known as *reciprocity failure*.

The response of the film is determined by measuring the optical density of the developed product. If the developed film transmits a fraction T of the light falling on it, then its *transmittance* is T and its density is defined by the equation

$$D = - \log T .\tag{3.3}$$

The minus sign serves only to make D a positive number, consistent with the idea that the darker the film, the more dense it is. Density is often defined in the equivalent form $D = \log(1/T)$, where $1/T$ is sometimes called *opacity*.

The characteristic curve of the film is obtained by plotting a D vs $\log \mathcal{E}$ curve (also known as an *H and D curve*, after its originators) as shown in Fig. 3.6. At very low exposures, the film is nearly transparent ($D \sim 0$). (There are some developed, unexposed grains that contribute to *fog density* and, in addition, light lost in the base material gives rise to a small *base density*.) Above a certain exposure, density begins to increase with increasing exposure. This region is the *toe* of the curve. Higher exposures render more of the available grains developable, and the curve goes through a *linear* region whose slope is known as *gamma* (γ). The response·saturates at the *shoulder* when all of the available grains are developed.

Most general-purpose photography is done on the toe and the lower linear portion of the characteristic curve. The average slope or *contrast index* is

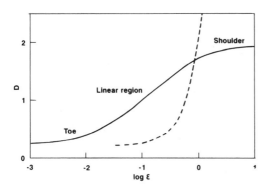

Fig. 3.6. *D* vs log $\mathcal{E}$ curves. *Solid curve, low-contrast pictorial film. Dashed curve, higher-contrast copying film*

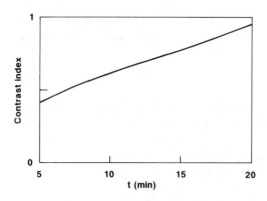

Fig. 3.7. Contrast index as a function of development time for a low-contrast pictorial film

somewhat less than gamma; either may be used as a measure of the relative *contrast* of the photograph. Higher contrast indices correspond to higher recorded contrast. For a given film and developer, contrast may be increased somewhat by increasing development time; curves of contrast index vs time (Fig. 3.7) are often available as guides.

In pictorial photography, the contrast index of the negative is usually slightly less than 1; the negative is printed so that the recorded contrast is roughly equal to 1, thereby faithfully reproducing the contrast of the original scene. *Line-copy films* and other special-purpose films may have much higher contrast; gamma may be limited to much less than 1 for particularly *widelatitude* exposures.

Most cameras have iris diaphragms that are calibrated in *stops*. Changing from one stop to the next changes the exposure by a factor of 2. Similarly, available *shutter speeds* (in fractions of a second) may be 1/250, 1/125, 1/60, . . . , so that each exposure time differs by a factor of 2 from the next.

To calibrate the lens aperture, we define the *relative aperture* or *F number* ϕ (sometimes called *focal ratio*) by the relation

$$\phi = f'/D \,, \tag{3.4}$$

where D is the diameter of the aperture stop. If two lenses have the same F number but different focal lengths and both image the same distant, extended object, then they both produce the same irradiance on the film plane. We discuss this fact in Sect. 4.17, but can easily see that, whereas the larger lens collects more light, it also has the greater focal length and therefore spreads the light over a proportionately larger area of film.

Suppose that the aperture of a certain lens is set at a given value of ϕ. To double the exposure we would have to increase the area of the aperture by a factor of 2. This is equivalent to decreasing ϕ by $\sqrt{2}$ or approximately 1.4. Therefore iris diaphragms are calibrated in multiples of 1.4, called *F stops*, such as 2.8, 4, 5.6, 8, 11, . . . , where the lower F number refers to the greater exposure. Changing from one F stop to the next changes the exposure by 2. F stops are written $F/2.8$, $F/4$, etc.; a lens whose greatest relative aperture is 2.8 is called an $F/2.8$ lens.

3.2.3 Resolving Power

In a great many photographic applications, resolution is limited by the granularity of the emulsion. For example, suppose the average spacing between grains in a certain film is 5 μm. Then, as with the eye, two points will be distinguishable if their images are separated by about twice the grain spacing. The *resolution limit RL* is thus about 10 μm. Photographic scientists more often speak of *resolving power RP*, the number of resolvable *lines per millimeter*, which is approximately

$$RP = 1/RL .\tag{3.5}$$

The resolving power in the example is about 100 lines/mm; most common films can resolve 50 to 100 lines/mm. *Copy films* are capable of two or three times this resolution, and certain glass *plates* used in spectroscopy or holography may resolve up to two or three thousand lines per millimeter. Processing can affect resolution only slightly.

3.2.4 Depth of Field

If a camera is focused sharply on a relatively distant object, the converging rays from a slightly nearer point are intercepted by the film before they reach a sharp focus, as in Fig. 3.8. The image on the film is a small disk called the *circle of confusion*. As long as the diameter of the circle of confusion is smaller than the resolution limit of the film, the nearer point is in acceptable focus. Blur will be evident only when the circle of confusion exceeds the resolution limit. The greatest acceptable amount of *defocusing* δ thus occurs when the circle of confusion is equal in size to the resolution limit. We may calculate δ by noting the similarity of the large and small triangles in Fig. 3.8,

$$\frac{D}{l' + \delta} = \frac{RL}{\delta} .\tag{3.6}$$

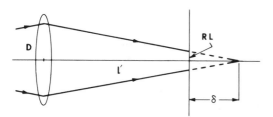

Fig. 3.8. Depth of focus

Assuming $\delta \ll l'$, we find

$$\delta = \phi(1 - m)/RP ,\tag{3.7}$$

where m is negative for a real, inverted image. $\phi(1 - m)$ is often called the *effective F number* and differs appreciably from ϕ only in *close-up photography*.

Because a more distant object would be focused in front of the film plane, the total *depth of focus* is equal to 2δ. *Depth of field* refers to the range of object distances that are imaged within a distance $\pm\delta$ of the film plane.

Depth of field could be calculated by direct application of the lens equation, but it is more easily deduced in the two important cases. When the camera is focused beyond a certain distance, called the *hyperfocal distance H*, depth of field extends to infinity. It is left as a problem to calculate the hyperfocal distance and show that depth of field extends from $H/2$ to ∞ when the camera is focused at H. The result shows that, for given F number, H is shortest with a short-focal-length lens.

The second important case occurs when the object is close compared with H. We may then apply the idea of longitudinal magnification and find that depth of field is approximately $2\delta/m^2$.

Problem. Show that the dependence of H on F number is given by

$$H = -f'^2 RP/\phi .\tag{3.8}$$

Explain why short focal-length lenses display greater depth of field than those with longer focal length.

Show further that the nearest point that is in sharp focus is approximately $H/2$ distant from the lens.

3.3 Projection Systems

It is useful to discuss projection systems (such as the slide projector) before bringing up instruments such as the microscope or telescope. Projection systems can be used to exemplify nearly all complicated imaging systems and to describe in very concrete terms an important concept known as *filling the aperture*.

Figure 3.9 illustrates a slide projector. The object, a transparency, is back lighted by a bright *projection lamp L*, and a greatly magnified image is projected on the screen S. Many of the rays that pass through the transparency will not ordinarily be intercepted by the entrance pupil of the *projection lens PL*. Because the projected image is quite large, it will be rather dim.

Many of the rays that leave the lamp will not pass through the transparency or the projection lens. To increase the efficiency of the optical system, the slide projector normally uses a *condensing lens CL* or *condenser* placed just ahead of the transparency. The condenser redirects many of the widely divergent rays toward the projection lens. Obviously, it works most efficiently when all of the

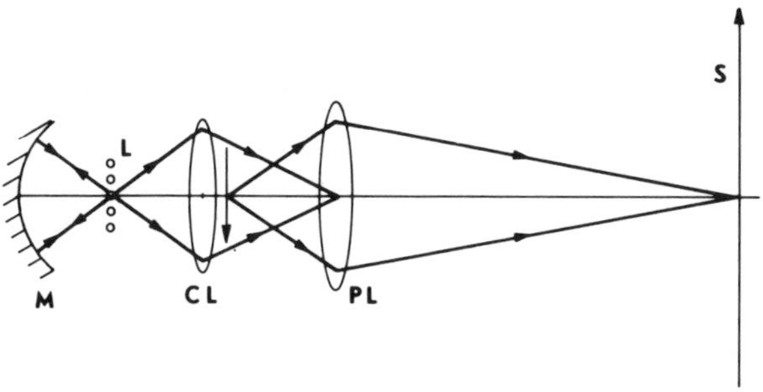

Fig. 3.9. Projection system

rays intercepted by the condenser are focused through the projection lens. That is, the condenser should project an image of the lamp filament into the entrance pupil of the projection lens (Sect. 3.7).

The image of the filament should be about the same size as the entrance pupil. If the filament image is too large, then light is lost by *vignetting*. On the other hand, if the image is too small, much of the projection lens is wasted because no light passes through it. The image of the filament then becomes the effective aperture stop, and resolution or picture clarity may suffer.

Because of the necessity of filling the aperture of the projection lens, projection lamps have several filaments, close together and parallel to one another. The lamp approximates a small, diffuse source. Often, either the projector or the lamp itself employs a concave mirror M to intercept rays traveling backward, away from the condenser. The mirror reflects an image of the filament back into the vicinity of the filament, so that these rays, too, will be directed by the condenser into the entrance pupil.

The condensing lens has no function other than to fill the aperture of the projection lens. It therefore need not be of high quality; condensers are often single-element lenses made of plastic or flame-polished glass. A condenser should be a relatively high-aperture lens, however; many condensing lenses have one aspheric surface to reduce spherical aberration and more effectively fill the entrance pupil.

A *condenser enlarger* is another system that works on the principle of the slide projector. In fact, an enlarger is no more than a special-purpose projection system. The relatively new *overhead projector* also operates on the same principle. The condenser is a large, flat, molded-plastic *Fresnel lens* located just below the writing surface. (A Fresnel lens is a series of molded, concentric-circular prisms designed to focus light exactly as a lens does, but with less thickness and therefore less mass.) The two-element projection lens incorporates a mirror to redirect the optical axis horizontally.

Most motion-picture projectors differ from slide projectors in that they project an image of the source into the plane of the film. This is possible primarily because the film is moving and will therefore not be burned by the hot image of the lamp filament. The condenser is used to increase the efficiency of the optical system; such a projection system is roughly equivalent to the slide projector. It is rarely used outside of motion-picture projectors, however; the slide projector remains the best example of a complicated projection system.

3.4 Hand Lens or Simple Magnifier

This instrument, sometimes called a simple microscope, is familiar and needs no introduction. Imagine that we wish to examine a small object closely. Its height is h, and we would normally scrutinize it at the near point d_v, where it subtends an angle h/d_v at the eye.

To magnify the object, we use a hand lens. Ideally, the object is located at the primary focal point F and projected to $-\infty$. The virtual image subtends $w = h/f'$ at the eye, as shown in Fig. 3.10. If w exceeds h/d_v, the image on the retina is proportionately larger with the lens than without. We therefore define the *magnifying power*

$$MP \equiv \frac{\text{angle subtended by virtual image}}{\text{angle subtended by object located at } d_v}. \tag{3.9}$$

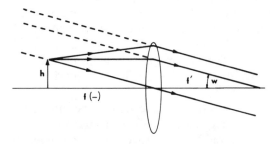

Fig. 3.10. Simple magnifier

(Do not confuse MP, which is dimensionless, with power in diopters.) Magnifying power is often used in place of magnification in systems where the image or object is likely to be located at ∞. In such cases, the magnification may be zero, but the magnifying power may not. In the case of the hand lens, we find immediately that

$$MP = d_v/f'. \tag{3.10a}$$

Because it is conventional to take d_v as 25 cm,

$$MP = 25/f'_{[cm]}. \tag{3.10b}$$

Simple lenses will serve well as hand lenses with magnifying powers possibly as great as 5. Eyepieces or special flat-field *comparators* conveniently provide magnifying powers of 10 or more.

The magnifying power of any hand lens (or eyepiece) can be increased by $+1$ (for example, from 5 to 6) by adjusting the position of the lens so that the image appears at the distance d_v from the eye. This is tiring, and little is gained. It is preferable to learn to use optical devices with the eye unaccommodated, that is with the muscles that control the lens completely relaxed and the lens focused a few meters away (not necessarily at infinity as was once believed).

Problem. An observer may well accommodate data his eye for d_v, rather than ∞, when he uses a hand lens. Show that the magnifying power is

$$MP = 1 + (d_v/f')$$ (3.11)

if the eye is assumed to be in contact with the hand lens. What if the eye is assumed to be located at F'? Explain why M depends on the position of the eye with respect to the lens.

3.5 Microscope

The microscope, or *compound microscope,* is best regarded as a two-stage instrument. The first lens, the *objective,* forms a magnified image of the object. Then, in effect, we examine that image with a good-quality magnifying glass called the *eyepiece.* The eyepiece projects the image to infinity.

Figure 3.11 shows a compound microscope. The distance $F_o'F_e$ between the secondary focal point of the objective and the primary focal point of the eyepiece is known as the *tube length g.* The objective projects an image with magnification

$$m = -g/f_o' ,$$ (3.12)

expressed in terms of g. The angular subtense of the image seen through the eyepiece is h'/f_e', or $-hg/f_o'f_e'$; dividing this value by h/d_v, we find that the total magnifying power of the microscope is

$$MP = -(g/f_o')\,MP_e .$$ (3.13)

where MP_e is the magnifying power of the eyepiece. The minus sign indicates merely that the object appears inverted. In microscopy this is generally of no consequence.

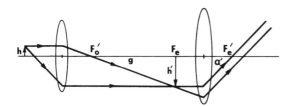

Fig. 3.11. Compound microscope

Many microscopes use $g = 160$ mm. The magnification stated on the barrel of an objective therefore assumes 160-mm tube length, and the objectives are designed to work best at this tube length. If the tube length differs from 160 mm, the value is usually stated on the barrel of the objective; the tube length of ∞ is becoming increasingly common. If a microscope objective is used at a tube length that differs from the specified tube length, spherical aberration may result, especially for higher-power objectives, say, those over $40 \times$.

Some manufacturers also define *mechanical tube length* or equivalent; this is usually the distance between the shoulder of the objective (the flat surface next to the threads) and the image, and is not standard.

The most common objectives have magnifications of 10, 20, or $40 \times$ when they are used with the standard tube length; see Table 3.1. Besides magnification, microscope objectives have on their barrel a second number called the *numerical aperture NA*. As we shall find in Sect. 3.8.3, the total magnifying power of the compound microscope should not greatly exceed the *useful magnifying power*

$$MP_u = 300\, NA \tag{3.14}$$

when the instrument is used visually. A typical 40-power objective may have a numerical aperture of 0.65. In this case MP_u is about 200, and the resolution limit of the instrument is less than 1 μm. Higher useful magnifying power is available with higher-power objectives, in particular *oil-immersion objectives* (Sect. 3.8.3), which may have numerical apertures as high as 1.6.

Table 3.1. Common microscope objectives

Magnification	Numerical aperture	Useful MP	RL[a]
4	0.15	45	2.2
10	0.25	75	1.3
20	0.50	150	0.7
40	0.65	195	0.5
60	0.80	240	0.4
100 (dry)	0.90	270	0.4
100 (oil)	1.25	375	0.3

[a] in micrometers and at 550 nm.

Most objectives are designed to be used with *cover slips* of specified thickness, usually 0.16–0.18 mm, to avoid introducing spherical aberration. Certain *metallurgical objectives*, on the other hand, are intended for looking at opaque surfaces and therefore are designed to be used without a cover slip.

A microscopic objective that is designed for use with a cover slip could be called a *biological objective*, to distinguish it from a metallurgical objective. If the cover slip is omitted, spherical aberration may be an important factor, particularly if the numerical aperture exceeds 0.5 or so. Sometimes, however, these objectives may be needed to examine a "dry" object, that is, one that is not immersed in a fluid and not in contact with a cover slip. In these cases, the cover slip may be mounted in air, just in front of the first element of the objective. Failure to include the cover slip will result in decreased image contrast and possibly errors when quantitative measurements, such as length, are important.

A *video microscope* can be little more than a microscope objective fixed to a closed-circuit video (television) camera. Viewing the output of the video camera on a monitor is easier and less tiring than looking into an eyepiece and can offer about equal image quality. In addition, sometimes the brightness or the contrast can be adjusted to reveal details that cannot be seen by the naked eye. Also, it is easy to use off-axis illumination to reveal otherwise faint phase images (Sect. 7.2), even in a conventional microscope.

The real power of video microscopy, however, is the ability to direct the video signal to a digital computer or a *frame digitizer*, which can capture or *digitize* the image for nearly simultaneous (real-time) processing or store it for future analysis (Sect. 7.4).

3.6 Scanning Confocal Microscope

The scanning confocal microscope is not new but is receiving renewed prominence, especially in biology. Such a microscope is sketched in Fig. 3.12. A point source, perhaps a laser, is focused with a high-quality microscope objective (not a condensing lens) onto the plane of the sample. A second objective lens re-images that source with the proper tube length onto a detector. Because the two objectives focus on the same plane, we call the microscope confocal. In some implementations, the detector has a pinhole before it.

To generate an image, the object itself may be translated, or *scanned*, across the image of the source. Where the object is transparent, for example, most of the light from the source is focused onto the detector. Where the object is opaque, the detector receives little light. The output of the detector as a function of time represents one line in the image. The object is then moved up or down (out of the

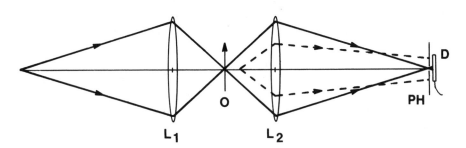

Fig. 3.12. Confocal microscope

plane of the page), and another line may be observed. A sequence of parallel, linear scans is called a *raster*, so we may say that the object is scanned in a raster pattern across the common focal plane of the two objectives. For some measurements, such as the width of a line or strip on an integrated circuit, the object need not be scanned in a raster because a scan of a single line gives enough information.

Unless the object can be scanned very fast (and in two dimensions), the image cannot be observed visually. It is therefore often preferable to devise a way to scan the image of a point source over a stationary object. Physically moving the object, however, may be more accurate for quantitative length measurements, but it is usually necessary to use digital image processing (Sect. 7.3) to extract the information.

One of the most-important advantages of scanning confocal microscopy over conventional microscopy is that it can be used for *optical sectioning*. This is a way of saying that out-of-focus images are invisible, so the scanning confocal microscope allows plane-by-plane imaging of thick objects such as biological specimens. The dashed lines in Fig. 3.12 show the paths of rays that originate at an out-of-focus object point. The source is approximately a point, and the object point is distant from the image of the source; therefore, it is not illuminated with high intensity, so its image is necessarily weak. In addition, if there is an aperture before the detector, the out-of-focus point is imaged well away from the plane of that aperture, so very little light passes to the detector. In consequence, object points that are located outside the depth of field of the objective are not just blurred but invisible. This is of immeasurable importance, say, for examining a small weak structure near a brighter structure or for examining relatively sparse structures like cells, without obstruction by the blurred images of defocused objects.

3.6.1 Nipkow Disk

A commercial scanning confocal microscope may scan the source (rather than the object) with a perforated disk called a *Nipkow disk*. One such scheme is shown in Fig. 3.13. A uniform, diffuse source S is focused by a beam splitter BS and a condensing lens CL onto a spinning disk D, the Nipkow disk. The disk is perforated with holes arranged in arcs of a spiral, as shown in the inset. Images of the holes are projected into the object plane by the microscope objective lens MO; their motion across the object plane forms the desired raster.

The objective lens directs light scattered or reflected by points on the object back through the holes in the Nipkow disk. From there, the light progresses through an eyepiece for direct viewing or to a video camera for viewing on a monitor or for image processing. In Fig. 3.13, an intermediate image is projected into the plane of the disk; the final image I in turn appears in the image plane of the relay lens (or eyepiece) RL. As long as the disk scans a complete image in less than about 1/60 s, this image can be viewed comfortably with minimal flicker.

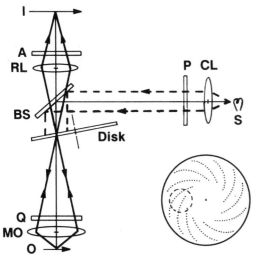

Fig. 3.13. Real-time confocal microscope with Nipkow disk. [After G. S. Kino, Intermediate Optics in Nipkow Disk Microscopes, in *Handbook of Biological Confocal Microscopy*, ed. by J. B. Pawley (Plenum, New York 1990)]

The disk in Fig. 3.13 passes only a very small fraction of the light incident on it. The source therefore must be very intense. In addition, the disk must be well polished and inclined at a slight angle to the incident light to eliminate from the image the light scattered by the disk. For further discrimination against scattered light, a polarizer *P* is placed after the source. The polarized light from the source passes twice through a quarter-wave plate *Q* (Sect. 9.2.2) before returning to the Nipkow disk. The two passes through the quarter-wave plate rotate the plane of polarization (Problem 9.4) by 180°, and the light is directed through a second polarizer, or analyzer, *A*, whose axis is rotated 90° to the first. Scattered light, however, retains the original polarization and is rejected by the second polarizer. Other schemes that use a Nipkow disk may instead use mirrors to direct the reflected light through a different set of holes from those used for illumination and in that way eliminate the light scattered by the disk.

A scanning confocal microscope that uses a Nipkow disk is almost necessarily used in reflection. Often this is an advantage, as when an opaque object is to be examined; at other times it is irrelevant or outweighed by the ability of the instrument to render out-of-focus planes invisible.

We defer further discussion of the scanning confocal microscope until Sect. 7.3.3.

3.7 Telescope

The essential difference between a telescope and a microscope is the location of the object. A telescope is used to view large objects a great distance away. As with the microscope, the telescope objective projects an image that is examined through the eyepiece.

Suppose the object to be very distant, but large enough to subtend angle α at the location of the telescope (Fig. 3.14). Viewed through the telescope, it subtends α'. As with the hand lens, we define the magnifying power of the telescope as

$$MP = \alpha'/\alpha .\qquad(3.15)$$

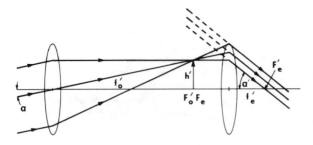

Fig. 3.14. Simple telescope

Geometry shows that, numerically, $\alpha = h'/f_o'$ and $\alpha' = h'/f_e'$. Thus,

$$MP = -f_o'/f_e' .\qquad(3.16)$$

where the minus sign is retained only because the image is inverted. If f_o' is made very much larger than f_e', a distant object can be made to appear very much bigger through a telescope than with the unaided eye.

3.7.1 Pupils and Stops

It is convenient to use the telescope to illustrate the importance of pupils and stops in an optical system. As with the camera, the opening that limits the total amount of light collected by the telescope is called the *aperture stop* or *limiting aperture*. In a properly designed telescope, the ring that holds the objective in place serves as the aperture stop and roughly coincides with the objective itself.

Now let us trace two bundles of rays through the telescope, as in Fig. 3.15. The ray *pr* that goes through the center of the aperture stop is known as the *principal ray*. The bundles that emerge from the eyepiece have a waist where the principal ray crosses the axis, slightly behind F_e'. The eyepiece projects an image of the aperture stop into the plane where the principal ray crosses the axis. This image is known as the *exit pupil* and lies in the plane of the waist. Careful examination of Fig. 3.15 reveals further that the exit pupil and the waist coincide exactly. All rays that enter the aperture stop and pass through the optical system also pass through the exit pupil. In the telescope, the aperture stop also serves as the *entrance pupil*.

Formally, the exit pupil is defined as the image of the aperture stop seen through all the optics beyond the aperture stop. The entrance pupil is the image

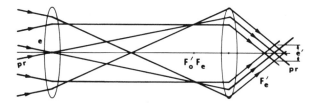

Fig. 3.15. Construction of telescope exit pupil

seen through the optics before the aperture stop. The entrance and exit pupils are also images of each other.

For best viewing, the eye should be located at or near the exit pupil. Otherwise, a great many rays will not enter the pupil of the eye, a phenomenon known as *vignetting*.

If we let e be the diameter of the entrance pupil and e', that of the exit pupil, we can easily show that

$$e/e' = MP .\qquad(3.17)$$

With a telescope, e is also the diameter of the aperture stop, so the diameter of the exit pupil is a fraction $1/MP$ of the diameter of the aperture stop.

Ideally, the exit pupil should not only coincide with the pupil of the eye, it should also be about the same size as the pupil of the eye. If the exit pupil is too large, vignetting inevitably occurs; in fact, the eye becomes the limiting aperture. Many of the rays that enter the objective are not allowed to enter the eye. Much of the objective is wasted, as we can easily see by projecting an image of the pupil of the eye into the plane of the objective. This image, and not the objective itself, is, in this case, the entrance pupil. Resolution and light-gathering ability of the telescope suffer when its exit pupil is too large. Nevertheless, some telescopes are designed with slightly larger exit pupils to give more freedom of eye movement.

We will find in Sect. 3.8.2 that if the magnifying power is so great that the exit pupil becomes much smaller than the pupil of the eye, the *useful magnifying power*

$$MP_u = 5D_{[cm]}\qquad(3.18)$$

has been exceeded. D is the objective diameter in centimeters. For given D, the magnifying power should not greatly exceed MP_u.

Earlier, we found that the pupil of the eye has a diameter of about 5 mm under average viewing conditions. For this reason, most *binoculars* are designed to have 5-mm exit pupils (although night glasses may have exit pupils as large as 8 mm). Thus, for example, 7-power binoculars will generally have a 35-mm objective. Such binoculars are specified as 7×35; the first number refers to the power and the second to the objective diameter in millimeters. Other common binoculars are 8×40 and 10×50, while night glasses may be 6×50. Small, lightweight binoculars, such as 8×25, have smaller exit pupils and are most useful during the day. Eight-power binoculars are the most powerful that can be hand-held comfortably.

3.7.2 Field Stop

In addition to the aperture stop, most telescopes have a *field stop* in the primary focal plane of the eyepiece. As with the camera, the field stop limits the angular acceptance or field of view to a certain value. It is located in an image plane, where the rays in a bundle pass through a point, to prevent partial vignetting of some bundles.

3.7.3 Terrestrial Telescopes

The image seen in the telescope is inverted. For astronomical purposes this is unimportant, but for terrestrial telescopes an erect image is preferable.

The simplest erecting telescope is the *Galilean telescope*, in which the eyepiece is a negative lens. Such telescopes are rarely used today, except for low-power opera glasses, but they inherently provide an erect image. *Prism binoculars* or *field glasses* are erecting telescopes that use two reflecting prisms to rotate the image so it is upright.

What is conventionally called a *terrestrial telescope* employs a third lens to project the image at unit magnification and thereby invert it (Fig. 3.16). A *periscope* is a telescope with many relays such as the erecting lens *EL* in Fig. 3.16.

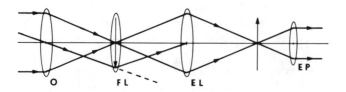

Fig. 3.16. Terrestrial or erecting telescope

The erecting lens or the eyepiece could become quite sizeable in a powerful terrestrial telescope with a large field. To prevent this, we place a *field lens FL* in or near the first image plane. The lens has little effect on the image, but redirects the divergent rays toward the axis, as shown also in Fig. 3.16.

The field lens is in many ways analogous to the condensing lens of the slide projector. It intercepts rays that would otherwise miss the erecting lens and directs them toward the axis. To perform its function properly, it must project an image of the aperture stop into the plane of the erecting lens. (Otherwise, vignetting will occur, as we noted in connection with the telescope eyepiece.) In addition, the image of the aperture stop should roughly fill the erecting lens, just as the image of the filament roughly fills the projection lens of a slide projector.

Incidentally, there is an image of the aperture stop, or a *pupil*, every time the principal ray crosses the axis. A diaphragm is nearly always placed in the plane of a pupil. The diaphragm is equal in size to the pupil and reduces scattered light without vignetting.

Finally, we must take similar care that the relay lens project an image of the field lens into the field stop associated with the eyepiece. The field lens and the field stop must be images of each other, as seen through the intervening optics. If they are not, then one or the other becomes the effective field stop, just as the size of the eye's pupil had to be considered in determining the exit pupil of a simple telescope.

3.8 Resolving Power of Optical Instruments

Physical or wave optics shows that, because of *diffraction*, the image of a very small point is not itself a point (Sect. 5.5). Rather, it is a small spot known as the *Airy disk*. The size of the spot depends on the relative aperture of the optical system. Because the image is not a point, two object points will not be distinguishable if their geometrical images fall within one Airy-disk radius of one another. For optics with circular symmetry, the *theoretical resolution limit RL'* in the image plane is

$$RL' = 1.22 \lambda l'/D , \tag{3.19}$$

where l' is the image distance, D the diameter of the aperture, and λ the wavelength of the light. Optical systems capable of achieving this resolution are said to be *diffraction limited*.

RL' is the separation between two geometrical-image points when the two points are just resolved. The separation between the actual points in the object plane is RL'/m, where m is the magnification of the system. Because two object points must be separated by at least RL'/m if they are to be resolved, RL'/m is called the resolution limit in the object plane. Because $m = l'/l$, we have

$$RL = 1.22 \lambda l/D \tag{3.20}$$

for the resolution limit in object space. Here the sign of l is not important and we take $RL > 0$.

3.8.1 Camera

The theoretical resolution limit of a camera is often more conveniently expressed by

$$RL' = 1.22 \lambda \phi(1 - m) , \tag{3.21}$$

where m is negative. For visible light, λ is about 0.55 μm, so RL' is about equal to the effective F number $\phi(1 - m)$ in micrometers. The reciprocal of RL' is known as the *theoretical resolving power* and is generally expressed in *lines per millimeter*.

In practice, most cameras never attain their theoretical resolving power for apertures much higher than $F/8$ ($\phi = 8$). At this relative aperture, the theoretical resolving power is perhaps 120 lines/mm. Few common films can achieve this resolution; as a result the film is most often the dominant factor. That is, common photographic objectives are rarely diffraction limited at the higher apertures (low F numbers). This is particularly so far off the lens axis and comes about because the optical designer must make many compromises to design a relatively inexpensive lens with the wide field required for general photography. He can afford to do so, in part, because of the limitations of the available films.

Lenses that are not diffraction limited are said to be *aberration limited*. Typical photographic objectives are thus aberration limited at F numbers lower than $F/11$ or so. As the aperture diameter is increased from $F/11$, they may show constant or slightly worsening resolution on the optical axis. Very high-aperture lenses ($F/1.4$) may be significantly poorer at their lowest F numbers than in their middle range. In addition, brightness and resolution may suffer at the edges of an image produced by a high-aperture lens that is "wide open".

3.8.2 Telescope

A telescope objective can theoretically resolve two points if their images are separated by $1.22 \lambda f_o'/D$ or more. This means that the angular separation of the two points must be greater than

$$\alpha_{\min} = 1.22 \lambda/D \, , \tag{3.22}$$

as indicated in Fig. 3.17. The eyepiece must have higher relative aperture (smaller F number) than the objective if it is to resolve the two images well. The eye then sees an apparent angular separation

$$\alpha'_{\min} = MP \cdot 1.22 \lambda/D \, , \tag{3.23}$$

where MP is the magnifying power of the telescope.

Unlike camera objectives, telescope objectives relatively rarely have apertures exceeding $F/11$. Reflecting objectives or achromatic doublets may therefore be very nearly diffraction limited close to the optical axis.

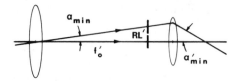

Fig. 3.17. Angular resolution limit of telescope

The angular resolution limit of the eye is about 0.3 mrad (1'). The eye will resolve two points only if α'_{min} has at least this value. When α'_{min} is just equal to 0.3 mrad, we speak of *useful magnification*, which is

$$MP_u = 5D_{[cm]} , \tag{3.24}$$

when λ is taken as 0.55 μm. To take full advantage of the telescope, MP must be at least MP_u and may be up to twice MP_u for comfortable viewing.

On the other hand, MP should not too greatly exceed MP_u. When MP is just MP_u, the eye fully resolves the disk-like images in the objective focal plane. Making MP larger (by increasing eyepiece power, for example) will not allow greater resolution because the eye already resolves the image in the objective focal plane. It will, however, increase the size of the point images on the retina. The light gathered by the objective from one point is spread over several receptors, rather than one. As a result, edges will not appear sharp, and image contrast, especially of fine detail, will decrease severely.

For this reason, magnifying power that greatly exceeds MP_u is known as *empty magnification* and should be avoided.

3.8.3 Microscope

A good microscope objective is designed to be diffraction limited when it is used at the proper tube length g. The object resolution limit is thus $1.22 \lambda \, l/D$. If we generalize to the case in which the object space has refractive index n, we must use the fact that the wavelength in the medium is equal to λ/n, where λ is the wavelength in air or vacuum. The resolution limit is thus $1.22\lambda \, l/nD$.

Microscopists usually rewrite this expression in terms of the *numerical aperture*, $NA = nD/2l$. $D/2l$ is just the half angle u subtended, in paraxial approximation, by the limiting aperture at the object. In terms of NA, the object resolution limit is

$$RL = 0.61 \lambda/NA . \tag{3.25}$$

(When paraxial approximation is avoided, the sine condition shows N to be

$$NA = n \sin u , \tag{3.26}$$

for a well corrected objective.)

With most objectives, $n = 1$, and $\sin u$ rarely exceeds 0.65 for a 40 $\times$ objective. Because λ is about 0.55 μm, resolution is limited to 0.5 μm or so. Certain *oil-immersion objectives* are designed to accommodate a drop of high-refractive-index oil held by surface tension between the objective and the microscope cover slip. These objectives may have 60 or 100 power and numerical apertures as high as 1.6. They must be used with the proper oil and, to avoid introducing aberration, with cover slips of specified refractive index and thickness.

The microscope, like the telescope, reaches *useful magnifying power* when the eye is just able to resolve all the detail that exists in the image plane of the

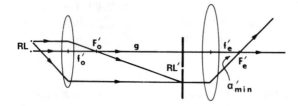

Fig. 3.18. Angular resolution limit of compound microscope

objective. If two points are separated by the resolution limit R, as in Fig. 3.18, then $MP = MP_u$ when $\alpha'_{\min}$ is equal to the resolution limit of the eye. The ray that passes through F'_e in Fig. 3.18 shows that

$$\alpha'_{\min} = RL'/f'_e \tag{3.27}$$

and the ray that passes through F'_o shows that

$$RL'/g = RL/f'_o . \tag{3.28}$$

But $RL = 0.61 \lambda / NA$, so $\alpha'_{\min}$ may be written, after some manipulation, as

$$\alpha'_{\min} = \frac{0.61 \lambda}{NA} \frac{MP}{d_v} , \tag{3.29}$$

where d_v is the shortest distance of distinct vision. Setting $\alpha'_{\min} = 0.3$ mrad and $\lambda = 0.55\ \mu$m, we find that

$$MP_u = 300\,NA . \tag{3.30}$$

As with the telescope, M should not exceed M_u by more than a factor of 2.

To calculate the depth of field of a microscope or any diffraction-limited system, see Problems 3.13 and 5.12.

3.8.4 Condensers

Usually, microscopes are employed with ordinary, white-light sources; an optical system is generally used to project the light from the source into the system. *Critical illumination* refers to imaging a diffuse source onto the plane of the sample. The source must be uniform so that its structure does not appear in the image seen through the microscope; usually tungsten lamps with flat, *ribbon filaments* are used for the purpose.

Köhler illumination is basically the same system as that shown in Fig. 3.9. In a microscope system, however, the source is not located in the plane labeled *L*; rather, there is an additional lens that projects an image of the source into that plane. An iris diaphragm located there allows adjustment of the size of the image that is projected into the aperture stop. As with the slide projector, this image should not underfill the aperture stop.

We shall see in Sects. 5.6 and 5.7 that the angular subtense of a source influences the *coherence* of the beam. Coherence somewhat influences the resolution limit of an optical system. The theory of *partial coherence* shows that the numerical aperture of the condenser should be roughly 1.5 times that of the objective. In that case, the resolution limit of the system is that given by (3.25). It is slightly poorer for other values of the condenser NA. As with the projector, the condenser need not have especially high optical quality.

When a laser is used as the source, the illumination system is not especially important because the light is highly coherent and the coherence cannot be adjusted. The resolution limit increases by about 30%, and the image is often accompanied by unsightly artifacts that result from interference (Sect. 5.7).

Problems

3.1 A certain eye has 5 D of myopia and 2 D of accommodation. Find the near and far points for that eye. What are the near and far points when the myopia is corrected with a -5 D lens? Will the patient require special reading glasses? (In cases like these, corrective lenses may be combined with weaker reading lenses in composite lenses called *bifocals*.)

3.2 (a) *Estimate* the change of position of the principal planes of the eye when spectacle lenses are used. Estimate the resulting change of magnification of the image on the retina. Why is the magnification change small when the patient wears contact lenses?

(b) Assume that a patient can tolerate a 10% difference of image size between his two eyes. What is the maximum acceptable difference of power between the left and right lenses?

(c) *Aphakia* is the lack of the crystalline lens (usually due to surgery to remove *cataracts*, or crystalline lenses that have become opaque). What is the approximate power of a *cataract lens* (a spectacle lens designed to correct aphakia)? What is the difference of magnification between the eyes of an *aphakic* with one normal eye and a cataract lens over the aphakic eye?

3.3 *Close-up Photography*. Two thin lenses in contact have magnification f_2'/f_1' when the object is located precisely at the focal point of lens 1.

(a) Explain why the lenses need not be precisely in contact.

(b) Suppose that the first lens has a focal length 50 mm and the second of 200 mm. Show that the magnification is in error by about 2% if the object is located 0.25 mm from F_1.

3.4 A camera has a 50-mm lens that is diffraction limited at $F/8$. It is used with a high-resolution film that can resolve 200 lines/mm. Find the useful magnifying power for photographing distant objects through a telescope that has an objective lens with diameter D. Explain why the result differs from the value $5 D_{[cm]}$, which applies to visual observations.

3.5 A certain film has a maximum density of about 4; the average density of the negative of a scene is less than 0.5. Estimate what fraction of the undeveloped silver originally in the emulsion remains after the photograph is developed. (The unused silver is eventually dissolved in the fixer and either discarded or salvaged.)

3.6 *Perspective Distortion.* A camera with a 50-mm (focal length) lens photographs a distant scene that has considerable depth. The photograph is enlarged by a factor of 10.

(a) Find the location from which each image in the enlargement will subtend an angle equal to that subtended by the original object. This location is called the *center of perspective.*

(b) Show that images of nearby objects will appear disproportionately large when the observation point is beyond the center of perspective. This phenomenon is known as *apparent perspective distortion* and is often seen in close-up photographs, where the camera was quite close to the subject.

3.7 We wish to locate the focal plane of an $F/11$ lens by focusing on an object whose distance is about 1000 times the focal length f' of the lens. Calculate the error in locating the focal plane. Assume that we can resolve about 100 lines/mm in the image plane. Compare the depth of focus with the focus error. This is the origin of the rule of thumb that 1000 focal lengths is "equal to" infinity.

3.8 A near-sighted scientist removes his spectacles and announces that he has just put on his magnifying glasses. Explain this peculiar remark and determine what magnifying power he has achieved if his spectacles have a power of $-5D$. Assume, if you like, that his range of accommodation is 4D.

3.9 Using his actual near point (not 25 cm) as a reference, find the magnifying power of a magnifying glass used by a *myope* (a patient who has myopia) whose near point d_v is much less than 25 cm. The focal length of the lens must be less than what value before the myope experiences a magnifying power appreciably greater than 1?

3.10 A certain scanning confocal microscope uses a $40 \times$, 0.65-NA objective with a tube length of 160 mm. The diameter of the pinhole that precedes the detector is 11 μm, and the depth of field of the objective lens is about 0.65 μm. (a) Use (2.46) to determine the distance between the primary focal point of the objective and a well-focused object point. (b) Suppose that the object is moved an additional 100 μm farther from the objective lens. At what distance from the secondary focal point is this object point brought into sharp focus? (c) What is the diameter of the resulting circle of confusion in the plane of the pinhole? (d) If this circle of confusion is uniformly illuminated, what fraction of the light passes through the pinhole? Because nearly all the light of a focused point passes through the pinhole, this result gives roughly the attenuation of an object point that is out of focus by only slightly more than the depth of field of the objective lens.

3.11 An object 1 m distant and 1 cm wide is viewed through a telescope that has a 50-mm (focal length) objective and a nominally $10 \times$ eyepiece. (a) What is the angular subtense of the object when it is viewed from the distance of 1 m with the naked eye? (b) What is the apparent angular subtense of the image seen through the eyepiece? Assume that the objective lens is now 1 m from the object. (c) What is the overall magnifying power of the telescope *referred to the object 1 m away*? [Answer: 2, not 1/2.]

3.12 (a) We outfit a microscope with a $40 \times$, 0.65-NA objective and view the end of an optical fiber with light whose wavelength is 550 nm. What power eyepiece should we use? What is the resolution limit? (b) An astronomical telescope has a 1-m (focal length) objective lens whose diameter is 7.5 cm. What power eyepiece should it have? Can it distinguish two stars separated by 10 μrad?

3.13 Use (3.7) to estimate the depth of field of a microscope objective in terms of its numerical aperture NA. Compare with Problem 5.12.

3.14 A camera uses a film that has resolving power $RP = 100$ lines/mm. At what focal length will the angular resolution limit of the camera equal that of the eye?

3.15 We wish to photograph distant objects through a telescope. For this purpose we choose a pair of 10×50 binoculars and a camera with a 50-mm, $F/2$ (maximum aperture) lens.

(a) Show how to take the photographs and sketch the instrumentation, including the correct position of the camera. Find the largest useful aperture diameter D of the camera lens.

(b) Assume that the camera is diffraction limited when set at diameter D. Explain why it is probably best to expose all the pictures at that aperture diameter. What if, as is more likely, the camera is not diffraction limited but is limited by the resolving power of the film?

(c) Suppose that the camera (without the telescope) requires an exposure time of t when it is set at diameter D. Find the exposure time when the camera is fixed to the binoculars. (In fact, the transmittance of the binoculars may be less than 1, so trial exposures will be necessary.)

3.16 A camera has a 35-mm (focal length) lens. It is used with a film that has a resolving power $RP = 30$ lines/mm, or resolution limit $RL = 1/30$ mm. The film, not the lens, limits the capabilities of the system. (a) Two points are separated by a small angle δ. What is the smallest value of δ that the camera can resolve (with this film in it)? Compare with the angular resolution of the eye (0.3 mrad). (b) The camera is set up to look through a telescope whose objective lens has diameter D as in Problem 3.15. What is the largest useful magnifying power of the telescope? Why does it differ from the value $5D_{cm}$ that pertains when the eye looks through the telescope?

3.17 For another application of a projection system, see Problem 11.1.

4. Light Sources and Detectors

In this chapter, we discuss radiometry (and photometry), blackbody sources, and line sources, and introduce several types of detector of optical radiation.

4.1 Radiometry and Photometry

The subject of *photometry* deals with the propagation and measurement of visible radiation. Photometry uses units that are based on the response of the human eye. For example, *luminous power* is identically zero at all wavelengths outside the visible spectrum. The subject of photometry has been greatly complicated by a profusion of units.

The more general subject is *radiometry*, which deals with propagation and measurement of any electromagnetic radiation, whether in the visible spectrum or not. The units of radiometry are watts and joules, for example. *Photometric units* are defined precisely analogously to these physical or *radiometric units*, and, in recent years, an effort has been made to define a consistent set of photometric units and terms. Because it is customary, we shall usually deal with photometric terminology. Nevertheless, the definitions and concepts involved are introduced first in terms of the familiar physical units.

4.1.1 Radiometric Units

To begin, consider a point source that radiates uniformly in all directions (Fig. 4.1). We measure the *radiant power* $d\phi$ emitted into a small solid angle $d\Omega$. (The solid angle $d\Omega$ subtended by a certain area element at the source is dA/r^2,

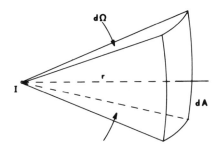

Fig. 4.1. Radiant intensity of a point source

where dA is the projection of the area element onto a sphere with radius r, centered about the source. A full sphere thus subtends a solid angle of 4π steradians.) The *radiant intensity* I of the point source is defined implicitly by

$$d\phi = I d\Omega . \tag{4.1}$$

Since the units of ϕ are watts, I has units of watts per steradian, which we may write $W \cdot sr^{-1}$.

The term, radiant intensity, is reserved for point sources; for a small plane source, we define *radiance L* by the equation

$$d^2\phi = L dS \cos \theta \, d\Omega , \tag{4.2}$$

where dS is the area of a differential element of the source. (We use the notation $d^2\phi$ as a reminder that two differential quantities appear on the right side of the equation.) θ is the angle indicated in Fig. 4.2 between the normal to dS and the line joining dS and dA. We insert a factor of $\cos \theta$ here because the source is inclined to dA and appears to have area $dS \cos \theta$. The quantity $(L dS \cos \theta)$ is analogous to the radiant intensity of a point source.

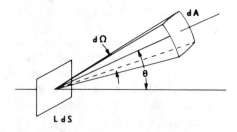

Fig. 4.2. Radiance of a small, plane area

The units of L are watts per square meter steradian, $W \cdot m^{-2} \cdot sr^{-1}$. For convenience, the unit $W \cdot cm^{-2} \cdot sr^{-1}$ is often used, even though the centimeter is not one of the mks or International System of units (SI).

Next, assume that a small element of area dA is irradiated by some power $d\phi$. ($d\phi$ is infinitesimal because dA is infinitesimal.) We define the *irradiance E* of the surface by the relation

$$d\phi = E dA , \tag{4.3}$$

independent of the inclination θ of the radiation to the surface (Fig. 4.3). E is the total power falling on a surface divided by the area of the surface, irrespective of the orientation of the surface. The units of E are watts per square meter, $W \cdot m^{-2}$. Again, $W \cdot cm^{-2}$ is often the preferred practical unit.

Several other quantities are less important and are derivatives of the quantities ϕ, L, E, and I. These include the total energy Q in joules or watt seconds and the energy density w in watt seconds per square meter, $W \cdot s \cdot m^{-2}$. Finally, *radiant emittance M* is used to describe the total power per unit area radiated

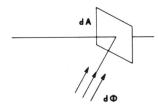

Fig. 4.3. Irradiance of a surface

from a surface. The units of M are watts per square meter; M is obtained by integrating radiance L over all possible angles.

When either of the quantities L or I depends on angle, we indicate this with a subscript θ. In addition, if we are interested in a single wavelength interval $\varDelta\lambda$ centered at wavelength λ, we use the subscript λ. Thus, for example, L_λ is known as *spectral radiance*, and its units are the units of radiance per unit wavelength interval, $W \cdot m^{-2} \cdot sr^{-1} \cdot nm^{-1}$. Similarly, we adopt the subscript v when we are working in a given frequency interval $\varDelta v$. The units of L_v are radiance per unit frequency interval, $W \cdot m^{-2} \cdot sr^{-1} \cdot Hz^{-1}$.

4.1.2 Photometric Units

Photometric units are defined in terms of the standard eye, whose response curve relates them to the physical or radiometric units. Photometric units are important, say, in illumination engineering, where the visual sensation of "brightness" is important.

The fundamental photometric unit is the *lumen* (lm). Just as radiant power is measured in watts, *luminous power* or *luminous flux* is measured in lumens. To circumvent the profusion of nomenclature that existed in the all-too-recent past, we use the same symbol ϕ for luminous power as for radiant power. When it is necessary to distinguish between luminous and radiant power, the subscripts v (for visual) and e (for energy) are used. Thus, ϕ_v is luminous power and ϕ_e, radiant power.

ϕ_e and ϕ_v are related to one another by the spectral response V_λ of the bright-adapted eye, shown in Fig. 3.2. V_λ is the *spectral sensitivity* or *luminous efficiency* of the eye and is normalized to 1 at its maximum value. This maximum occurs in green light at about 555 nm. In a small wavelength interval at this wavelength only

$$1 \text{ watt} = 673 \text{ lumens} , \tag{4.4}$$

by definition. At all other wavelengths, the visual sensation produced by 1 W is equal to 673 lumens multiplied by V_λ; V_λ is always less than 1. (The term, *luminous efficacy*, is reserved for the ratio of luminous power to radiant power at a given wavelength. The luminous efficacy is thus equal to $(673 \text{ lm} \cdot W^{-1}) \cdot V_\lambda$.)

The quantities derivable from luminous power ϕ_v are precisely analogous to the radiometric quantities. The *luminous intensity* I_v of a point source is defined exactly as radiant intensity; the units of I_v are therefore $lm \cdot sr^{-1}$. Similarly the

Table 4.1. Important radiometric and photometric quantities and their units

Symbol (SI Units)	Photometric Unit	Radiometric Unit	Definition
ϕ	luminous power (flux), lm	radiant power, W	
I	luminous intensity, $\mathrm{lm \cdot sr^{-1}}$ (cd)	radiant intensity, $\mathrm{W \cdot sr^{-1}}$	power radiated by point source into unit solid angle
L	luminance $\mathrm{lm \cdot m^{-2} \cdot sr^{-1}}$ ($\mathrm{cd \cdot m^{-2}}$)	radiance $\mathrm{W \cdot m^{-2} \cdot sr^{-1}}$	power radiated from unit area into unit solid angle
E	illuminance, $\mathrm{lm \cdot m^{-2}}$ (lux)	irradiance $\mathrm{W \cdot m^{-2}}$	total power falling on a unit area
Q	luminous energy $\mathrm{lm \cdot s}$	radiant energy J ($\mathrm{W \cdot s}$)	
M		radiant emittance (radiant exitance), $\mathrm{W \cdot m^{-2}}$	total power radiated (in all directions) from a unit area

luminance L_v of a point source (formerly called *brightness*) is analogous with radiance L_e, and its units are $\mathrm{lm \cdot m^{-2} \cdot sr^{-1}}$. The luminous power per unit area incident on a surface is called *illuminance* H_v (formerly *illumination*) and has units of $\mathrm{lm \cdot m^{-2}}$. Finally, the lumen per steradian is known as the *candela* and abbreviated cd; the lumen per square meter is called the *lux*.

Table 4.1 relates the important radiometric and photometric quantities and their units. We can compute values in, say, photometric units from those in radiometric units by using the response V_λ of the eye. For example, if we know the irradiance $E_{\lambda e}$ at a given wavelength λ, then the corresponding value of illuminance $E_{\lambda v}$ is just

$$E_{\lambda v} = (673\ \mathrm{lm \cdot W^{-1}}) V_\lambda E_{\lambda e} .\tag{4.5}$$

If the illuminating source is not monochromatic, then $E_{\lambda e}$ is interpreted as the irradiance produced by the source in a narrow band centered at λ. In thiat case, the total illuminance E_v is found by integration to be

$$E_v = (673\ \mathrm{lm \cdot W^{-1}}) \int V_\lambda E_{\lambda e}\, d\lambda ,\tag{4.6}$$

where the limits of integration can generally be taken to be 400 to 700 nm.

4.1.3 Point Source

The power radiated by a point source with intensity I into solid angle $d\Omega$ is

$$d\phi = I d\Omega .\tag{4.7}$$

The total power radiated by a uniform point source ($I_\theta = I$) is thus

$$\phi = 4\pi I \ . \tag{4.8}$$

Suppose, as in Fig. 4.1, that we irradiate (or illuminate) an area dA with this point source The area is located a distance r from the source and subtends a solid angle $(dA \cos\theta)/r^2$, where we let the normal to dA be inclined at angle θ to the line that joins dA and the source. The power that falls on dA is thus

$$d\phi = I\frac{dA}{r^2} \cos\theta \ . \tag{4.9}$$

We may find the irradiance of the surface from the definition

$$d\phi = EdA \ , \tag{4.10}$$

where $d\phi$ is the total power falling on the area element. Equating these last two expressions, we find that

$$E = \frac{I}{r^2} \cos\theta \tag{4.11}$$

is the irradiance brought about by a point source. This is well known as the *inverse-square law*. We consider a source to be a point source only when it is so small or distant that the inverse-square law applies.

4.1.4 Extended Source

We generally idealize this case and assume the radiance of a source to be independent of angle; that is,

$$L_\theta = L \ . \tag{4.12}$$

Such sources are known in optics as *Lambertian sources* and appear equally bright at all angles.

We will shortly need an expression for the total power radiated by a small Lambertian source. We may find this by using the construction of Fig. 4.4. The

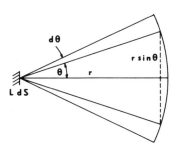

Fig. 4.4. Cross-section of a plane source radiating into a thin conical shell

power radiated into a solid angle $d\Omega$ is given by (4.2). Assuming the source to be the surface of an opaque body, we seek the power radiated into a hemisphere (not a sphere). We look first at the power radiated into the thin conical shell whose cross section is indicated in Fig. 4.4. We find the solid angle corresponding to that shell by constructing a hemisphere with radius r. The shell intersects the sphere and defines an annulus on the surface of the sphere. The width of the annulus is $r\,d\theta$, and its radius is $r\sin\theta$. Its area dA is accordingly

$$dA = 2\pi(r\sin\theta)r\,d\theta ,\qquad(4.13)$$

and the solid angle it subtends, therefore,

$$d\Omega = 2\pi\sin\theta\,d\theta .\qquad(4.14)$$

The power radiated into $d\Omega$ is therefore

$$d^2\phi = (L\,dS)\,2\pi\sin\theta\cos\theta\,d\theta .\qquad(4.15)$$

We may determine the total power $d\phi$ radiated into a cone by integrating this expression from 0 to the half-angle θ_0 of the cone,

$$d\phi = 2\pi L\,dS \int_0^{\theta_0} \sin\theta\cos\theta\,d\theta .\qquad(4.16)$$

The result is

$$d\phi = \pi L\,dS\,\sin^2\theta_0 .\qquad(4.17)$$

The total power radiated into the hemisphere is

$$d\phi = \pi L\,dS ,\qquad(4.18)$$

because $\theta_0 = 90°$ for this case.

4.1.5 Diffuse Reflector

Suppose irradiance E falls on a small area dS. Then the total power falling on the surface is

$$d\phi_i = E\,dS .\qquad(4.19)$$

Suppose that the surface scatters a fraction k of this power in such a way that its radiance is not a function of angle. Then,

$$d\phi_s = k\,d\phi_i ,\qquad(4.20)$$

where $d\phi_s$ is the total scattered power. Such a surface is called a *Lambertian reflector*. It is immaterial whether the surface radiates power or scatters it. The important fact is that power leaves the surface in all directions. We may therefore apply the result of the last section and write that

$$d\phi_s = \pi L_s \, dS \,, \tag{4.21}$$

where L_s is the apparent radiance of the surface.

Combining the last three equations, we find that

$$L_s = kE/\pi \,. \tag{4.22}$$

L_s is the radiance (or luminance) of a Lambertian reflector with irradiance E striking it.

Only a few diffusing surfaces make good Lambertian reflectors in the visible portion of the spectrum. Such surfaces look nearly white if k is about 1 and appear equally bright at any viewing angle. Pressed blocks of MgO, $BaSO_4$, or Teflon powder and certain flat paints make the best diffusers. Milk, snow, and heavy, white unglazed paper are also excellent diffusers.

Certain diffusers may scatter light into a full sphere, like *ground glass*. A few, such as milky-white *opal glass*, approximate Lambertian reflectors; their apparent radiance is

$$L_s' = kE/2\pi \,. \tag{4.23}$$

Ground glass itself is a rather poor diffuser and scatters most of the light falling on it into a narrow cone centered about the direction of the incident light. Sometimes this is desirable, because then, in the forward direction, the radiance of the surface greatly exceeds L_s'.

4.1.6 Integrating Sphere

An *integrating sphere* is a hollow sphere whose inside surface is coated with an approximately Lambertian reflector. It may be used to measure the total power of a beam whose irradiance distribution is arbitrary or of a source whose radiance distribution is arbitrary.

Suppose there is a source located anywhere inside the sphere. The source may be self luminous, or it may be a scatterer illuminated through a hole in the sphere. A detector (or a piece of opal glass with a detector behind it) is located elsewhere on the inner surface of the sphere. A small stop prevents propagation directly from the source to the detector; the detector is illuminated only by diffuse reflections from the inside of the sphere.

We call the area of the detector dA and calculate the radiant power falling on dA. Let dS be an arbitrary element of area on the inside surface of the sphere. Light shines directly from the source onto dS. The apparent radiance of dS is L, so the radiant power falling onto dA is

$$d^2\phi_1 = L \cos \alpha \, dS \, d\Omega \,, \tag{4.24}$$

where α is the angle shown in Fig. 4.5 and $d\Omega$ is the solid angle subtended by dA

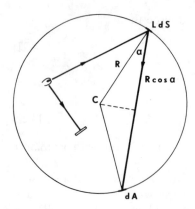

Fig. 4.5. Integrating sphere

at dS. The distance between dA and dS is $2R \cos \alpha$, so the equation reduces to

$$d^2\phi_1 = L \, dS \, dA/4R^2 . \tag{4.25}$$

Several cosines have canceled as a result of the spherical geometry; the equation is independent of α and therefore independent of the position of dS. If we use (4.22) and integrate with respect to S, we find that

$$d\phi_1 = (k \, dA/4\pi R^2) \int E \, dS , \tag{4.26}$$

where E is the irradiance from the source.

The integral of E over the entire sphere is just the total radiant power ϕ_s emitted by the source. Therefore, $d\phi_1$ may be rewritten in terms of ϕ_s as

$$d\phi_1 = k \, \phi_s \, dA/(4\pi R^2) . \tag{4.27}$$

Some of the power scattered by the inner surface of the sphere is scattered a second, third, and more times before falling onto the detector. Precisely similar reasoning reveals that $d\phi_2 = k \, d\phi_1$, $d\phi_3 = k \, d\phi_2$, and so on, where $d\phi_i$ is the power scattered onto the detector after i reflections off the inside of the sphere. Therefore, the total power that falls onto the detector is

$$d\phi = (\phi_s \, dA/4\pi R^2)(k + k^2 + k^3 + \cdots) , \tag{4.28}$$

or, when the infinite series is summed,

$$d\phi = \frac{k}{1-k} \frac{\phi_s}{4\pi R^2} dA . \tag{4.29}$$

The detector and the openings in the sphere must occupy only a small fraction of the sphere's area, or else the analysis breaks down. Likewise, the source must be small compared to the area of the sphere, so that the stop that separates the source from the detector will not obscure too large a solid angle. Even if k is as large as 98%, the efficiency $d\phi/\phi_s$ of the sphere is apt to be only a few percent. Still, because it allows measurement of arbitrary irradiance distributions, the integrating sphere has proven to be a useful instrument for evaluating diffuse sources and scatterers. With laser sources, even the low efficiency of the integrating sphere is not a great drawback.

4.1.7 Image Illuminance

Consider a system, such as a camera, in which a lens projects a real image of a bright, extended object. The image distance is l', and the image is off axis by θ (Fig. 4.6). dS is a small area of the lens, and dS' a small portion of the image plane.

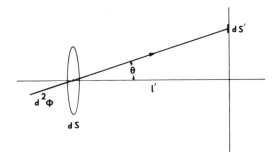

Fig. 4.6. Image illuminance (cosine-fourth law)

An observer stationed at dS' sees light originating at the lens, giving it an apparent luminance L given by

$$d^2\phi = L \, dS \cos\theta \, d\Omega \,, \tag{4.30}$$

where $d^2\phi$ is the total luminous power passing through the lens (dS) and directed toward dS'. The solid angle $d\Omega$ subtended by dS' is

$$d\Omega = \frac{dS' \cos\theta}{(l'/\cos\theta)^2} \,, \tag{4.31}$$

which yields

$$d^2\phi = L \, dS \, dS' \frac{\cos^4\theta}{l'^2} \,. \tag{4.32}$$

If the aperture of the lens is relatively small compared with l', we may assume θ to be approximately constant over the aperture. Integrating over the aperture (really, the exit pupil) thus yields

$$d\phi = L \, dS' \frac{\cos^4 \theta}{l'^2} \frac{\pi D^2}{4} , \tag{4.33}$$

where D is the diameter of the aperture.

The illuminance of dS' follows from the definition (4.3), and we write it as

$$E = \frac{\pi L}{4} \frac{\cos^4 \theta}{(l'/D)^2} . \tag{4.34}$$

The illuminance does not depend on object distance or other geometric parameters, but only on the source luminance. The angular dependence is severe, in that illuminance decreases as $\cos^4 \theta$. This is known as the *cosine-fourth law*. l'/D is the *effective F number*, which has been discussed in connection with the basic camera (Sect. 3.2).

Nearly all optical systems suffer from the cosine-fourth law. One way of circumventing this law in wide-angle optics is to employ a curved image "plane". The image surface lies on a circular cylinder whose radius is about equal to the image distance. The illuminance then follows a cosine law. By way of example, suppose a camera requires a 90° field, or a 45° half-field. Because $\cos^4 45° = 1/4$, a flat film plane loses two F stops between the center and the edge of the format. With a curved film surface, the loss is only 70% or about half of one F stop.

In the optical system of the eye, the retinal illuminance determines the sensation of brightness. Because l' is fixed, we find that the illuminance E_r at the fovea depends on only the object luminance and the diameter D of the pupil,

$$E_r \propto LD^2 . \tag{4.34a}$$

When the object is a point source, the image illuminance has a different dependence on F number and diameter. In Fig. 4.7, we show a point source with luminous intensity I illuminating an element dS of a lens. We treat only the case where the source lies on the axis of the lens. The total power falling on dS is

$$d\phi = I \, dS/l^2 . \tag{4.35}$$

As before, we assume that the lens is small enough compared with l' that we can

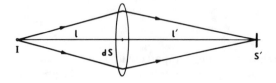

Fig. 4.7. Image illuminance resulting from a point object

ignore angular dependences. Thus, the power collected by the lens is

$$\phi = I\,\pi D^2/4l^2 \;,\tag{4.36}$$

where D is the lens diameter.

Diffraction theory shows that the image of a point is a small disk whose radius is about $\lambda l'/D$, where λ is the wavelength of the light (Sect. 5.7). Thus, the light collected by a perfect lens is focused into an area S' approximately

$$S' = \pi\left(\frac{\lambda l'}{D}\right)^2 .\tag{4.37}$$

The average illuminance E in that area is just ϕ/S', or

$$E = \frac{I}{4l^2}\frac{D^2}{(\lambda l'/D)^2} \;.\tag{4.38}$$

There are two important points. First, the illuminance does, in this case, depend on the distance of the object. Second, the illuminance depends not only on the F number but has an additional factor of D^2 that was not present in the case of an extended object. This is because the size of the image of a point decreases as the aperture of a perfect lens increases, whereas the size of the image of an extended source remains constant. This is one reason why astronomical telescopes are so large (even though their resolution is limited by atmospheric effects). As the aperture is increased, for given focal length, the illuminance of star images increases ideally as D^4, whereas the illuminance of the diffuse background of the sky increases only as D^2. Thus, the stars can be made to stand out against the sky and very dim stars can be photographed.

4.1.8 Image Luminance

Previously we were concerned with the illuminance of an image projected onto a screen or film. Here we seek the luminance of the *aerial image* that is viewed directly and not projected on a screen.

Consider an element dS of a diffuse source with luminance L (Fig. 4.8). The power radiated into an element dA of a lens is

$$d^2\phi = L\,dS\,dA/l^2 \;.\tag{4.39}$$

For convenience, we take $\cos\theta = 1$.

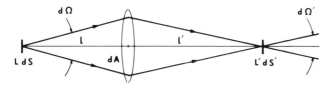

Fig. 4.8. Luminance of aerial image

All of the power collected by dA is directed through dS' (the image of dS). The apparent luminance L' is given implicitly by

$$d^2\phi' = L'dS'\,d\Omega' \,, \tag{4.40}$$

where $d\Omega'$ is the solid angle indicated to the right of dS'. Because $d\Omega'$ is the same as the solid angle subtended by the lens,

$$d\Omega' = dA/l'^2 \,. \tag{4.41}$$

(The "opposite angles are equal" theorem of plane geometry applies as well to solid angles.) If we assume that the lens does not attenuate the beam, then $d^2\phi$ and $d^2\phi'$ are identical; thus

$$L\,dS\frac{dA}{l^2} = L'\,dA\frac{dS'}{l'^2} \,. \tag{4.42}$$

The lens-area element dA is common to both sides and cancels. The image luminance L' is therefore

$$L' = L\frac{dS}{dS'}\left(\frac{l'}{l}\right)^2 \,. \tag{4.43}$$

To relate dS and dS', we use the Lagrange invariant hnu. For generality, suppose that the lens has media with index n on its left and n' on its right. The half-angle subtended by the lens at dS is

$$u = D/2l \,, \tag{4.44a}$$

and, similarly,

$$u' = D/2l' \,. \tag{4.44b}$$

The square of the Lagrange invariant shows that

$$h^2n^2\left(\frac{D}{2l}\right)^2 = h'^2n'^2\left(\frac{D}{2l'}\right)^2 \,, \tag{4.45}$$

from which the factors of $D/2$ cancel. If, now, we associate h and h' with the dimensions of the area elements dS and dS', we find that

$$\frac{dS}{dS'}\left(\frac{l'}{l}\right)^2 = \left(\frac{n'}{n}\right)^2 \,, \tag{4.46}$$

whether we take circular or square area elements.

Thus, the apparent luminance of the image is

$$L' = (n'/n)^2 L \,. \tag{4.47}$$

For a lens in air, $n' = n$, and

$$L' = L .\tag{4.48}$$

This important result can be stated in a general way. The apparent luminance of an extended object is unchanged by any optical system, except for attenuation by the components. (The theorem is, of course, not true for systems that have diffusing elements.) The most important consequence of this fact is that it is impossible to make an aerial image appear brighter by changing to a higher-aperture lens, for example by shortening the focal length. Although the lens condenses the light into a smaller image, it also increases the solid angle of the rays that emerge from the image. The two effects cancel to yield the same luminance in all cases. This result is known as the *radiance theorem* or the *conservation of radiance*.

4.2 Light Sources

Here we discuss specifically classical or *thermal sources* and postpone laser sources to a chapter of their own (Chap. 8). Thermal sources are so named because they radiate electromagnetic power in direct relation to their temperature. For our purposes, we divide them into two classes: *blackbody radiators* and *line sources*. The former are opaque bodies or hot, dense gases that radiate at virtually all wavelengths; line sources, as their name implies, radiate on the whole at discrete wavelengths only.

4.2.1 Blackbodies

The basic physics of blackbodies is taken up in most modern physics texts; here we discuss only the results that are important to applied optics. We begin by noting that hot, opaque objects, dense gases and other materials behave as blackbodies and radiate power according to *Stefan's law*,

$$M = \varepsilon \sigma T^4 .\tag{4.49}$$

M is the radiant emittance (total power radiated per unit area) of the surface of an opaque object whose temperature in thermal equilibrium is T. σ is a universal constant, the Stefan-Boltzmann constant,

$$\sigma = 5.67 \times 10^{-8} \ \mathrm{W \cdot m^{-2} \cdot K^{-4}} .\tag{4.50}$$

(In practical units, $\sigma = .5.67 \times 10^{-12} \ \mathrm{W \cdot cm^{-2} \cdot K^{-4}}$.) ε is known as the *emissivity* of the surface and varies between 0 and 1. For a true blackbody, ε is precisely 1. The emissivity of most real materials is less than 1 and varies slightly with temperature and wavelength. Many blackbodies are good Lambertian sources; their radiance is therefore $\varepsilon \sigma T^4/\pi$.

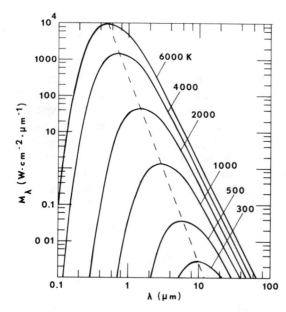

Fig. 4.9 Spectral radiant emittance of blackbodies that have various temperatures

Figure 4.9 shows the spectra of blackbodies at various temperatures. $M_\lambda(T)$ is the radiant emittance per unit wavelength interval at temperature T. Several things are apparent from the figure. A hot blackbody radiates more energy at every wavelength than a cooler one. In addition, most of the radiation is in the infrared portion of the spectrum for blackbodies at temperatures less than a few thousand kelvins. (An object can be heated to a dull-red incandescence at about 500 K or so; what the eye sees is the short-wavelength tail of the blackbody radiation.) Comparatively cool bodies radiate entirely in the infrared. For example, bodies at typical ambient temperatures, say 300 K, have maximum radiant emittance in the neighborhood of 10 μm. The atmosphere is largely transparent in this region of the spectrum; as a result, the 8 to 14 μm wavelength region is important in certain applications, such as remote sensing of crop growth, water pollution, and so on.

The shift of the blackbody spectrum toward shorter wavelengths at higher temperature is described by *Wien's displacement law*,

$$\lambda_m T = 2898 \ \mu\text{m} \cdot \text{K} , \tag{4.51}$$

where λ_m is the wavelength at which the blackbody has its maximum emittance. For rule-of-thumb calculations, it is convenient to remember that the sun as seen through the atmosphere is approximately a blackbody with λ_m about 480 nm (slightly less than the wavelength of maximum sensitivity of the eye) and a surface temperature of 6000 K.

The curves in Fig. 4.9 are accurately described by *Planck's law*, which is usually written in terms of frequency $\nu \ (= c/\lambda)$ and states that the spectral radiant emittance of a true blackbody is

$$M_v(T) = \frac{2\pi v^4}{c^3} \frac{hv}{e^{hv/kT} - 1}.$$ (4.52)

The quantity h is *Planck's constant,*

$$h = 6.624 \times 10^{-34} \, J \cdot s .$$ (4.53)

To derive his law, Planck had to assume that matter radiated discrete quantities of electromagnetic radiation called *quanta.* The energy of each quantum is hv. The success of this assumption provided the foundation for the modern quantum theory of atomic structure.

Figure 4.9 is a log-log plot of Planck's law; in this representation, all of the curves have the same shape. The dashed line connects the peaks of each curve. To find $W_\lambda(T)$ for any temperature, it is necessary only to find λ_m from Wien's displacement law. We then locate the position of the maximum value of $W_\lambda(T)$ along the dashed line. The remainder of the curve may be plotted by interpolating between two existing curves or otherwise making use of the fact that the curves are all identical.

Planck's law was derived under the assumption of thermal equilibrium. Conservation-of-energy arguments show that when a blackbody is in thermal equilibrium with its surroundings, the *absorptance* α, the fraction of incident power absorbed, must be equal to ε; that is,

$$\varepsilon = \alpha .$$ (4.54)

For a true blackbody, $\varepsilon = 1$. Therefore, since $\alpha = 1$, the blackbody theoretically absorbs all the radiation falling on it. No radiation is reflected, transmitted, or scattered; this is the origin of the name, blackbody.

The condition $\alpha = 1$ also gives us a clue as to how to construct a good blackbody radiator. Such blackbody emitters are needed to calibrate detectors or standard lamps. The best blackbody is a hole in a cavity, such as those shown in Fig. 4.10. If the interior is smooth and highly reflecting, any radiation that enters the opening will be reflected many times before finding its way back out. At each reflection there is some absorption; whatever radiation finally does

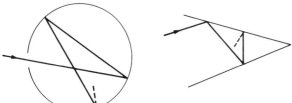

Fig. 4.10. Practical blackbodies

make its way out is only a small fraction of the incident radiation. (The cavity walls are not rough because rough surfaces always scatter radiation, so some of the incident radiation would be scattered directly out of the cavity.) The opening to the cavity thus closely approximates an ideal blackbody.

Tungsten lamps and other sources of continuous spectra are sometimes called *graybodies*, provided that their emissivities are nearly independent of wavelength. At 3000 K, tungsten has an average emissivity of about 0.5.

Glass and other materials that we normally consider transparent may be nearly opaque in the infrared. Such materials may transmit or reflect radiation, as well as absorb and emit. For example, the absorptance of glass for the thermal radiation from a 300-K blackbody is 0.88. If energy is to be conserved, about 12% of the radiation falling on the glass must be partially transmitted, reflected, or both. Such considerations lead us to conclude that

$$\alpha + R + T = 1 \tag{4.55}$$

for a real body for which $\alpha \neq 1$. R and T are the total reflectance and transmittance, including any radiation that may be scattered diffusely.

For an opaque body, $T = 0$. Therefore,

$$\alpha = 1 - R . \tag{4.56}$$

Highly reflecting surfaces are poor blackbodies, whether they are matte or shiny.

If ε varies with wavelength, the preceding three equations hold true at each wavelength; that is, $\varepsilon_\lambda = \alpha_\lambda$ and so on.

Besides tungsten, practical sources of infrared or blackbody radiation include the commercially available Globar and Nernst glower. The former can be operated at about 1500 K and has a fairly uniform emissivity as a function of temperature. The Nernst glower may be run as hot as 2000 K.

The carbon arc, an electric discharge in air between two carbon electrodes, reaches a temperature of the order of 6000 K. High-pressure gas arcs, such as xenon arcs in quartz envelopes, exhibit spectral radiances in the visible and ultraviolet equivalent to that from blackbodies with temperatures in excess of 6500 K. Other arc lamps include high-pressure mercury and sodium lamps, which are among the most efficient in generating visible light. The sun has already been mentioned as a blackbody with a temperature about 6000 K. Outside the earth's atmosphere, the irradiance due to the sun is about 735 mW·cm^{-2}; the atmosphere attenuates the radiation by at least 75%, depending on the solar elevation, and the spectrum only faintly resembles a 6000-K blackbody.

At low temperatures, say 250–350 K, many nonmetals behave as blackbodies with emissivities as high as 0.8 or more. Most clean, metallic surfaces have comparatively low emissivities in this temperature range. The outdoor terrain has an emissivity that averages about 0.35; the emissivity of snow is 0.95. The daytime sky is roughly equivalent to a blackbody at ambient temperature with an emissivity approaching 1 at the horizon and falling to a low value at the zenith. The average emissivity of the daytime sky is about 0.7, less in high-altitude or low-humidity environments, where there is less absorption resulting from the presence of water vapor and carbon dioxide in the air. Clouds seen from below are approximately blackbodies 1 kelvin or so below ambient temperature. The nighttime sky is often assumed to be a blackbody with an effective temperature of 190 K.

4.2.2 Color Temperature and Brightness Temperature

The *color temperature* of a graybody is the actual temperature of a blackbody that has the same color as the graybody. Color temperature therefore provides an estimate of the true temperature of a graybody. The estimate is low, except, possibly, in the case that the emissivity of the graybody is significantly greater at short wavelengths than at long wavelengths. If the graybody is truly gray (that is, if its emissivity is independent of wavelength), then color temperature is the same as true temperature.

At temperatures of 800–1000 K, a graybody is dull red. At about 1200 K, it is bright red, turning to yellowish red at 1400 K and to the near white of an incandescent bulb by 2000 K. Between 3000 and 5000 K, a graybody is intense white; it turns to pale blue between 8000 and 10 000 K.

The *brightness temperature* of a graybody is the true temperature of a blackbody that has the same radiance at a given wavelength, usually about 650 nm. Brightness temperature is measured by making a visual comparison between the brightness of a tungsten filament and the unknown graybody; the instrument used to make such comparisons is known as a *radiation pyrometer*. Because the graybody always emits less radiation at a given wavelength than a blackbody at the same temperature, brightness temperature is always less than true temperature.

4.2.3 Line Sources

Incandescent gases in which there is little interaction between excited atoms, ions, or molecules are good examples of sources of *line radiation*. Neon lights and low-pressure sodium and mercury lamps are such sources (Table 4.2).

To understand these sources, it is necessary to have some conception of the quantum theory of atomic structure. We begin with a description of the *Bohr picture* of the atom; details of this theory and of the more correct quantum theory may be found in any modern physics text.

In the Bohr picture (which strictly speaking applies only to hydrogen), the atom consists of a positively charged nucleus surrounded by orbiting negative electrons called *bound electrons*. Our primary concern is the outermost electron. This electron may circle the nucleus in any one of a number of discrete orbits; orbits other than these are not allowed. The total energy of the electron in any orbit is the sum of its electrostatic potential energy and its kinetic energy; the

Table 4.2. Important spectral lines[a]

Wavelength [nm]	Element[a]	Wavelength [nm]	Element[a]
768.2	K	471.3	He
670.8	Li	486.1	H (F)
667.8	He	467.8	Cd
656.3	H (C)	447.1	He
643.8	Cd	443.8	He
589.6, 589.0	Na (D)	438.9	He
587.6	He	435.8	Hg
579.1	Hg	434.0	H
577.0	Hg	430.8	Fe (G)
546.1	Hg	410.2	H
527.0	Fe (E)	407.8	Hg
508.6	Cd	404.7	Hg
504.8	He	396.8	Ca (H)
501.6	He	393.4	Ca (K)
492.2	He	365.0	Hg
491.6	Hg	253.7	Hg
480.0	Cd		

[a] Letters in parentheses are the designations given to the lines by Fraunhofer. The most important laser wavelengths are given in Table 7.1.

total energy is least when the electron is in its smallest orbit. The energy of an atom is the sum of the energies of its electrons.

The outermost electron may be shifted from its smallest orbit to another by addition of the proper amount of energy. When this happens, we say that the atom has been raised from its *ground energy level* or *ground state* to a *higher energy level* or *excited state*. Similarly, by absorbing or emitting precisely the right amount of energy, the atom can jump from one energy level to any other. Such jumps are called *transitions*. Because the outermost electron is confined to a discrete number of orbits, the atom has only a discrete set of energy levels and therefore a discrete number of transitions. Further, some transitions are comparatively unlikely and are called *forbidden* transitions.

If the electron absorbs enough energy, it may be set completely *free* of the atom. The atom acquires a positive charge and is known as an *ion*. When one or more of the remaining electrons is raised from its ground state, the ion exhibits energy levels different from those of the original atom.

In a gas-discharge lamp, an electric current is passed through a partly ionized gas. Occasionally, a free electron strikes an atom and raises it to a higher energy level. The atom soon falls to a lower energy level, frequently by the mechanism of emitting enough electromagnetic radiation to ensure that energy is conserved. This is called *spontaneous emission*.

The frequency of the radiation is determined by the difference in energy between the two levels. If the energy difference is ΔE, then the frequency v of the emitted radiation is such that

$$hv = \varDelta E .\tag{4.57}$$

As long as its atoms are in thermal equilibrium with its surroundings, an intense line radiator can never exceed the spectral emittance $M_\lambda(T)$ of a black-body at the same temperature as the line source. That is, the emissivity of a line source cannot exceed 1. This rule is broken only by laser sources, the atoms of which are not in thermal equilibrium with their surroundings.

The spectral lines emitted by an isolated, stationary atom are extremely sharp; for most practical purposes the atom can be said to emit *monochromatic radiation* (that is, one frequency or wavelength only).

In a real gas, the atoms are not at all stationary, but move with comparatively high velocities, depending on the temperature. Because some of the atoms are moving toward the observer, some away from him, and others in other directions, the freqency of the radiation is shifted by different amounts by the Doppler effect. As a result, the very nearly monochromatic radiation of the isolated atom is almost never observed.

When a gas is dense, the atoms' collide with electrons and with one another more frequently than when the gas is more rarified. Further, in a high-temperature or a high-current discharge lamp, atoms are continually influenced by the electric fields of electrons and ions as they pass by. The effect of both the collisions and the electric fields is to broaden the spectrum still further. Therefore, high-pressure gas-discharge lamps emit spectra that are more like *bands* than like lines.

In addition, a *continuum* of radiation results from the presence of many free electrons in hot, dense gases. This continuum comes about, in part, from *recombination radiation*, or the capture of free electrons by ions. Recombination radiation is continuous because the free electron (as opposed to the bound electron) is not restricted to discrete values of energy, but rather may acquire any value. Figure 4.11 illustrates why this results in a continuum of radiation:

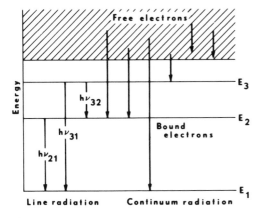

Fig. 4.11 Energy levels of a typical line radiator

the tails of the arrows are not restricted, thus allowing a wide range of energy to be emitted.

Another factor contributes to continuum radiation. In classical electrodynamics, accelerated charges are known to radiate. As the free electron (which is in constant motion) passes ions or other free electrons, it is accelerated by their electric fields. Even if the free electron is not captured, the radiation resulting from this acceleration will add to the continuum.

When the gas is hot enough and dense enough, the continuum becomes intense, and the spectral lines become so broad that they begin to overlap. When, this happens, the spectrum becomes more like that of a blackbody radiator than like that of a line source. The hotter and denser the gas, the more closely it approximates a blackbody.

A rather similar situation prevails in certain solids or liquids, though at lower temperatures. Frequently, a solid may contain certain atoms or ions with energy-level structures like that of Fig. 4.11. However, because the density of a solid or a liquid is very much greater than that of a gas, the energy levels are virtually always broadened considerably. As a result, the spectra of solids, even at room temperature, consist of bands tens of nanometers or more wide.

4.2.4 Light-Emitting Diodes (LEDs)

The *light-emitting diode* or LED is a relatively new type of light source and has seen considerable application in alphanumeric and other displays. Because of its small size and low power requirements, it also has great value in optical communications and computers.

Basically, the LED consists of a *junction* between heavily doped p- and n-type semiconductors, such as gallium arsenide. An n-type semiconductor has many, highly mobile electrons, whereas a p-type material has less mobile, positive *holes*. When two such materials are joined, the mobile charges reorient themselves until the energy-level structure of Fig. 4.12 results. The structure is characterized by two *bands*, one above and one below the *forbidden gap* labeled E_g. Neither electrons nor holes are allowed to have energies that fall within the gap (see also Sect. 4.3.1).

We fix our attention on the more mobile electrons. At low temperatures, virtually all of the electrons have energies below the level (the Fermi level) E_f

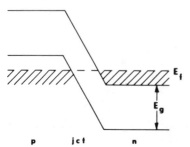

Fig. 4.12. Energy-level structure of the light-emitting diode

indicated by the dashed line. The shaded regions below E_f indicate the presence of electrons. Electrons above the forbidden gap are mobile and are known as *conduction electrons*. The unshaded region in the p-type material just above E_f indicates the absence of electrons or the presence of holes.

The conduction electrons cannot occupy the holes in the p-type material, because they are physically separated by the width of the junction. If an external voltage V is applied, the conduction electrons can be attracted toward the p-type materal. If the p-type material is made positive and if the voltage drop across the junction is large enough, the conduction electrons will be *injected* into or across the junction. Examination of Fig. 4.12 shows that the electron needs to acquire, approximately, energy E_g to cross the potential barrier into the p-type material. Therefore,

$$eV \gtrsim E_g \tag{4.58}$$

is required. The electron is now free to undergo a transition across the forbidden gap and to combine with a hole. In suitable semiconductor materials, this gives rise to the emission of light.

LEDs are commonly available in the near-ir and red regions of the spectrum; they have been made with wavelengths as short as that of green light.

4.3 Detectors

Detectors of light (including uv and ir radiation) fall into two classes, *thermal* and *quantum* detectors. Quantum detectors depend upon the absorption of a quantum of radiant energy; in essence, they measure the rate at which quanta interact with the detector. In a thermal detector, on the other hand, the radiation heats the detector element, and the ensuing temperature rise is measured.

In addition, it is sometimes convenient to classify detectors as those which record or detect an image and those which do not. Image-recording detectors may be either thermal or quantum detectors, but the vast majority are quantum detectors.

4.3.1 Quantum Detectors

These detectors fall into two general subclasses. Both work as a result of the *photoelectric effect*, the excitation of an electron by a quantum of electro-magnetic energy. In the *external* photoelectric effect, the incident light excites the electron to the point that it completely escapes from an irradiated surface, much as sufficient energy can free a bound electron from an atom. The *internal* photoelectric effect, of which *photoconductivity* is one example, refers to the case in which the energy of the quantum is insufficient to free the electron from the detector surface, but is sufficient to excite the electron to the point where it changes the electrical conductivity of the material.

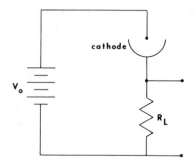

Fig. 4.13. Vacuum photodiode

Detectors whose operation is based on the external photoelectric effect are called *phototubes*. The simplest phototube, the *vacuum photodiode*, consists of two electrodes, an anode and a *photocathode* in an evacuated glass envelope. The photocathode is coated with a material to which the electrons are only weakly bound. It is held at a negative potential V_0 with respect to the anode (Fig. 4.13). When radiation with sufficiently high quantum energy strikes the surface of the photocathode, electrons are released and current flows through an ammeter or series load resistance R_L. (The phototube is a current source.)

The least expensive and perhaps most common vacuum photodiode is made of a half-cylinder photocathode surrounding an anode that is just a wire running along the axis of the cylinder. It has a comparatively fast response and can measure pulses with about 10-ns duration. In the *biplanar diode*, the electrodes are closely spaced planes with comparatively high voltage between them. Such diodes minimize the transit time of the electrons and thereby realize response times as short as a fraction of a nanosecond.

For increased sensitivity, some phototubes are filled with a gas at low pressure. If the electric potential between the anode and the cathode is great enough, an electron emitted by the cathode is accelerated to the point where it has enough energy to strike and ionize a gas atom. This creates a second free electron, known as a *secondary electron*. Secondary electrons may go on to create other secondary electrons and produce an amplification factor of 5 or 10. Unfortunately, the gas-filled phototube is comparatively slow, having a response time of the order of a millisecond.

The *multiplier phototube* or *photomultiplier* is a vacuum photodiode with a very high-gain amplifier included within the tube. The amplifier is constructed of up to a dozen electrodes called *dynodes*. The dynodes are coated with a material that will give rise to secondary electrons, much as the gas in the gas-filled phototube. The potential difference between each succeeding dynode is typically about 100 V; the potential drop across an entire tube ranges from about 500 V to several kilovolts.

The electrons emitted by the cathode are electrostatically focused onto the first dynode, where they give rise to secondary electrons. These are in turn focused onto the succeeding dynodes until they reach the anode. The voltage must be extremely well regulated, because the number of secondary electrons is a sensitive function of voltage.

The overall gain in a photomultiplier tube may be so high (10^{10} or more) that care must be taken that the illumination and the corresponding anode current remain sufficiently small that the tube not become saturated.

The speed of response of the photomultiplier, like that of the vacuum photodiode, is limited by the transit time of electrons traveling from one electrode to the next. Because there are many such electrodes, the response time of the photomultiplier is greater than that of the vacuum photodiode. Nevertheless, well designed photomultiplier tubes have response times close to 10 ns.

Typical photocathodes are silver-oxygen-cesium and other mixtures. Their response is good throughout much of the visible and uv. In the uv, phototubes are often limited by the transmittance of the envelope. In the ir, the external photoelectric effect is comparatively useless because the energy per quantum is insufficient to free electrons from the photosensitive surface with great efficiency. Nevertheless, biplanar diodes are in use at $1\,\mu$m, where certain applications that require the fast response times of these detectors are possible because of the high irradiance produced by lasers.

Photoconductive detectors are semiconductors. Such materials are characterized by an energy-level structure such as that seen in Fig. 4.14. A pure or *intrinsic* semiconductor displays a *valence band* or continuous set of energy levels in which the electrons are bound to the solid and not able to move freely, somewhat as bound electrons in an atom are not free to move about in a gas. There is, in addition, a range of energies called the *forbidden gap*; electrons in the solid may not take on values of energy within the forbidden gap. A *conduction band* lies above the forbidden gap. Electrons with energies in the conduction band are free to move about in the solid. They are, however, bound to the solid as a whole and cannot escape the surface as in vacuum photodiodes.

At room temperature, most of the electrons in an intrinsic semiconductor lie in the valence band. The semiconductor has a high electrical resistance. If the material is exposed to intense-enough radiation whose quantum energy is greater than the *band gap* between the valence and conduction bands, then enough electrons may be excited to the conduction band to decrease the electrical resistance of the material. In addition, positive holes created in the

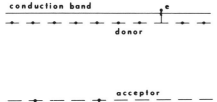

conduction band

donor

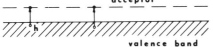

acceptor

valence band

Fig. 4.14 Energy-level structure of an impurity semiconductor

valence band by the departure of the electrons contribute to the conductivity of the material.

The band gap of many semiconductors is greater than the quantum energy of ir radiation. To reduce the energy necessary to bring about electrical conduction, the semiconductor may be *doped* with a small amount of impurity. The impurity creates energy levels within the forbidden gap, as shown in Fig. 4.14. Depending upon the impurity, the levels may be *donor levels*, which provide electrons to the conduction band, or *acceptor levels*, which provide holes to the valence band (that is, accept electrons from the valence band). Impurity semiconductors that have free electrons in the conduction band are called *n type*; those with holes in the valence band are called *p type*. The locations of the impurity levels with respect to the valence or conduction bands determine the minimum quantum energy that can be measured with a doped-semiconductor detector.

The impurity levels also provide a significant number of electrons or holes in the respective bands due to thermal effects at room temperature. These electrons or holes may mask any effect that results from irradiation of the detector. Therefore, in many cases, the detector has to be cooled, typically to liquid-nitrogen temperature, on 77 K. Thermal effects become increasingly important as we move farther into the ir, because of the decreasing quantum energy and the relative weakness of most ir sources.

The most common photoconductive detectors are probably lead sulfide and cadmium sulfide, both of which may be used at room temperature. Cadmium sulfide is restricted to the visible, but lead sulfide has high response as far into the ir as 3 or 4 μm. Other important photoconductors are germanium, especially doped with gold or mercury, indium antimonide, lead telluride, and mercury cadmium telluride. Common photoconductors have response to a few micrometers; photoconductors have been made with response to several hundred micrometers.

The response time of photoconductors is generally longer than that of vacuum photodiodes. It is determined by the rate at which the electrons and holes recombine, and varies greatly from semiconductor to semiconductor. The fastest semiconductors have time constants of a fraction of a microsecond; on the other hand the response time of some cadmium-sulfide detectors can be nearly a tenth of a second.

The photoconductive detector is generally used with a load resistor R_L and a battery or voltage source, all in series. The circuit is that of Fig 4.13, with the photoconductor in place of the vacuum photodiode. When there is no radiation falling on the detector, the voltage across the load resistor is

$$V_L = \frac{R_L}{R + R_L}, \tag{4.59}$$

where R is the *dark resistance* of the photoconductor. When the photoconductor is irradiated, its resistance decreases by ΔR. The corresponding change ΔV_L of the voltage across the load resistance is measured as the output of the device.

Another type of detector that is based on the internal photoelectric effect is the *photovoltaic detector*. Such detectors are made of junctions between two semiconductors, one of which is doped with acceptors, the other with donors. Because the holes and electrons are mobile, the charges orient themselves as in Fig. 4.12, which shows the band structure of the light-emitting diode. Because the conduction electrons and the holes are drawn to opposite sides of the junction, there is a strong electric field in the region of the junction. In addition, there are neither electrons nor holes within the junction region, because they combine with one another when they are not physically separated.

When the junction is irradiated, valence-band electrons may be excited to the conduction band, thereby creating electron-hole pairs. However, because of the strong electric field in the junction region, the electrons and the holes created there are accelerated in opposite directions and prevented from recombining with one another. A current is created because of the velocity imparted to these charged particles. If we connect an ammeter across the junction, we will detect the current as long as the irradiation is continued. This is the photovoltaic effect.

The most common photovoltaic-effect detectors are silicon and selenium cells, also called *solar cells*. The silicon cell in particular can convert a large fraction of incident power to electrical power. Gallium arsenide and its relatives can be used as fast photovoltaic detectors; see also Sect. 12.1.

Frequently, a photodetector is made by *reverse bias* of a *pn* junction; that is, the positive side of a battery or other voltage source is connected to the *n*-type material, the negative side to the *p*-type material. A detector operated in this way is called a *photodiode* and is sometimes said to be operated in the photo-conductive mode (as opposed to photovoltaic mode).

The region near the junction of a *pn* diode is known as the *depletion layer* because there are virtually no electrons or holes there. The depletion layer is bounded on both sides by regions of comparatively high space charge. Only carriers created in or near the depletion layer are effective in contributing to the photocurrent; carriers created in the bulk of the material are unlikely to be accelerated by the electric field in the junction before they recombine with carriers of the other type. The diode is designed so that the depletion layer is located as near to the surface of the detector as possible. The optical absorption and the width of the depletion layer are chosen so that most of the carriers are created in the depletion layer. These carriers are swept apart rapidly by the field in the depletion layer. The speed of response of the diode is determined by the transit time of carriers across the depletion layer and by the capacitance of the depletion layer itself.

Reverse-biasing the diode increases the width of the depletion layer. It therefore ensures that more carriers are created in the depletion layer and reduces the capacitance of the depletion layer. Reverse-biasing increases both speed and sensitivity of a photodiode.

Sometimes the width of the depletion layer is controlled by fabricating a *pin diode*, where the *i* stands for "intrinsic". That is, a comparatively thick layer of high-resistivity material is located between the *p* and the *n* materials. This creates a thick depletion layer.

When the reverse bias is great enough, multiplication of the photoelectrons will occur because of secondary emission. A diode designed to multiply small signals in this way is called an *avalanche photodiode*. Avalanche photodiodes are fast and sensitive, but they are comparatively *noisy* at very low irradiance because the number of secondary electrons generated per primary electron is uncertain.

Some photodiodes are available in small packages that also contain high-gain *operational amplifiers*. These devices are capable of detecting very low levels of radiation and can sometimes function as solid-state substitutes for photomultiplier tubes. They are usually more stable than photomultiplier tubes and require lower voltages (typically 6–15 V) with less precise regulation.

Other quantum detectors include the eye, photographic film, and the video camera (Sect. 7.4.1). These are imaging detectors, unlike the detectors we have just discussed.

4.3.2 Thermal Detectors

This class of detector depends on the heating of the detector element by the incident radiation. If the mass of the element is small enough, then the temperature rise that results from even a small amount of radiation will be large enough to be measured.

The earliest infrared detectors were *radiation thermocouples*. These devices are still widely used in infrared spectroscopy, and a newer variety is essential in measuring accurately the output of high-power lasers at all wavelengths.

The radiation thermocouple consists of a blackened detector surface, usually a small, thin chip of material coated with gold black. This is an evaported film of gold that is nearly uniformly black at all wavelengths from the uv well into the ir (including the important 8 to 14 μm region).

To measure the temperature rise that results from the incoming radiation, the detector is fixed to a small thermocouple. This device, made of very fine wires to lessen heat conduction from the detector, is merely a junction between two dissimilar metals. A voltage drop always exists at such a junction; its actual value depends on the temperature of the junction. Therefore, measuring the voltage drop across the junction is a means of measuring the temperature of the junction.

Practical radiation thermocouples generally have a second junction in series with that fixed to the detector. This junction is shielded from the radiation and is inserted so its voltage opposes that of the other junction. It is known as the *reference* or *cold junction*. The junction fixed to the detector is called the *hot junction*. The primary function of the cold junction is to make the output of the radiation thermocouple directly proportional to the increase ΔT of the temperature of the detector owing to the incident radiation. Without the cold junction, the device would indicate the absolute temperature T of the detector; because ΔT is always much less than T, taking measurements would involve observing very small voltage differences. In addition, small changes in the ambient temperature would be indistinguishable from the effects of radiation.

When the cold junction is in place, the output of the thermocouple is 0 when the irradiance is 0, irrespective of small changes of the ambient temperature. In addition, the output of the radiation thermocouple is now a nearly linear function of irradiance.

A *radiation thermopile* is a somewhat more sensitive device that is made up of several hot and cold junctions in series, and all fixed to the same detector.

Radiation thermocouples and thermopiles are comparatively slow devices and are almost always used at dc or rates of change of irradiance corresponding to a few hertz. Their time constant is determined by the thermal characteristics of the detector element; a thermocouple can always be made slightly faster by decreasing its area and therefore its sensitivity.

Massive radiation thermopiles, whose detector elements are often stainless steel cones, are used to measure the output power of high-power lasers (or, in the case of pulsed lasers, the energy of a single pulse).

Thermocouples and thermopiles must be calibrated with a standard source, such as a blackbody at 500 K. (Laser devices are calibrated with joule heating by a known amount of electrical power.) Once calibrated, a thermopile may be used as a secondary standard for calibrating other sources.

A *bolometer* is a thermal detector whose electrical resistance changes with temperature. Most of the discussion of the thermocouple applies to the bolometer as well. Unlike the thermocouple, however, the bolometer does not generate a voltage of its own, but must be connected to an external voltage source. A practical bolometer contains a matched pair of elements, one of which is the detector, and one of which serves a function analogous to the cold junction of the thermocouple. The device is connected to some kind of electrical bridge network, which measures the difference in resistance between the two elements. The output of a metallic bolometer is linearly related to irradiance.

A bolometer with a semiconductor element rather than a metal element is called a *thermistor*. Thermistors are about 10 times more sensitive than metal bolometers.

If a small flake of thermistor material is fixed with good thermal contact to a comparatively massive heat sink, then the thermistor will cool rapidly after any incident radiation is abruptly cut off. A bolometer with a fairly short response time can be made in this manner. Sensitivity is sacrificed because of large overall mass of the device. Often speed is more important than sensitivity; response times as short as 10 ms are possible.

A *pyroelectric detector* is made from a material that has an internal electrical polarization. The material may be a crystal or a plastic. Certain crystals naturally have an internal polarization. The plastic, on the other hand, is polarized by placing it, at a high temperature, in an electric field. When the temperature is reduced and the plastic therefore hardened, an electrical polarization is retained by the material.

When a pyroelectric material is heated, the electrical polarization changes slightly. The change can be detected as a displacement current.

A pyroelectric detector is made by blackening the surface of a pyroelectric material. The material is placed between two electrodes; any change of the irradiance of the blackened surface is accompanied by a change of the temperature of the material, and therefore by a current in the external circuit. The detector responds only to changes of irradiance; it must be used with pulsed sources or sources whose radiation is modulated, as by mechanical chopping.

Like other thermal detectors, which often have blackened surfaces, the pyroelectric detector is sensitive throughout most of the visible and ir spectra.

4.3.3 Detector Performance Parameters

Several parameters are needed to characterize the performance of any detector. We have already alluded to one or two; here we take them up in more detail.

The performance parameters we shall consider are the output of the detector for given irradiance, the spectral sensitivity, the minimum detectable signal, and the modulation-frequency response.

The *responsivity* $\mathcal{R}$ of a detector is the ratio of the detector's output to the input. The precise definition depends upon the application. In infrared detectors, responsivity is generally volts or amperes per watt (in more practical units, microvolts or microamperes per microwatt); that is, it is the ratio of output voltage or current to input radiant power. For detectors that operate primarily in the visible spectrum, responsivity is sometimes expressed as amperes per lumen, where a particular incandescent tungsten source is assumed. Responsivity at a given wavelength is called *spectral responsivity* $\mathcal{R}_\lambda$.

One of the properties that determine the responsivity of a quantum detector is *quantum efficiency*, the number of photoelectrons created per incident quantum. The quantum efficiency of most photocathodes is low, less than 10%, whereas silicon detectors have been made with quantum efficiencies approaching 100%.

The responsivity of nearly all detectors depends on wavelength. This dependence on wavelength is generally called the *spectral sensitivity* or, more properly, *spectral responsivity* of the detector. Spectral responsivity is the second important detector parameter. Quantum detectors, especially, have a comparatively limited wavelength range of spectral responsivity, so the choice of detector often depends on the nature of the source.

Often the source is pulsed or its irradiance is modulated. Frequently the modulation is accomplished by mechanical chopping with a toothed wheel spun by a synchronous motor. The output of the detector is observed with a frequency-sensitive amplifier. This technique may allow a relatively weak source to be detected in the presence of a bright surrounding that is not chopped and therefore not detected. It is therefore often important to ask how the responsivity of the detector varies with chopping frequency.

At low frequencies, the detector output will closely follow the variation of irradiance. At higher frequencies, the output will not follow the input nearly so well. In thermal detectors, this is because the rate of change of the temperature of the detector element is limited by the thermal mass of the element. In quantum detectors, several factors may limit the speed of response. For example, in vacuum photodiodes, the important factor is the transit time of the electrons from one electrode to the next. In photoconductors, the carriers have a finite lifetime, and this is one factor that determines the rate at which the output of the detector can be varied.

Both the rise and decay of many detectors are exponential with time. They behave as low-pass filters; their responsivity is accordingly

$$\mathscr{R}(f) = \frac{\mathscr{R}_0}{(1 + 4\pi^2 f^2 \tau^2)^{1/2}}, \tag{4.60}$$

where $\mathscr{R}_0$ is the responsivity at zero frequency. τ is the *response time* or *time constant* of the detector. The frequency $f_c = 1/2\pi\tau$ is often called the *cutoff frequency*. $\mathscr{R}(f)$ is nearly constant from dc to f_c.

A detector with a cutoff frequency f_c or a time constant τ will give a true representation of pulsed sources only when the pulse duration is long compared with τ.

In certain applications, we will need to detect an extremely weak signal. This is especially true in infrared systems, although it can also apply to optical communications and various applications in the visible.

All electronic systems suffer from *noise*. Noise comes about because of the discrete nature of the electric charge, because of thermally generated carriers, and because of other effects. If we assume sufficiently sensitive and noise-free electronics, then the minimum radiant power that we can detect is that which will bring about an output large enough that it can be distinguished from noise in the detector itself. It is conventional to define that minimum as the power that produces an output exactly equal to the noise output of the detector. (In other words, the signal-to-noise ratio must exceed 1 for a signal to be deemed detectable.)

For many common sources of noise, the noise power is proportional to the bandwidth Δf of the detection electronics. Such noise is known as *white noise*. Most detectors are incorporated into circuits that display either the current or the voltage developed by the detector. Because current and voltage are proportional to the square root of power, noise current and noise voltage are proportional to $(\Delta f)^{1/2}$. Therefore, the minimum detectable radiant power within an electrical bandwidth Δf increases in direct proportion with $(\Delta f)^{1/2}$.

The *noise equivalent power* (NEP) is the minimum detectable power at a given electrical frequency and within a given bandwidth Δf. (Some authors and manufacturers define NEP as the minimum detectable power per unit bandwidth. In that case, NEP is not the minimum detectable power in a given bandwidth; rather, the minimum detectable power in a bandwidth Δf is equal to the product of NEP and $(\Delta f)^{1/2}$. When it is defined in this fashion, NEP has the units of watts per (hertz)$^{1/2}$, pronounced "watts per root hertz". The use of the term NEP for least detectable power per unit bandwidth is therefore something of a misnomer, because the units of power are watts.)

NEP is specified at a given wavelength, chopping frequency, and bandwidth, or, alternatively, for a given-temperature blackbody source, chopping frequency, and bandwidth. Many ir detectors are described by their NEP measured using a 500-K blackbody, a chopping frequency of 90 or 900 Hz, and a bandwidth of 1 Hz. Often, NEP is written NEP (500, 90, 1), for example, to indicate the blackbody temperature, chopping frequency, and bandwidth. In most cases, the temperature of the detector must be specified as well. When the detector is intended for use with a specific laser, the NEP is determined at the wavelength of that laser, and that wavelength (rather than a blackbody temperature) is used in the specification of the detector.

Other figures of merit are often used to characterize detector performance. The most common is D^*, called *dee star* or *specific detectivity*. D^* is the reciprocal of NEP, normalized to unit area and unit bandwidth; D^* and NEP are related by the equation

$$D^* = A^{1/2} \Delta f^{1/2} / \text{NEP} . \qquad (4.61)$$

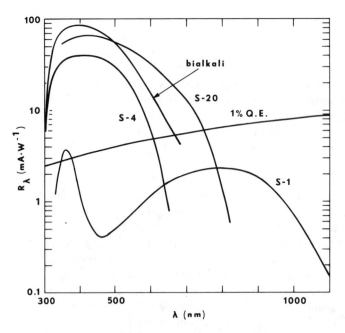

Fig. 4.15. Spectral responsivity of several photocathodes. [After *Electro-Optics Handbook* (RCA Corporation, Harrison, NJ 1968)]

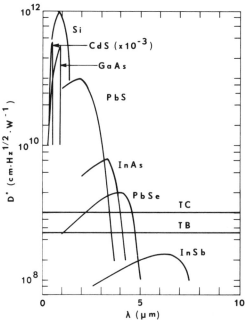

Fig. 4.16. Specific detectivity of room-temperature detectors that operate in the visible and near infrared. TC, thermocouple. TB, thermistor bolometer. [After *Electro-Optics Handbook* (RCA Corporation, Harrison, NJ 1968); P. W. Kruse et al., *Elements of Infrared Technology* (Wiley, New York 1962); W. L. Wolfe (ed.): *Handbook of Military Infrared Technology* (US Government Printing Office, Washington, DC 1965)]

D^* is useful in comparing one detector material or generic type with another, whereas NEP compares a specific detector with another. NEP, along with responsivity, time constant, and spectral responsivity, is therefore more often the desirable parameter when specifying a detector.

Figure 4.15 shows the spectral responsivity $\mathscr{R}_\lambda$ of some common photocathodes; Fig. 4.16 shows D^* for common room-temperature detectors.

Problems

4.1 Suppose that the radiance of a source is proportional to $\cos^m \theta$. Such a source roughly approximates a semiconductor laser, for which m may be as much as 15 or 20. Derive a relation analogous to (4.17) for the power radiated into a cone whose half angle is θ_0.

4.2 Calculate the irradiance E' falling on a small area element dA located close to a Lambertian diffuser that is irradiated with irradiance E. For convenience, assume the diffuser to be circular and to subtend angle θ_0 at dA. [Answer: $E \sin^2 \theta_0$.]

4.3 Two scientists (who really ought to know better!) attempt to bring out, or enhance, the edges in a photographic image by making a "sandwich" of a

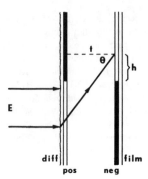

Fig. 4.17. Edge enhancement of contact printing

positive, a negative, and a diffuser, and contact printing the sandwich onto a third piece of film (Fig. 4.17). Use the result of Problem 4.2 to *estimate* the irradiance as a factor of h. If the film has a latitude of one-half F stop, roughly what will be the width of the exposed line?

4.4 An aperture with a diameter of 0.7 mm is illuminated with a tungsten lamp; its radiance is equal to that of the lamp, or about $10 \text{ W} \cdot \text{cm}^2 \cdot \mu\text{m}^{-1}$ when $\lambda = 0.85 \ \mu\text{m}$. Its image is projected with magnification $m = 1/40$ onto an optical fiber. The numerical aperture of the lens is 0.14, and all the light incident on the fiber is coupled into the core of the fiber (see Chap. 10).

(a) The light is filtered with a bandpass filter with $\Delta\lambda = 10$ nm. How much power is coupled into the fiber? [Answer: approximately 0.2 μW.]

(b) The light is mechanically chopped with frequency $f = 100$ Hz. A detector claims NEP $= 10^{-11}$ W when the bandwidth is 1 Hz. An electronic circuit with a bandwidth of 10 Hz amplifies the output of the detector. (i) What is the NEP of the system? (ii) The poorest fiber to be tested has a loss of 20 dB, or a transmittance of 0.01. Will the detector be adequate?

4.5 More-precise treatment of $\cos^4$ law.

(a) A Lambertian source with area dA and radiance L is located at an angle θ off the axis of a thin lens. The area of the source is dS, and the lens projects an image with area dS'. Show that the irradiance E on the area element dS' obeys the $\cos^4$ law.

(b) Compare your result with (4.34). What is the apparent radiance of the lens? [Hint: If the magnification of the image is m, what is dS'/dS?]

4.6 (a) Assume, for the moment, that all the photoelectrons emitted from the cathode of a vacuum photodiode are emitted with zero energy. Considering one such electron, show that a current $i(t)$ flows during the time that the electron is in flight and that the magnitude of the current is proportional to $(V/d)t$, where V is the voltage drop across the photodiode, d is the spacing between the electrodes, and t is time. Find the time τ that the electron is in flight for a fast biplanar diode, for which $d = 2$ mm and $V = 3$ kV. If our assumptions are correct, τ is the speed of response of the photodiode to an extremely short optical pulse, and $i(t)$ describes the output current waveform.

(b) Now assume that the photoelectrons are emitted with a range of energies of no more than 1 eV or so. (Visible-light quanta have energies of 1-2 eV; the energy that binds an electron to the photocathode is a fraction of 1 eV.) Show that the initial energy is negligible and therefore that variations of transit time are unimportant compared to transit time itself.

(c) Perform similar calculations for a more common photodiode, for which V is of the order of 100 V and d is of the order 1 cm. [Answers: 100 ps and 3 ns.]

4.7   Refer to (4.59) and prove that the maximum value of ΔV (for given radiant power) is attained when $R_L = R$. That is, the maximum response is attained when the load resistance is equal to the dark resistance of the photoconductor. (Assume that the change ΔR in R is small compared with R itself.) Suppose the photoconductor is to be used in ambient light, where its working resistance is R'. What is the optimum load resistance?

This theorem is not true when ΔR is a large fraction of R.

4.8   Assume that a quantum detector absorbs all the radiation falling onto it and that each quantum of light releases one photoelectron. Use simple physical arguments to show that the responsivity of the detector is approximately 0.5 A/W in the visible region of the spectrum, where the quantum energy is approximately 2 eV.

4.9   A certain opaque source does not obey Lambert's law ($L_\theta = L$) but rather has $L_\theta = 1/\cos\theta$. Show that the power radiated into a cone is

$$\frac{d\phi_{\text{cone}}}{d\phi_{\text{hemisphere}}} = \frac{d\Omega}{2\pi} ,$$

where the axis of the cone is perpendicular to the surface.

5. Wave Optics

In this chapter we discuss certain optical phenomena for which geometric or ray optics is insufficient. Primarily *interference* and *diffraction*, these phenomena arise because of the *wave nature of light* and often cause sharp departures from the rectilinear propagation assumed by geometric optics. For one thing, diffraction is responsible for limiting the theoretical resolution limit of a lens to a finite value. This is incomprehensible on the basis of ray optics.

The necessary background for this chapter is derived with as few assumptions as possible; some familiarity with wave motion is helpful. We cover mainly interference of light, far-field diffraction and just enough near-field diffraction to allow a clear understanding of holography. We shall later devote space to the important interferometric instruments and treat multiple-reflection interference in sufficient detail to apply to laser resonators.

5.1 Waves

The simplest waves are described by trigonometric functions such as sines, cosines, and complex exponential functions. A traveling wave on a string (Fig. 5.1) may be described by the equation

$$y = a \cos \frac{2\pi}{\lambda}(x - vt) , \tag{5.1}$$

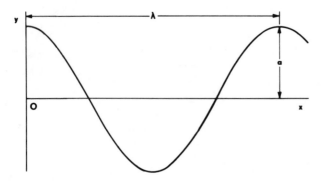

Fig. 5.1. Traveling wave

where x is the position along the string and y, the displacement of the string from equilibrium. a is the *amplitude* of the wave. The wave has crests when the cosine is 1; examining the string at $t = 0$, we can easily see that λ is the distance between crests, or the *wavelength*.

The argument of the cosine is known as the *phase* of the wave. Suppose that we consider a constant value ϕ_0 of the phase, so

$$\frac{2\pi}{\lambda}(x - vt) = \phi_0 , \tag{5.2}$$

and differentiate both sides of this equation with respect to t. The result is

$$v = dx/dt . \tag{5.3}$$

v is the velocity with which a point of *constant phase* (a crest, for example) propagates along the string; it is known as the *phase velocity* of the wave. Electromagnetic waves are two-dimensional, and v is the propagation velocity of a *plane of constant phase* in that case.

By examining the wave at one point, for convenience $x = 0$, we may find the *frequency* v of the wave. The *period* τ is the time required for the argument to change by 2π, and v is the number of periods per second, or the reciprocal of the period. Thus,

$$v = v/\lambda . \tag{5.4}$$

This relation is more commonly written as

$$v\lambda = v . \tag{5.5}$$

For compactness, we define two new quantities, *wavenumber k*,

$$k = 2\pi/\lambda , \tag{5.6}$$

and *angular frequency ω*,

$$\omega = 2\pi v . \tag{5.7}$$

(In spectroscopy, wavenumber is $1/\lambda$, when λ is expressed in centimeters.) The equation for the wave is then

$$y = a\cos(kx - \omega t) . \tag{5.8}$$

In terms of ω and k, the phase velocity is ω/k.

Finally, the wave need not have its maximum amplitude when $t = 0$ nor at $x = 0$. We account for this by writing

$$y = a\cos(kx - \omega t + \phi) , \tag{5.9}$$

where ϕ is a constant known as the *relative phase* of the wave.

5.1.1 Electromagnetic Waves

Light is a transverse, electromagnetic wave characterized by time-varying electric and magnetic fields. The fields propagate hand in hand; it is usually sufficient to consider either one and ignore the other. It is conventional to retain the electric field, largely because its interaction with matter is in most cases far stronger than that of the magnetic field.

A transverse wave, like the wave on a plucked string, vibrates at right angles to the direction of propagation. Such a wave must be described with vector notation, because its vibration has a specific direction associated with it. For example, the wave may vibrate horizontally, vertically, or in any other direction; or it may vibrate in a complicated combination of horizontal and vertical oscillations. Such effects are called *polarization* effects (see Chap. 9). A wave that vibrates in a single plane (horizontal, for example) is said to be *plane-polarized*.

Fortunately, it is not generally necessary to retain the vector nature of the field unless polarization effects are specifically known to be important. This is not the case with most studies of diffraction or interference. Thus, we will generally be able to describe light waves with the *scalar equation*,

$$E(x, t) = A \cos(kx - \omega t + \phi) , \tag{5.10}$$

where $E(x, t)$ is the electric field strength, A the amplitude, and x the *direction of propagation*.

The speed of light is almost exactly

$$c = 3.00 \times 10^8 \text{ m/s} \tag{5.11}$$

and the average wavelength of visible light

$$\lambda = 0.55 \ \mu\text{m} . \tag{5.12}$$

Because $v\lambda = c$, the frequency of visible light is approximately

$$v = 6 \times 10^{14} \text{ Hz} . \tag{5.13}$$

Detectors that are able to respond directly to electric fields at these frequencies do not exist. There are detectors that respond to radiant power, however; these are known as *square-law detectors*. For these detectors, the important quantity is not field amplitude A, but its square, intensity I,

$$I = A^2. \tag{5.14}$$

(We should properly use the term irradiance, but intensity is still conventional where only relative values are required.)

5.1.2 Complex-Exponential Functions

It is much more convenient to employ complex-exponential functions, rather than the trigonometric functions and their cumbersome formulas. For our purposes, complex-exponential functions are *defined* by the relation,

$$e^{i\alpha} = \cos\alpha + i\sin\alpha \,, \tag{5.15}$$

where $i = \sqrt{-1}$. The *complex conjugate* is found by replacing i with $-i$. The electric field is written

$$E(x, t) = Ae^{-i(kx - \omega t + \phi)} \,, \tag{5.16}$$

where it is understood that only the real part of E represents the physical wave. The intensity is defined as the *absolute square* of the field,

$$I(x, t) = E^*(x, t)E(x, t) \,, \tag{5.17}$$

where * denotes complex conjugation. $I(x, t)$ is always real. In a medium whose refractive index is equal to n, the intensity is

$$I = nE^*E \,, \tag{5.18}$$

according to a result of electromagnetic theory.

5.2 Superposition of Waves

Consider two waves, derived from the same source, but characterized by a *phase difference* ϕ. They may be written as

$$E_1 = Ae^{-i(kx - \omega t)} \quad \text{and} \tag{5.19a}$$

$$E_2 = Ae^{-i(kx - \omega t + \phi)} \,. \tag{5.19b}$$

For convenience, we allow them to have the same amplitude. If the waves are superposed, the resultant electric field is

$$E = Ae^{-i(kx - \omega t)}(1 + e^{-i\phi}) \,. \tag{5.20}$$

Before calculating the intensity, we rewrite $(1 + e^{-i\phi})$ by removing a factor of $e^{-i\phi/2}$ from both terms,

$$1 + e^{-i\phi} = e^{-i\phi/2}(e^{-i\phi/2} + e^{i\phi/2}) \quad \text{or} \tag{5.21a}$$

$$1 + e^{-i\phi} = 2e^{-i\phi/2}\cos(\phi/2) \,. \tag{5.21b}$$

This technique allows us to write E as a product of real functions and complex-exponential functions only. Because

$$e^{i\alpha} \cdot e^{-i\alpha} = 1 \,, \tag{5.22}$$

we may immediately write

$$I = 4A^2 \cos^2(\phi/2) , \tag{5.23}$$

where A^2 is the intensity of each beam, separately. We shall use this result several times to describe cos^2 *fringes*.

The intensity of the superposed beams may vary between 0 and twice the sum $2A^2$ of the intensities of the individual beams. The exact value at any point in space or time depends on the relative phase ϕ. In particular,

$$I = 4A^2 \quad \text{when } \phi = 2m\pi , \tag{5.24a}$$

and

$$I = 0 \quad \text{when } \phi = (2m + 1)\pi , \tag{5.24b}$$

where m is any integer or 0.

Because energy must be conserved, we realize that we cannot achieve *constructive interference* without finding *destructive interference* elsewhere. Interference of two uniform waves may therefore bring about quite complicated distributions of energy.

5.2.1 Group Velocity

Now consider two waves that have slightly different values of k and ω. The total electric field that results from these two waves propagating collinearly is

$$E = e^{-i(kx - \omega t)}(1 + e^{-i(\Delta kx - \Delta \omega t)}) , \tag{5.25}$$

where $k + \Delta k$ and $\omega + \Delta \omega$ are the wavenumber and angular frequency of the second wave. This equation has the same form as (5.20), with $\phi = \Delta kx - \Delta \omega t$. Therefore, we may use (5.21b) to write the total electric field as

$$E \propto e^{-i(kx - \omega t)} \cos[(\Delta kx - \Delta \omega t)/2] . \tag{5.26}$$

The cosine factor is an envelope function that modulates the traveling waves. It is itself a traveling wave and has velocity

$$v_g = \Delta \omega / \Delta k , \tag{5.27}$$

according to (5.1–5.3). v_g is known as the *group velocity* of the wave; in the limit as Δk and $\Delta \omega$ approach 0, the group velocity approaches the derivative,

$$v_g = d\omega/dk . \tag{5.28}$$

This is the velocity at which a plane of constant phase (such as the peak) *of the envelope* propagates. It need not be the same as the phase velocity. Equation (5.28) turns out to be correct even when the wave is not restricted to two discrete frequencies but, more generally, consists of a relatively narrow range of frequencies. In that case, the envelope often appears as a pulse, rather than the cosine function. The group velocity is then the velocity at which the peak of that pulse propagates.

5.2.2 Group Index of Refraction

Consider a medium in which the index of refraction $n(k)$ is a function of wavenumber k. The phase velocity of a wave is $c/n(k)$. Therefore,

$$\omega = ck/n(k) . \tag{5.29}$$

The group velocity may be found by differentiating ω with respect to k,

$$v_\mathrm{g} = \frac{c}{n}\left(1 - \frac{k}{n}\frac{dn}{dk}\right) . \tag{5.30}$$

In a *dispersive medium*, that is, one in which n is a function of wavelength or wavenumber, dn/dk is not 0; the phase velocity c/n is then unequal to the group velocity.

We may define the *group index* n_g by the relation that

$$v_\mathrm{g} = c/n_\mathrm{g}, \quad \text{or} \tag{5.31}$$

$$n_\mathrm{g} = n/\left(1 - \frac{k}{n}\frac{dn}{dk}\right) . \tag{5.32}$$

Rewriting in terms of wavelength λ and using the approximation that, for small x, $1/(1 + x) \cong 1 - x$, we find after a little manipulation that

$$n_\mathrm{g} = n - \lambda(dn/d\lambda) . \tag{5.33}$$

Most commonly, $dn/d\lambda$ is negative, so the group index is greater than the index of refraction. (Regions of the spectrum where $dn/d\lambda$ is positive are regions of high absorption and need not concern us.)

The group velocity may not exceed the velocity of light; however, there are conditions under which the phase velocity can do so. This is possible because energy is not propagated with the phase velocity, but with the group velocity.

To visualize this state of affairs, we turn to Fig. 5.2. In this figure, a pulse consists of a wave with only a few wavelengths and a finite lateral extent. This pulse travels at the angle θ to the horizontal line. We take the phase and group velocities to be equal; that is, the medium is dispersionless.

An observer who positions detectors along the horizontal axis will see a pulse of energy travel by with velocity component $c \cos \theta$, as expected; this is the horizontal component of the group velocity. However, the horizontal distance between the planes of constant phase is $\lambda/\cos \theta$, which is larger than the wavelength. If these planes are to propagate without changing their shape, the

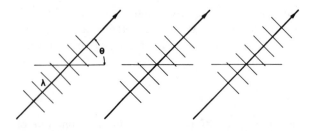

Fig. 5.2 Phase and group velocities

horizontal component of the phase velocity must be $c/\cos\theta$, a value larger than c. This component is faster than the pulse itself but does not carry energy ahead of the pulse because its amplitude falls to 0 as soon as it reaches the leading edge of the pulse. This is not a contrived example; it accurately describes the phase velocity of a mode in an optical fiber and shows how v (but not v_g) can exceed c without violating relativity.

5.3 Interference by Division of Wavefront

The *wavefront* refers to the maxima (or other planes of constant phase) as they propagate. The wavefront is normal to the *direction of propagation*. One way of bringing about interference is by dividing the wavefronts into two or more segments and recombining the segments elsewhere.

5.3.1 Double-Slit Interference

Suppose a monochromatic *plane wave* (a collimated beam, or a beam with plane wavefronts) is incident on the opaque screen shown in Fig. 5.3. Two infinitesi-

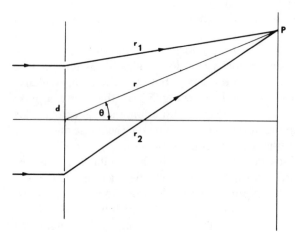

Fig. 5.3. Double-slit interference

mal slits a distance d apart have been cut into the screen. Each slit behaves as a point source, radiating in all directions. We set up an observing screen a great distance L away from the slits. Light from both slits falls on this screen. The electric field at a point P is the sum of the fields originating from each slit

$$E = A(e^{-ikr_1} + e^{-ikr_2})e^{i\omega t}, \tag{5.34}$$

where A is the amplitude of the waves at the viewing screen and r_1, r_2 are the respective distances of the slits from P. Because the factor $e^{i\omega t}$ is common to all terms and will vanish from the intensity, we shall hereafter drop it.

If L is sufficiently large, r_1 and r_2 are effectively parallel and differ only by $d \sin \theta$. Thus

$$E = Ae^{-ik\gamma_1}(1 + e^{-ikd \sin \theta}) . \tag{5.35}$$

The phase difference between the two waves is

$$\phi = kd \sin \theta , \tag{5.36}$$

and we can immediately write

$$I = 4A^2 \cos^2\left(\frac{\pi}{\lambda} d \sin \theta\right) \tag{5.37}$$

from the earlier treatment of superposition. $d \sin \theta$ is called the *optical path difference* (OPD) between the two waves. For small angles, $\sin \theta = x/L$, and the *interference pattern* has a $\cos^2$ variation with x. Maxima occur whenever the argument of the cosine is an integral multiple of π, or where

$$\text{OPD} = m\lambda \quad \text{(constructive interference)} . \tag{5.38a}$$

This result is generally true and comes about because the waves have a relative phase of a multiple of 2π whenever the optical path difference between them is an integral multiple of the wavelength.

Similarly, minima (in this case, zeroes) occur whenever

$$\text{OPD} = (m + 1/2)\lambda \quad \text{(destructive interference)} . \tag{5.38b}$$

When this relation holds, the waves arrive at the observing screen exactly 180° out of phase. If the waves have equal amplitudes they cancel each other precisely.

In fact, the $\cos^2$ fringes do not extend infinitely far from the axis. This is so far at least two reasons, (a) the light is not purely monochromatic, and (b) the slits are not infinitesimal in width.

The first relates to the *coherence* of the light, which we shall discuss later. This effect brings about a superposition of many double-slit patterns, one for each wavelength, so to speak. Each wavelength brings about slightly different fringe pattern from the rest, and at large angles θ the patterns do not coincide

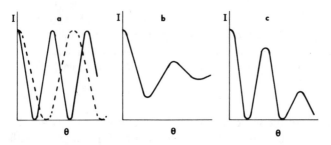

Fig. 5.4. (a) Superposition of two incoherent, double-slit interference patterns. (b) Double-slit pattern with light that is not monochromatic. (c) Double-slit pattern with finite slits

exactly (Fig. 5.4a). This results in the washing out and eventual disappearance of the fringes, as shown in Fig. 5.4b.

The second effect has to do with diffraction, which we also discuss later. In the derivation, we assumed each slit to radiate uniformly in all directions. This is valid only for zero slit width. A finite slit radiates primarily into a cone whose axis is the direction of the incident light. For this reason, the intensity of the pattern falls nearly to zero for large θ. With a good, monochromatic source, this is usually the important effect and is shown in Fig. 5.4c.

5.3.2 Multiple-Slit Interference

If we generalize from two slits to many (Fig. 5.5), we find that the OPD between rays coming from adjacent slits is $d \sin \theta$. Thus, the OPD between the first and the jth slit is $(j - 1)d \sin \theta$. The total electric field at a point on the distant

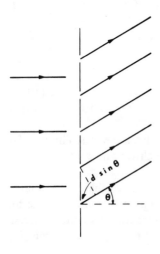

Fig. 5.5. Multiple-slit interference

observation screen is a sum not of two terms, but of many,

$$E = A e^{-i k r_1} [1 + e^{-i\phi} + e^{-2i\phi} + \ldots + e^{-(N-1)i\phi}] , \tag{5.39}$$

where N is the number of slits and $\phi = kd \sin \theta$ as before.

The term in brackets is a geometric series whose common ratio is $e^{-i\phi}$. The sum of the terms in the series may be found by a well known formula to be

$$\text{series sum} = \frac{1 - e^{-iN\phi}}{1 - e^{-i\phi}} . \tag{5.40}$$

We use the same technique as before: factor $e^{-iN\phi/2}$ from the numerator and $e^{-i\phi/2}$ from the denominator, and rewrite the sum as

$$\text{series sum} = e^{-i(N-1)\phi/2} \frac{\sin N\phi/2}{\sin \phi/2} . \tag{5.41}$$

Thus, the intensity of the interference pattern is

$$I(\theta) = A^2 \frac{\sin^2 N\phi/2}{\sin^2 \phi/2} = A^2 \frac{\sin^2 (\pi/\lambda \, Nd \sin \theta)}{\sin^2 (\pi/\lambda \, d \sin \theta)} . \tag{5.42}$$

At certain values of θ, the denominator vanishes. Fortunately, the numerator vanishes at (among others) the same values of θ. The indeterminate form $0/0$ must be evaluated by studying the limit of $I(\theta)$ as θ approaches one of these values. The evaluation is particularly simple as θ approaches 0, where the sine is replaced by its argument. Thus,

$$\lim_{\theta \to 0} I(\theta) = A^2 N^2 . \tag{5.43}$$

The denominator is 0 at other values of θ and intuition shows that $I(\theta)$ approaches $N^2 A^2$ in all such cases as well.

If N is a fairly large number, $I(\theta)$ is large at these angles. Conservation of energy requires that $I(\theta)$ be relatively small at all other angles, and direct calculation will bear this out.

A typical interference pattern is sketched in Fig. 5.6. The sharp peaks are known as *principal maxima* and appear only when

$$\frac{\pi}{\lambda} d \sin \theta = m\pi; \quad m = 0, \pm 1, \pm 2, \ldots , \tag{5.44}$$

or when

$$m\lambda = d \sin \theta . \tag{5.45}$$

This is known as the *grating equation*, and m is known as the *order number* or *order*.

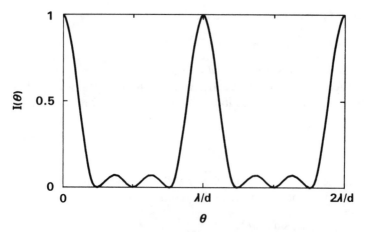

Fig. 5.6. Four-slit interference pattern

The smaller peaks are called *secondary maxima* and appear because of the oscillatory nature of the numerator of $I(\theta)$. When $N \gg 1$, the secondary maxima are relatively insignificant, and the intensity appears to be 0 at all angles where the grating equation is not satisfied. At all angles that satisfy the grating equation, the intensity is $N^2 A^2$; it falls rapidly to 0 at other angles.

5.4 Interference by Division of Amplitude

Interference devices based on division of amplitude use a partial reflector to divide the wavefront into two or more parts. These parts are recombined to observe the interference pattern.

5.4.1 Two-Beam Interference

Here we consider two parallel faces whose reflectances are equal and small compared with 1. We observe the reflected wave as a function of θ, as in Fig. 5.7. Because the reflectance is small, we need not consider multiple reflections. Hence, we again have a superposition of two waves.

To calculate the form of the interference pattern, we need the OPD between the two reflected waves. It is generally found by extending line CB to E. The OPD is the excess distance traveled by the second ray. It is therefore equal to $AB + BF$. Because AB and BE are equal, the OPD is the length of line EF. Thus, by trigonometry,

$$\text{OPD} = 2d\cos\theta ,\qquad(5.46)$$

where d is the separation between the reflecting surfaces.

The phase difference between the waves is therefore $2kd\cos\theta$. Because the reflected waves have very nearly equal amplitudes A, the total reflected field is

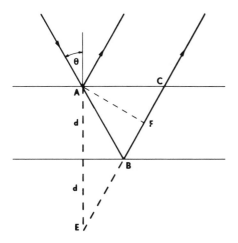

Fig. 5.7. Two parallel surfaces with low reflectances

found by superposition to be

$$I_R = 4A^2\cos^2(kd\cos\theta) \, ,\tag{5.47}$$

which are again $\cos^2$ fringes. Energy is conserved, so the transmitted intensity I_T can be found by subtracting I_R from the incident intensity I_0. As in previous examples, the reflected intensity has a maximum whenever the OPD is an integral multiple of the wavelength,

$$m\lambda = 2d\cos\theta \quad \text{(constructive interference)} \, .\tag{5.48a}$$

Similarly,

$$(m + 1/2)\lambda = 2d\cos\theta \quad \text{(destructive interference)} \, .\tag{5.48b}$$

In this derivation, we have tacitly assumed either that the reflected waves suffered no *phase change on reflection*, or that the phase change was the same at each surface. In fact, neither is generally true; the phase change on reflection can be important when the spacing between the surfaces is less than a few wavelengths.

For example, suppose the surfaces of Fig. 5.7 were the upper and lower faces of a block of glass, index $n \sim 1.5$. The intensity reflected from each surface is about 4% near normal incidence, so the two-beam approximation works fairly well. Electromagnetic theory shows, however, that the beam reflected from the low-to-high-index interface undergoes a phase change of π on reflection, whereas the other beam does not. This phase change is exactly equivalent to an additional half-wavelength of optical path and therefore *reverses* the conditions (5.48) for destructive and constructive interference. With metallic and other mirrors, the phase change need not be either 0 or π and causes a shift of the reflected interference pattern.

Problem. Show in the general case of a slab whose index n is greater than 1 that $2d \cos \theta$ must be replaced by $2nd \cos \theta'$, where θ' is the angle of refraction (inside the glass). The quantity nd is known as the *optical thickness* of the slab; optical thickness is a concept that is often important in interference experiments.

5.4.2 Multiple-Reflection Interference

Here we have surfaces with relatively high reflectance and cannot ignore the effect of multiple reflections. When the number of reflections is large, we shall find sharp transmitted fringes, just as in the case of multiple-slit interference.

Consider the situation of Fig. 5.8, where the viewing screen is a great distance away. The surfaces each reflect a fraction r of the incident amplitude A and transmit a fraction t. That is, the *amplitude reflectance* is r and the *amplitude transmittance* is t. To calculate the intensity on the screen at angle θ, we require the amplitude and relative phase of each transmitted wave. The first wave has passed through two surfaces and has been twice attenuated by t. Its amplitude is therefore At^2, and we define its relative phase at the observation point as 0.

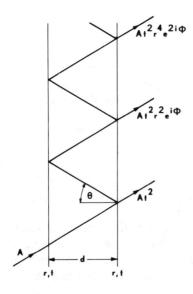

Fig. 5.8. Interference by multiple reflections

The second transmitted wave in addition suffers two reflections and a phase change

$$\phi = 2kd \cos \theta ; \tag{5.49}$$

the third, another two reflections and another phase change; and so on. The total field on the viewing screen is thus

$$E = At^2(1 + r^2 e^{-i\phi} + r^4 e^{-2i\phi} + \ldots) . \tag{5.50}$$

The sum of the geometric series in parentheses is

$$\text{series sum} = 1/(1 - r^2 e^{-i\phi}) . \qquad (5.51)$$

The transmitted intensity I is the absolute square E^*E of the transmitted amplitude. We may write I in terms of the (intensity) *reflectance* R and the (intensity) *transmittance* T. Because intensity is the square of amplitude, $R = r^2$ and $T = t^2$. The transmitted intensity is written after some manipulation

$$\frac{I_T}{I_0} = \left(\frac{T}{1 - R}\right)^2 \frac{1}{1 + F\sin^2\phi/2}, \quad \text{where} \qquad (5.52)$$

$$F = 4R/(1 - R)^2. \qquad (5.53)$$

Here, T is not set equal to $(1 - R)$ to allow the possibility of absorption or scattering loss in the mirrors themselves. The difference between T and $(1 - R)$ may well be important in instruments with metal mirrors or dielectric mirrors with reflectances near 100%. The factor $T/(1 - R)$ is a constant, and we shall for convenience drop it hereafter and assume perfect reflectors.

The transmittance I_T/I_0 is equal to 1 whenever $\sin^2\phi/2 = 0$. If R is relatively near 1, then F is large compared to 1. Thus, the transmittance falls rapidly to a small value as the sine deviates from 0.

Figure 5.9 is a sketch of I_T/I_0 vs ϕ for various values of R. Maxima occur when $\phi/2 = m\pi$, or when

$$m\lambda = 2d\cos\theta, \quad m = 0, 1, 2, \dots . \qquad (5.54)$$

Again, m is the *order number* or order of interference. For a fixed value of d, we will observe transmission at fixed values of θ. If the plates are illuminated with a range of angles, this will appear as a series of bright rings. On the other hand, if we remain at $\theta = 0$ and vary d, we will observe transmission at specific values of d only.

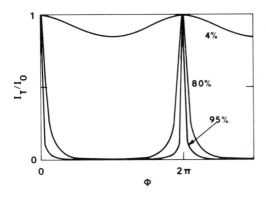

Fig. 5.9. Multiple-reflection interference patterns

5.5 Diffraction

Although the distinction is sometimes blurry, we shall say that *diffraction* occurs when light interacts with a single aperture. Interference occurs when several beams interact. If a screen has several apertures, we can say that each aperture causes a spreading of the beam by diffraction. Far from the screen the beams overlap. This results in an *interference pattern*.

Diffraction is observed *whenever* a beam of light is restricted by an opening or by a sharp edge. Diffraction is very often important even when the opening is many orders of magnitude larger than the wavelength of light. However, diffraction is most noticeable when the opening is only somewhat larger than the wavelength.

We can account for diffraction, or at least rationalize its existence, by *Huygens's construction*. Today, we interpret Huygens's construction as a statement that each point (or infinitesimal area) on a propagating wavefront itself radiates a small spherical wavelet. The wavelets propagate a short (really, infinitesimal) distance, and their resultant gives rise to a "new" wavefront. The new wavefront represents merely the position of the original wavefront after it has propagated a short distance.

More specifically, Huygens's construction is shown in Fig. 5.10. The wavefront in this case is a part of a plane wave that has just been allowed to pass through an aperture. A few points are shown radiating spherical wavelets. Both experience and electromagnetic theory indicate that the wavelets are radiated primarily in the direction of propagation. They are thus shown as semicircles rather than full circles.

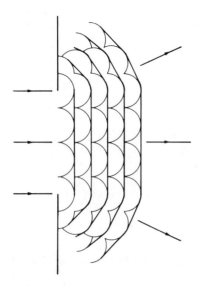

Fig. 5.10. Huygens's construction

The spherical wavelets combine to produce a wavefront lying along their common tangent. The new wavefront is nearly plane and nearly identical with the original wavefront. At the edges, however, it develops some curvature owing to the radiation of the end points away from the axis. Succeeding wavefronts take on more and more curvature, as shown, and eventually the wavefront becomes spherical. We then speak of a *diverging wave*.

Double-slit interference occurs because diffraction allows the light from the individual slits to interact. Close to the slits, where diffraction is not always noticeable, interference is not observed. Only the geometrical shadow of the slits is seen. Far enough from the slits, when the divergence due to diffraction is appreciable, the diffracted beams begin to overlap. Only beyond this point is interference important.

Sufficiently far from the diffracting aperture, we can assume that the rays from the two slits to the point of observation are parallel. This is the simplest case, known as *Fraunhofer diffraction* or *far-field diffraction*.

For most diffracting screens, the observing plane would have to be prohibitively distant to allow observation of Fraunhofer diffraction. The approximation is in fact precise only at an infinite distance from the diffracting screen. Fraunhofer diffraction is nevertheless the important case. This is so because the far-field approximation applies in the focal plane of a lens. One way to see this is to recognize that the diffraction pattern, in effect, lies at infinity. A lens projects an image of that pattern into its focal plane.

The fact can also be seen from Fig. 5.11. Rays leaving the diffracting screen at angle θ contribute to the intensity at a single point on the distant observing screen. The lens brings these rays to a point in its focal plane, where they contribute to the intensity of that point.

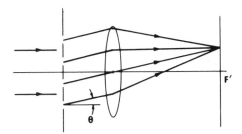

Fig. 5.11 Fraunhofer diffraction in the focal plane of a lens

In paraxial approximation, all paths from the lens to the point are equal, so no unwanted path lengths are introduced by the lens. The same is true of a well corrected lens, and a true Fraunhofer pattern is observed only with well corrected (diffraction-limited, see Sect. 3.8) optics.

Finally, we have been tacitly assuming that the diffracting screen is illuminated with plane waves. If this is not so and it is illuminated with spherical waves originating from a nearby point source, the pattern at infinity is not a Fraunhofer pattern. It is nevertheless possible to observe Fraunhofer diffraction with a well corrected lens; it can be shown that the Fraunhofer pattern lies in the plane into which the lens projects the image of the point source, no matter what the location of the source. Illuminating with collimated light is just a special case.

5.5.1 Single-Slit Diffraction

This is shown in one dimension in Fig. 5.12. We appeal to Huygens's construction and assume that each element ds of the slit radiates a spherical wavelet. The observing screen is located a distance L away from the aperture, and we seek the intensity of the light diffracted at angle θ to the axis.

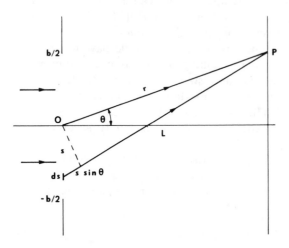

Fig. 5.12. Fraunhofer diffraction by a single opening

The center O of the aperture is located a distance r from the observation point P. The OPD between the paths from θ and from the element ds (at s) is $s \sin \theta$, in Fraunhofer approximation.

The electric field at P arising from the element is

$$dE = A \frac{e^{-ik(r + s\sin\theta)}}{r} ds . \tag{5.55}$$

Here, A is the amplitude of the incident wave, assumed constant across the aperture. We obtain the r in the denominator by realizing that the element is essentially a point source. The intensity from the point source obeys the inverse-square law, so the amplitude falls off as $1/r$. We drop $s \sin \theta$ from the denominator because it is small compared with r. We cannot, however, drop it from the phase term $k(r + s \sin \theta)$ because very small changes of $s \sin \theta$ cause pronounced changes of the phase of the wavelet relative to that of another wavelet.

The total field at P is the sum of the fields due to individual elements. If the dimension of the slit is b and its center, $s = 0$, this is just the integral

$$E(\theta) = A\frac{e^{-ikr}}{r} \int_{-b/2}^{b/2} e^{-(ik\sin\theta)s}\, ds \, , \tag{5.56}$$

where constant terms have been removed from the integral. The integrand is of the form $\exp(as)$, so the integral is easily evaluated:

$$E(\theta) = A\frac{e^{-ikr}}{r}\frac{2\sin[(kb\sin\theta)/2]}{ik\sin\theta} \, . \tag{5.57}$$

If we multiply both numerator and denominator by b, and define

$$\beta = \frac{1}{2}kb\sin\theta \, , \tag{5.58}$$

we may write

$$E(\theta) = \frac{Ab}{r}e^{-ikr}\left(\frac{\sin\beta}{\beta}\right), \quad \text{or} \tag{5.59a}$$

$$I(\theta) = \frac{I_0 b^2}{r^2}\left(\frac{\sin\beta}{\beta}\right)^2 . \tag{5.59b}$$

More proper analysis, based on electromagnetic theory and a two-dimensional integration would include an additional factor of i/λ in the expression (5.59a) for $E(\theta)$, but the important part is the variable $(\sin\beta)/\beta$.

Figure 5.13 shows $I(\theta)$ vs θ for a single slit, normalized to 1. The *principal maximum* occurs when θ approaches 0, and $(\sin\beta)/\beta$ becomes 1. The diffracted intensity is 0 at angles (except 0) for which $\sin\beta = 0$. The first such zero occurs at angle

$$\theta_1 = \lambda/b \, , \tag{5.60}$$

where θ is assumed small.

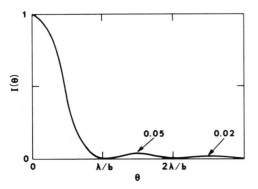

Fig. 5.13. Single-slit diffraction pattern

If the viewing screen is the focal plane of a lens, then the first minimum is located a distance

$$RL = \lambda f'/b \tag{5.61}$$

from the center of the pattern, which extends in the direction perpendicular to the edges of the aperture. Over 80% of the diffracted light falls within $2\lambda f'/b$ of the center of the pattern, and the first *secondary maximum* is only about 5% as intense as the principal maximum.

Similar analysis can be carried out with a circular aperture in two dimensions. The result is similar, except that the pattern is a disk, known as the Airy disk, with radius defined by the first zero as

$$RL = 1.22 \lambda f'/D \;, \tag{5.62}$$

where D is the diameter of the aperture. It is the finite size of the Airy disk that limits the theoretical resolving power of any optical system.

Problem. Calculate the Fraunhofer-diffraction pattern of a slit whose center is located a distance s_0 away from the axis of the system. Show that the result is identical with (5.59a) multiplied by a complex-exponential function, $\exp(-iks_0 \sin \theta)$. Show further that the intensity is identical with (5.59b) and is centered about the angle $\theta = 0$.

This result applies only to Fraunhofer diffraction and therefore presumes that $s_0 \ll L, r$. The argument $ks_0 \sin \theta$ of the complex-exponential function is a *phase factor* that results from the shift of the aperture.

5.5.2 Interference by Finite Slits

Earlier, we noted that division-of-wavefront interference occurs because light is diffracted by the individual apertures. This implies, for example, that the interference pattern should vanish in those directions in which the diffracted intensity is 0. The pattern should be strongest where the diffracted intensity is greatest. If the slits are identical, this implies that the diffraction pattern with finite slits should be given by

$$\text{(interference pattern)} \times \text{(diffraction pattern of single slit)} \;, \tag{5.63}$$

where "interference pattern" refers to the pattern derived with infinitesimal slits. It is possible to verify this relation by direct integration over an aperture consisting of several finite slits.

The significance is mainly for multiple-slit interference. As we shall see in Chap. 6, a *diffraction grating* may well have slits whose widths are about equal to their spacing. Figure 5.14 shows the diffraction pattern in such a case. The dashed line is the diffraction pattern of a single slit, and the various orders of interference are indicated as peaks. Zero-order diffraction is of no interest, but the first and higher orders are weak because light is diffracted into their directions. Occasionally a principal maximum will fall so close to the diffraction minimum that it is barely detectable. In this case, we speak of *missing orders*.

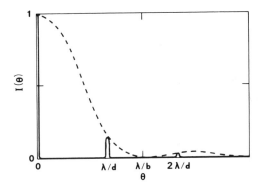

Fig. 5.14. Multiple-slit interference with finite slits

5.5.3 Fresnel Diffraction

Fresnel diffraction by an aperture refers to the general case, in which either the aperture is not illuminated by a collimated beam or the diffracting screen is not distant compared with the size of the aperture. Diffraction by a single straight edge is always Fresnel diffraction.

In this section, we discuss only enough Fresnel diffraction to allow understanding of the *zone plate*, an imaging device that is interesting partly because of its similarity to a hologram.

We begin with *Fresnel's construction*, Fig. 5.15. A point source P illuminates a large aperture centered at O, a distance a away. We seek the intensity at P', a' way from O.

Consider a point Q in the aperture, located where $OQ = s$. Light travels from Q to P' because of diffraction (see Huygens's construction). We take $s \ll a, a'$. PQ differs from PO by δ, whereas $P'Q$ differs from $P'O$ by δ'. δ is related to a and s by

$$a^2 + s^2 = (a + \delta)^2 . \tag{5.64}$$

Because s is small, δ is small. We may thus expand the square and drop the term

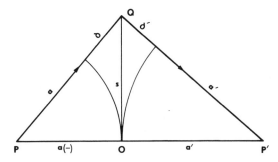

Fig. 5.15. Fresnel's construction

δ^2, giving

$$\delta \cong s^2/2a ,\tag{5.65}$$

which is known as the *sag formula*. (The *sag* is actually the perpendicular distance from Q to the circle, but for small angles the sag is nearly equal to δ.) The sag formula similarly relates δ', a', and s. The total path difference $\Delta(=\delta' + \delta)$ between PP' and PQP' is thus

$$\Delta = \frac{s^2}{2a'} + \frac{s^2}{2a} .\tag{5.66}$$

Fresnel's construction consists of defining radii $s_1, s_2, s_3, \ldots,$ such that $\Delta_1 = \lambda/2$, $\Delta_2 = 2\lambda/2$, $\Delta_3 = 3\lambda/2, \ldots,$ or, more generally, $\Delta_m = m\lambda/2$. The annuli defined by two successive radii are *Fresnel half-period zones*, named so because the field arising at P' from one annulus is, on the whole, π out of phase with the field from the neighboring annulus.

The radii are easily found from the formula

$$\frac{m\lambda}{2} = \frac{s_m^2}{2}\left(\frac{1}{a'} + \frac{1}{a}\right)\tag{5.67}$$

to be proportional to square roots of integers,

$$s_m \propto \sqrt{m} .\tag{5.68}$$

The area of any annulus is

$$S_m = \pi s_m^2 - \pi s_{m-1}^2 ,\tag{5.69}$$

which can be written

$$S_m = \pi \frac{aa'}{a' + a}\lambda ,\tag{5.70}$$

independent of m or s. For given geometry, the areas of all the Fresnel zones are equal. Thus, each zone contributes approximately the same amplitude as the rest to the field at P'.

Suppose the aperture to be a circle with N zones. Because the alternate zones contribute fields that are out of phase by π, the total amplitude at P' is

$$A_{P'} = A_1 - A_2 + A_3 - A_4 + \ldots \pm A_N .\tag{5.71}$$

The sign of A_N depends on whether N is odd or even. If N is odd, alternate zones cancel, but one zone is unpaired. In that case, $A_{P'}$ is about equal to the amplitude A_1 contributed by the first zone alone. On the other hand, if N is even, $A_{P'}$ is nearly 0. Therefore, the intensity $I_{P'}$ at P' varies between 0 and I_1 $(= A_1^2)$, depending on the precise geometry.

Fig. 5.16. Fresnel zone plate. [Figure courtesy of M.W. Farn, MIT Lincoln Laboratory]

We may construct a *Fresnel zone plate* by blocking either the even-numbered or the odd-numbered zones (Fig. 5.16). Suppose we have chosen to block the even zones. Then,

$$A_{P'} = A_1 + A_3 + \ldots + A_N , \tag{5.72}$$

which is approximately

$$A_{P'} = N A_1 / 2 , \tag{5.73}$$

where $N/2$ is the number of unobstructed zones. Squaring, we find

$$I_{P'} = N^2 I_1 / 4 , \tag{5.74}$$

an increase of $(N/2)^2$ over the contribution from the first zone alone. ($N/2$ is the total number of clear zones.)

Finally, we may write, from the equation for the radii s_m,

$$\frac{1}{a'} + \frac{1}{a} = \frac{m\lambda}{s_m^2} , \tag{5.75}$$

or, since $s_m \propto \sqrt{m}$,

$$\frac{1}{a'} + \frac{1}{a} = \frac{1}{s_1^2/\lambda} . \tag{5.76}$$

Now we change the sign of a because it is an algebraically negative quantity (see Chap. 1) and find that

$$\frac{1}{a'} - \frac{1}{a} = \frac{1}{s_1^2/\lambda} . \tag{5.77}$$

This is the lens equation, with focal length

$$f' = s_1^2/\lambda \tag{5.78}$$

and shows that P' is the image of P. A zone plate is thus an imaging device whose focal length depends on wavelength and on the geometry of the zone plate. Fainter images correspond to $1/3, 1/5, \ldots$ of f' because, for example, at $f'/3$ a single Fresnel zone includes exactly three rings. This gives rise, less efficiently, to a secondary focus.

Finally, we could have used precisely the same arguments, had we chosen P' to the left of O, rather than to the right. This indicates that each zone plate has a series of negative foci in addition to the positive foci. Because of this and because of zero-order or undiffracted light, the zone plate is only about 1/10 as efficient as a lens in bringing light to the primary focus. In addition, a zone-plate image of an extended object will have low contrast because of glare caused by the additional diffracted orders.

5.5.4 Far and Near Field

We have defined Fraunhofer or far-field diffraction as that which is observed whenever the source and observation plane are very distant from the diffracting screen. If the diffracting screen is close to either the source or the observation plane, Fresnel or near-field diffraction may be observed. We are now in a position to distinguish more precisely between these two cases.

Consider a diffracting screen whose greatest overall dimension is $2s$. If the source is located at ∞, then the diffracting screen will fall within a single Fresnel zone when the observation point is located a distance s^2/λ beyond the screen. If the observation point is moved closer than s^2/λ the screen will occupy more than one Fresnel zone. This is known as the region of *near-field diffraction*.

In the same way, if the observation point is moved beyond s^2/λ, the screen occupies less than one Fresnel zone. This is the region of *far-field* or *Fraunhofer* diffraction. In this region, all points on the screen are equidistant from the observation point, to an accuracy of less than $\lambda/4$. This corresponds closely to our assumptions when we derived the diffraction pattern of a slit.

The simple pinhole camera can be used to illustrate near- and far-field diffraction. The pinhole camera is a small hole punched in an opaque screen, with a viewing screen located at a distance f' beyond the hole. When the hole is very large, the image of a distant point is the geometrical shadow of the opaque screen. The diameter of the image is thus equal to the diameter of the pinhole. The limit of resolution in the image plane is about 1.5 times larger. When the hole is small, Fraunhofer diffraction applies, and the limit of resolution is 0.61 $\lambda f'/s$, where s is the *radius* of the pinhole. For intermediate-size holes, Fresnel diffraction applies, and little can be said without detailed calculation.

If we express the resolution limit in units of s and the focal length in units of s^2/λ, the previous arguments result in the solid lines in Fig. 5.17. The horizontal line corresponds to geometric optics (large hole) and the slanted line to Fraunhofer diffraction (small hole).

Experimental measurements, indicated by dots, show that Fraunhofer diffraction accurately describes the pinhole camera for focal lengths f' greater than s^2/λ. Thus, Fresnel diffraction corresponds to $f' \lesssim s^2/\lambda$.

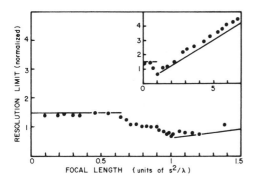

Fig. 5.17. Resolution limit of pinhole camera, illustrating regions of near- and far-field diffraction. Inset: Same data on scale of 0–7. [After M. Young: Am. J. Phys. **40**, 715 (1972) and Appl. Opt. **10**, 2763 (1971)]

When the hole is very large, the outermost Fresnel zones may become very small compared with the minute imperfections in the edges of the pinhole. When this happens, diffraction effects are washed out by the irregularity of the diffracting screen, and geometric optics adequately describes the propagation of light through the aperture. Thus, geometric optics is adequate when $f' \ll s^2/\lambda$.

We may extend these arguments to the case of a relatively nearby source by looking back to the treatment of the Fresnel zone plate. If the source is located a distance a before the screen, and the observation point a distance a' beyond, we can still define the quantity

$$\frac{1}{f'} = \frac{1}{a'} - \frac{1}{a} \tag{5.79}$$

for the system. As before, whenever $f' > s^2/\lambda$, Fraunhofer diffraction applies; whenever $f' < s^2/\lambda$, Fresnel diffraction applies; and whenever $f' \ll s^2/\lambda$, diffraction effects may be unimportant.

In sum, light does not acquire a beam divergence $\sim \lambda/D$ immediately after passing through an aperture (unless the size of the aperture is comparable to the wavelength). Rather, it first propagates through the aperture and casts a

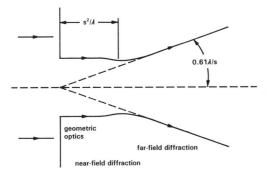

Fig. 5.18. Propagation of collimated beam past an aperture. [After M. Young, Imaging without lenses or mirrors, Phys. Teacher, 648–655 (Dec. 1989)]

geometrical shadow near the aperture. Farther away, it displays a near-field diffraction pattern and, in the case of a circular opening, actually comes to a weak focus a distance s^2/λ beyond the aperture. Only past s^2/λ do we observe a beam divergence $1.22\lambda/D$. Well beyond s^2/λ a true Fraunhofer pattern is observed (Fig. 5.18).

Finally, we stress the point that these arguments do not apply only to the pinhole camera or to the case of the circular aperture, but to any diffracting screen whose greatest overall dimension is $2s$.

5.5.5 Babinet's Principle

Consider a diffracting screen that is made up of a number of separate, not necessarily identical slits. The Fraunhofer-diffraction pattern E_1 produced by that screen is described by integrating (5.55) over the clear portions of the diffracting screen.

Now define the *complementary screen* as that screen that is opaque wherever the original screen is transparent, and vice versa. In the language of photography, the complementary screen is the negative of the original screen. The diffraction pattern E_2 of the complementary screen is described by integrating (5.55) over the clear portions of the complementary screen or, alternatively, over the opaque portions of the original screen.

If we form the sum $E_1 + E_2$, we find that it is equal to the amplitude E_0 of the unobstructed wave, because $E_1 + E_2$ represents integration over the entire plane.

This is *Babinet's principle*; it has its greatest utility in the calculation of Fraunhofer-diffraction patterns. In that case, $E_0 = 0$ everywhere except where $\theta = 0$. Therefore, $E_2 = -E_1$. If we square E_1 and E_2 to find the intensities, we find that $I_1 = I_2$. The Fraunhofer-diffraction patterns of complementary screens are identical, except for a small region near the axis of the system.

5.5.6 Fermat's Principle

We have just seen that a zone plate focuses light by ensuring that the optical paths through different points on the zone plate either are equal or differ by multiples of one wavelength. This is an opportune time to examine the imaging properties of lenses from the same point of view.

To simplify the discussion, we use Fig. 5.19 for a "len" with the object point at ∞. The optical path from the vertex O of the "len" to the secondary focal point F' is nf', where, according to the "len" equation, $f' = nR/(n-1)$. We wish to compare nf' with the corresponding optical path length from Q' through Q to F'. According to the sag formula, $QQ' = y^2/2R$, so the total optical path length $QQ'F'$ is

$$(y^2/2R) + nd , \tag{5.80}$$

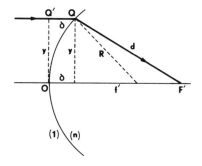

Fig. 5.19. Imaging by a "len". Fermat's principle

where d is related to R and y through the equation

$$d^2 = y^2 + [f' - (y^2/2R)]^2 . \tag{5.81}$$

If we use the "len" equation to write R in terms of f', drop terms of order y^4, rewrite d as

$$d \cong f'[1 - y^2/f'^2(n - 1)]^{1/2} , \tag{5.82}$$

and use the binomial expansion,

$$(1 - x)^{1/2} \cong 1 - x/2 , \tag{5.83}$$

when $x \ll 1$, we find that the optical path length between Q' and F' is just equal to nf', irrespective of the value of y. That is, in the paraxial approximation, all optical paths from the object point to the image point are equal.

Since a lens is merely a sequence of spherical refracting surfaces, the statement is true in general. Thus, in a sense, all imaging is a result of interference, just as in the more obvious case of the zone plate. In fact, when the paraxial approximation is invalid and the optical path depends slightly on y, aberrations result. Aberrations are sometimes measured by determining the *wavefront aberration*, or the deviation of the wavefront emerging from the exit pupil from a perfect sphere centered at the image point. When the wavefront aberration is less than $\lambda/4$, the image is very nearly diffraction limited.

Certain optical systems project images by means of an index of refraction that is a smoothly varying function of position. These systems are often most easily analyzed by requiring that all optical paths from object point to image point be equal. We shall later determine the properties of an optical fiber with such a development.

Finally, this discussion is a special case of *Fermat's principle*, which states that, if a light ray propagates between two points, then the optical path it follows is an extremum; that is, it is either the minimum or the maximum possible optical path between the two points. Many systems that employ a varying index of refraction are most easily analyzed through application of Fermat's principle.

5.6 Coherence

Until now, we have almost always assumed light to be completely *coherent*, in the sense that any interference experiment resulted in high-quality interference fringes. In general, this is not the case, except with certain laser sources; the light from most sources is said to be *incoherent* or *partially coherent*.

When conditions are such that the light is incoherent, it is not possible to detect interference effects. A discussion of wave optics is incomplete without considering the conditions that must exist for an interference experiment to be performed successfully.

Light sources are today put into one of two categories, laser sources and *thermal sources*. A typical thermal source is a gas-discharge lamp. In such a lamp, light is emitted by excited atoms that are, in general, unrelated to each other. Each atom emits relatively short bursts or *wave packets*. If an atom is excited several times, it can emit several consecutive wave packets. These packets are generally far apart (compared with their duration) and are emitted randomly in time. The packets emitted by a single atom therefore bear no constant phase relation with each other.

Suppose we try to perform interference by division of amplitude (Fig. 5.20) with the packets emitted by a single atom. The wave reflected from the second surface is delayed with respect to the first because of the finite speed of light. If the delay is greater than the duration of the wave packet, the two reflected packets will not reach the detector simultaneously. There will therefore be no interference pattern, and we would compute the intensity at the viewing screen by adding the intensities (not amplitudes) of the reflected waves. The light is said to be incoherent for the purpose of this experiment.

This statement would be true even if the wave packets were emitted so rapidly that several packets entered the apparatus at the same time. (In this case, the light would undergo rapid amplitude and phase fluctuations.) Because the packets are emitted at random, they bear no definite phase relation. We would

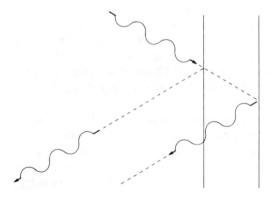

Fig. 5.20. Time coherence: interference of waves that have finite duration

sometimes detect a maximum and sometimes a minimum for any given OPD. Over the long term, we would observe constant intensity and would regard the light as incoherent.

Similarly, the waves from one atom bear no definite relation with the waves from any other atom. By precisely the same reasoning, we conclude that the light emitted by one atom is incoherent with that emitted by any other atom. Superposition of the waves from different atoms or different points on the same source is therefore described by adding intensities, not amplitudes.

In the following sections, we make these arguments more quantitative and find the conditions under which a given thermal source may be considered coherent.

5.6.1 Temporal Coherence

We begin by examining the interference resulting from two parallel reflecting surfaces separated by d and approximately perpendicular to the incident light. We found earlier that constructive interference would occur for values of d such that $m\lambda = 2d$. We take the reflectance low enough that $\cos^2$ fringes result.

Suppose now that we regard λ as only one wavelength of a nearly monochromatic beam whose *spectral width* is $\Delta\lambda$, where $\Delta\lambda \ll \lambda$. Each infinitesimal wavelength interval within $\Delta\lambda$ gives rise to its own interference pattern; we assume these wavelengths to be incoherent with one another. The net interference pattern is therefore the sum of a great number of $\cos^2$ patterns.

At some optical path difference $2d$, one extreme wavelength will exhibit a maximum, whereas the other extreme exhibits a minimum. This is so because

$$m\lambda \neq m(\lambda + \Delta\lambda) , \tag{5.84}$$

when $m \gg 1$.

At this value of $2d$, therefore, the fringes will be completely washed out, and the intensity will be constant.

If we call λ the extreme wavelength that has a maximum at $2d$, then

$$m\lambda = \text{OPD} \tag{5.85}$$

as above. If the other extreme $\lambda + \Delta\lambda$ has a minimum for the same value of OPD, then

$$(m - \tfrac{1}{2})(\lambda + \Delta\lambda) = \text{OPD} . \tag{5.86}$$

If we subtract one equation from the other, we find

$$\tfrac{1}{2}(\lambda + \Delta\lambda) = m\,\Delta\lambda . \tag{5.87}$$

We may drop $\Delta\lambda$ from the left side, and use (5.85) to find

$$\text{OPD} \cong \lambda^2/2\Delta\lambda . \tag{5.88}$$

We will thus observe fringes only if OPD is less than this value. We therefore define *coherence length* l_c as

$$l_c = \lambda^2/2\Delta\lambda \ . \tag{5.89}$$

We may rewrite this expression in terms of frequency. We start with the relation

$$v\lambda = c \tag{5.90}$$

and differentiate both sides, with c constant. The result is best expressed

$$\Delta v/v = \Delta\lambda/\lambda \tag{5.91}$$

with no regard to the signs of Δv and $\Delta\lambda$. The coherence length is thus

$$l_c = c/2\Delta v \ . \tag{5.92}$$

This suggests we define *coherence time* t_c by

$$l_c = ct_c \ , \tag{5.93}$$

so

$$t_c = 1/2\Delta v \ . \tag{5.94}$$

A more precise analysis integrates the individual $\cos^2$ fringes over the proper lineshape and shows that

$$t_c = 1/\Delta\omega \tag{5.95}$$

where $\Delta\omega = 2\pi\Delta v$ refers to the full width of the line at the half-intensity points.

We may interpret this result by suggesting that the continuous bursts or wave packets emitted by the atoms have a duration about equal to t_c. The phase will vary abruptly between one packet and another, later one, so interference can be observed only when the OPD is less than the length of a single packet.

When the OPD is close to 0, the fringes will have high contrast. The contrast gradually diminishes after the OPD exceeds l_c. When the fringes have high contrast, the light is said to be *highly coherent*. Light may be assumed to be highly coherent when

$$\text{OPD} < l_c \tag{5.96}$$

The light is incoherent when the OPD greatly exceeds l_c and *partially coherent* for intermediate values of OPD.

5.6.2 Spatial Coherence

Here we begin with a double-slit experiment. We consider an extended, monochromatic source that subtends a small angle $\Delta\phi$ at the diffracting screen. As in the case of temporal coherence, each infinitesimal portion of the source gives rise to a $\cos^2$ pattern. We assume that the patterns add incoherently because they must be derived from different radiating atoms within the source.

The infinitesimal source on the axis produces interference maxima at angles θ such that $m\lambda = d \sin \theta$, and the first maximum off the axis occurs at angle λ/d in small-angle approximation. A second infinitesmal source off the axis by $\Delta\phi$ produces the same interference pattern shifted by $\Delta\phi$. The patterns will effectively nullify each other when a minimum of one occurs at the same angle as a maximum of the other, which happens when

$$\Delta\phi = \lambda/2d \ . \tag{5.97}$$

If the source subtends an angle greater than $\lambda/2d$, the light reaching the screen is incoherent. If we let δ be the source dimension and L its distance from the screen, we may write

$$d = \lambda L/2\delta \ . \tag{5.98}$$

The value of d roughly delineates the *coherence area* of a source whose dimension is δ and whose distance from the screen is L. (The coherence area itself is equal to $\pi d^2/4$ with circular symmetry or d^2 with rectangular symmetry.) As with temporal coherence, the fringes have good contrast when the slits are close together, and the contrast diminishes as the slit separation approaches $\lambda L/2\delta$. Beyond this value the light is, essentially, incoherent. More-rigorous analysis shows that the region of high coherence extends to a slit separation of

$$d_c = 0.16\,\lambda L/\delta \ , \tag{5.99}$$

and the value d_c is usually said to delineate the coherence area of the source. The light reaching the two slits is nearly incoherent when d exceeds $\lambda L/2\delta$ and partially coherent for intermediate values of d.

5.6.3 Coherence of Thermal Sources

We therefore see that any thermal source can be made to achieve any desired degree of spatial or temporal coherence by limiting its extent with a pinhole or its wavelength range with a filter or monochromator. Both operations are costly in the sense that the coherent source is made extremely dim.

A *single-mode laser*, on the other hand, emits a beam that is almost completely coherent in space and time. This is one factor that has made the laser such an important advance in optics. (See "Coherence of Laser Sources", Sect. 8.3.)

5.6.4 Coherence of Microscope Illumination

In a microscope, the spatial coherence area may be adjusted by adjusting the numerical aperture of the condensing lens (Sect. 3.8.4). The condensing lens is filled with light, and, for analyzing the coherence of the light in the plane of the object, we may regard the condensing lens as the source. As we have just seen (Sect. 5.6.2), if the source subtends a large angle, it may be considered a spatially incoherent source; if it subtends a small angle, it is spatially coherent. The angular subtense of the condensing lens is, approximately, twice its numerical aperture. Therefore, if the numerical aperture *of the condensing lens* is small, the illumination in the microscope is spatially coherent or, at least, has a high degree of spatial coherence.

More-rigorous analysis shows, in fact, that the light may be considered coherent only if the numerical aperture of the condensing lens is less than one-tenth that of the objective lens (Problem 5.19). This is so because, then, the dimension d_c of the coherence area is greater than the resolution limit of the objective lens. Within each resolution limit, the lens effectively sees a single coherent source. Because the intensity at any point is influenced almost entirely by the illumination at other points within a few resolution limits, light whose coherence area is greater than a few resolution limits in diameter is effectively coherent.

Conversely, the illumination is largely incoherent – that is, has a small coherence area as defined by (5.99) – if the numerical aperture of the condensing lens is much greater than that of the objective. This is so because, now, d_c is much less than the resolution limit of the objective. Within each resolution limit, the lens now sees an array of many sources that are incoherent with each other; this is exactly what we mean when we say that a source is incoherent. Since the numerical aperture of a high-power objective typically exceeds 0.5, the condenser's numerical aperture usually cannot greatly enough exceed that of the objective, and the light in most microscopes is only partially coherent.

For certain quantitative measurements, such as an absolute measurement of the width of a line on an integrated circuit, the coherence of the illumination must be precisely controlled so that the edges of the line may be located precisely; for this application, the illumination is usually made highly coherent, so that the position of the edge may be associated with a particular intensity; see Sect. 7.3.3. Reducing the numerical aperture of the condensing lens reduces the irradiance greatly, however, so it is often preferable to use partially coherent illumination and account for the resulting error by acquiring an accurate calibration standard. In addition, the resolution limit of a microscope (or any lens system) is least when the light is nearly incoherent (Sect. 5.7).

5.7 Theoretical Resolution Limit

Any treatment of the theoretical resolution limit naturally follows from the discussion of diffraction by an aperture. We have deliberately postponed this topic, however, until after our discussion of coherence. In this section, we shall consider resolution not only in systems with incoherent illumination, but also in systems using the highly coherent radiation that may be emitted by a laser.

5.7.1 Two-Point Resolution

Consider the optical system of Fig. 5.21. The object consists of two incoherent points of equal radiometric intensity. The optical system is *diffraction limited* in that the images of the points are almost exactly Fraunhofer-diffraction patterns of the aperture.

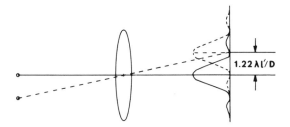

1.22 λl′/D

Fig. 5.21. Two-point resolution

When the points are very far apart, the diffraction patterns are likewise far apart, and we can resolve the points clearly. As the points are moved together, the diffraction patterns, having finite extent, begin to overlap. When the overlap is sufficiently great, the points become *unresolved*.

We now address the problem of finding the *resolution limit* of the two images. This refers to the smallest separation between the two diffraction patterns that will allow us to distinguish the two points. Usually, we apply the *Rayleigh criterion*, which states that the points are just resolved when the maximum of one diffraction pattern coincides with the first minimum of the other. This situation is shown in Fig. 5.22.

When the object is illuminated incoherently, the total intensity at any point is the sum of the individual contributions. When the points are separated as

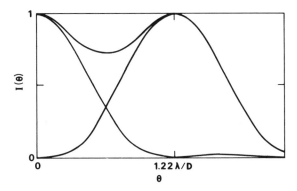

$I(\theta)$

0   1.22 λ/D

θ

Fig. 5.22. Rayleigh criterion

specified by the Rayleigh criterion, the intensity dips about 20% between the two maxima. When they are closer, the intensity dips less than 20%, and we say the points are not resolved.

Thus, two image points are just resolved when their separation (more precisely, that of their geometrical images) is equal to the radius of the Airy disk. That is,

$$RL' = 1.22 \lambda f'/D \qquad (5.100)$$

is the resolution limit for lenses with circular symmetry. RL' is often known as the *Rayleigh limit*. It is extremely difficult to resolve points or lines finer than the Rayleigh limit.

We have discussed the effect of the theoretical resolution limit on optical systems in Chap. 3.

5.7.2 Coherent Illumination

Resolution in coherent light is complicated by the fact that we must consider two cases, separately. First, let us coherently illuminate an object that does not diffuse or scatter light. Such an object must be viewed either in transmission or by direct reflection; otherwise light from the object does not enter the eye. This is the case known as *specular illumination*.

As before, we consider a two-point object. Because the points are illuminated coherently, we must add the amplitudes of the diffraction patterns, rather than the intensities, and then square to find the total intensity. We shall assume the object points to have very nearly the same phase, because they are close together and the object is smooth.

Figure 5.23 shows the intensity diffraction pattern calculated for two coherently illuminated points separated by 1.3, 1.6 and 1.9 times λ/D, subject to the assumption that they are illuminated with the same relative phase. In the first case, there is virtually no dip at all between the geometric-image points, so the points are not resolved. We therefore move them apart until there is a significant dip. Following the result of the calculation for incoherent points, we may call the

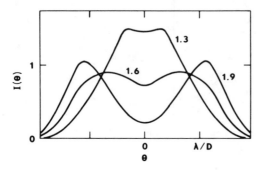

Fig. 5.23. Diffraction images of two adjacent, coherently illuminated points. Separations between the geometric images are 1.3, 1.6 and 1.9 times λ/D

separation for which the intensity dips 20% the *coherent resolution limit* RL'_c. Figure 5.23 shows that the diffraction pattern has a 20% dip when the image points are separated by

$$RL'_c = 1.6\lambda f'/D \tag{5.101}$$

for the case of circular symmetry.

RL'_c is slightly greater than the incoherent resolution limit RL'. In a way, this comes about because the amplitude diffraction pattern displays a secondary maximum 20% as strong as the principal maximum, as opposed to 4% for the intensity diffraction pattern. Thus, the radius of the diffraction pattern (and hence of the image of a point) is effectively larger in coherent light than in incoherent.

5.7.3 Diffused, Coherent Illumination

When the object is rough and diffuses or scatters light like ground glass, the situation is changed by the existence of a *speckle pattern*. The speckle pattern arises because the object is locally very rough and causes a random phase to be superimposed on the amplitude distribution of the coherent wave. The speckle pattern is a sort of diffraction pattern of a random diffracting screen and is thus a random distribution of intensity.

One of the most useful and straightforward ways to describe the speckle pattern is to think of a relatively coarse diffraction grating. The grating is so coarse that the principal maxima are separated by small angles.

Suppose we take two such gratings and locate them in the same plane, but align the grooves perpendicular to one another. If we illuminate the gratings coherently and project the diffracted light into the focal plane of a lens, we will observe a rectangular array of bright spots, corresponding to the principal maxima of the gratings. If the gratings are coarse enough, the spots are very close together, and the light diffracted from the gratings may be used for diffuse illumination. A screen made in this way is called a *nonrandom diffuser*.

Later on (Sect. 6.1) we shall calculate the chromatic resolving power of a diffraction grating. We will find there that the principal maximum that is diffracted in a given direction θ is, in fact, diffracted into a small range of angles surrounding θ. The angular width or divergence $\Delta\theta$ of the diffracted light is

$$\Delta\theta = \lambda/Nd\cos\theta , \tag{5.102}$$

where N is the total number of rulings on the grating and d is the distance between rulings.

Similarly, the light scattered by a (square) nonrandom diffuser is diffracted into cones whose angular width is $\Delta\theta$. If we focus the scattered light with a lens, we will therefore observe spots whose size R' is roughly the product of $\Delta\theta$ and the focal length f',

$$RL' = \lambda f'/Nd\cos\theta . \tag{5.103}$$

Now assume for simplicity that the diffuser is placed directly before the lens. Then the product Nd, which is the total width of the grating, is to all intents equal to the diameter D of the lens. Thus, the spots in the focal plane are characterized by a dimension

$$RL' = \lambda f'/D \cos \theta \ . \tag{5.104}$$

The result is independent of the period d of the gratings. Further, when $\cos \theta \sim 1$, the spots are about equal to the resolution limit of the lens, assuming rectangular symmetry. For circular symmetry, we would expect the spot sizes to be about equal to the Rayleigh limit, and, as above, dependent only on the F number of the lens and the wavelength of light.

Finally, the intensity in the focal plane is the Fraunhofer-diffraction pattern of the diffuser. The size of the spots does not depend on the location of the screen with respect to the lens, but is generally equal to RL'.

We are now in a position to discuss the *speckle pattern* that appears in the image of a random diffuser. To do so, we must conceptually construct the random diffuser from the nonrandom diffuser. The nonrandom diffuser is made of crossed gratings. This is equivalent to a rectangular array of holes. To randomize the diffuser, we merely place a small *phase plate* before each hole. These plates are thin, transparent lamellae whose optical thicknesses are chosen randomly to randomize the phases of the waves emerging from the openings. Their use so radically alters the wavefront emerging from the diffuser that the regular array of spots in the focal plane of the lens is transformed to an irregular, random array of bright and dark patches known as *speckles*.

However greatly the spots may be distorted, the average size of the speckles remains equal to the resolution limit of the lens, and this is true no matter what the details of the random diffuser. For finite conjugates, f' must be replaced with l' in the preceding equations.

A second approach to the properties of a random diffuser may be helpful. We begin by assuming that a lens projects a real image of a coherently illuminated diffuser. We consider the intensity at a given point, say the center of a bright speckle. The speckle is bright because a great number of waves from neighboring object points happen to interfere constructively at that point. The image of any object point is a diffraction pattern that is significant only over a region whose radius is about equal to the Rayleigh limit. For coherent light, we may take the extent of the diffraction pattern to be about RL'_c. We therefore conclude that only geometrical-image points within a radius about RL'_c can contribute significantly to the intensity at the center of the speckle.

If the intensity at a point is great, it is because the relative phases of the waves arriving at neighboring points were all about the same. The intensity will be large at another nearby point because many of the same waves contribute to the intensity at that point as well. On average, therefore, the intensity will be great over an area whose radius is about RL'_c. The average speckle size is about equal to that of the Airy disk or a little larger.

Because of the random phases, the speckles are not round, but very irregular. A great many are much larger than the Airy disk, and, as a result, the presence of the speckles degrades resolution considerably. In addition, the speckles give the image an unsightly appearance.

Figure 5.24 compares the imagery in incoherent, specular-coherent, and diffuse-coherent illumination. All photographs were exposed with the same relative aperture or F number. The series on the left shows a standard, three-bar target, whereas that on the right depicts a target made from an optometrist's tumbling-E chart. The latter target was chosen to explore the difficulty of recognizing a pattern, rather than simply resolving bars. Examination of such photographs reveals that in certain cases the presence of speckles effectively increases the resolution limit to 5 times the Rayleigh limit or more.

Sometimes, coherent light will be shone onto a diffuser and scattered onto a second surface, such as a detector, with no intervening lens. Then the diameter of the speckle is approximately $1.6 \, \lambda l'/D$, where l' is the distance from diffuser to detector and D is the diameter of the illuminated area of the diffuser. If the

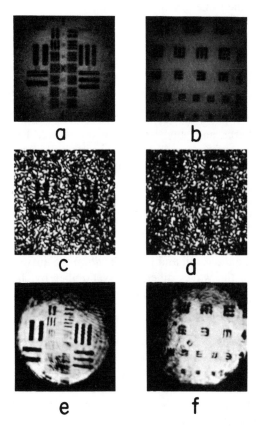

a

b

c

d

e

f

Fig. 5.24. a-f. Imaging in coherent and incoherent light. *Top*, incoherent illumination. *Center*, diffused, coherent illumination. *Bottom*, specular, coherent illumination. [After M. Young et al.: J. Opt. Soc. Am. **60**, 137 (1970)]

detector is not at least an order of magnitude larger than the speckles, substantial noise may result from relative motion of the components or changes of the structure of the laser beam.

5.7.4 Quasi-Thermal Source

When the intensity or monochromaticity of a laser is required, but the light must be spatially incoherent, a laser may be shone onto a moving diffuser. Typically, a small portion of a rotating ground-glass screen is illuminated by the laser and is in turn used as the source for the optical system. Speckles are formed by the diffuse reflection from the ground glass, but they are not stationary because of the motion of the glass. As long as the speckles move fast enough, the images seen, for example, in Fig. 5.24c and d become averages of many different speckle patterns. Here, fast enough means that the speckle pattern must shift by a few times the average speckle size within one integration time of the detector – about 30 ms for the human eye or a typical video camera. Then, the imagery becomes identical to partially coherent imaging and depends only on the diameter of the illuminated area of the ground glass. A source made by spinning a diffuser is often called a *quasi-thermal source* because its imagery is identical to that produced by a thermal source with the same dimensions.

Problems

5.1 *Wavelength and Spatial Coherence.* Consider a double-slit experiment performed with an infinitesimal source and slits, but where the source has wavelengths λ_1 and λ_2. Show that interference will not be observed in order m when $\Delta\lambda/\lambda = (1/2m)$ where $\Delta\lambda = \lambda_2 - \lambda_1$. Under what conditions would we say the source was completely incoherent?

5.2 Suppose that two waves have the same amplitude A, but that they differ in angular frequency by the amount $\Delta\omega$. They are in phase at time $t = 0$. Using complex-exponential notation, find the intensity of the two waves at time t. Explain the result physically. Examine the result and just *write down* the answer to the analogous problem, in which the waves differ in wavenumber by Δk. That is, deduce the intensity resulting from waves with different wavenumbers, assuming that they were in phase where $x = 0$.

5.3 (a) A *Fresnel biprism* is an isosceles prism with one obtuse angle that is nearly 180° and two acute angles that are only a few degrees. A distant point source illuminates the base of a Fresnel biprism at normal incidence. Find the two-beam interference pattern that falls onto a distant screen in terms of the distance L between the source and the screen and the acute angle α and index of refraction n of the prism. (Use the approximation that $\sin\theta = \theta$ for small angles.)

(b) *Lloyd's Mirror*. A point source illuminates a horizontal glass plate at nearly grazing incidence, so the reflectance is nearly 1. Find the intensity as a function of height h along a vertical screen a distance L from the point source. Assume that $h \ll L$.

5.4 Show analytically that (5.43) is true at all values of θ for which $m\lambda = d \sin \theta$.

5.5 Calculate the intensity (as a function of phase difference ϕ) of two beams whose amplitudes A_1 and A_2 are small but not equal. Express the result as $\cos^2$ fringes. (It is about as easy to use trigonometric formulas as to do the problem any other way.) As a check, show that your result reduces to the right result when the reflectances are equal. What are the maximum and minimum values of the overall intensity?

5.6 (a) Calculate the maxima and minima of a coating whose optical thickness is nd and whose index of refraction n is less than the index of refraction n_g of the glass substrate.

(b) *Sketch* the reflectance as a function of the parameter $\lambda/4nd$. [Note: A half-wave layer is equivalent to one of zero thickness in that the reflectance of such a layer is equal to that of the bare glass. A half-wave layer is sometimes called an *absentee layer*.]

(c) *Sketch* the reflectance of a layer that has $n > n_g$.

5.7 (a) Consider the Fraunhofer-diffraction pattern of a single slit. Show that (the optical path between the center of the slit and the first minimum) differs from (the optical path between one edge of the slit and the first minimum) by one-half wavelength. Explain.

(b) A circular hole with radius s is cut into an opaque screen. Collimated light falls onto the screen at right angles to it. A viewing screen is located a distance D^2/λ ($= 4s^2/\lambda$) beyond the hole. Consider a ray that originates on the edge of the hole and intersects the viewing screen on the axis of symmetry of the opaque screen. Show that the OPD between that ray and the axial ray is $\lambda/4$. What would be the OPD in the case of true Fraunhofer diffraction? What is the significance of the distance D^2/λ? (It may be prudent to use D^2/λ rather than s^2/λ for the far-field distance when the exact intensity profile rather than width or general appearance of a diffraction pattern is needed.)

5.8 Show by direct integration that the interference pattern produced by a grating with identical, finite slits is equal to the interference pattern produced by a grating with infinitesimal slits multiplied by the diffraction pattern of a single slit.

5.9 (a) What is the radius s_m of the largest Fresnel zone plate that can be made on photographic film whose resolution limit is RL? The focal length of the zone plate is f'.

(b) What is the limit of resolution of a lens with the same radius? Comment.

5.10   A diffracting screen consists of five 10-μm slits separated from one another by 20 μm. The overall dimension of the screen is therefore 130 μm. Make a rough sketch of the diffraction pattern 200 μm from the screen, assuming that the screen is illuminated at normal incidence with collimated light whose wavelength is 500 nm. [Hint: Calculate s^2/λ for the individual slits, as well as for the screen as a whole.]

5.11   A large slit with width B contains a hair with width b; the hair lies parallel to the edges of the slit. Making sure to label appropriate intensities and widths, *sketch* the resulting Fraunhofer-diffraction pattern. Draw a similar sketch for a dot with diameter d inside a circular aperture with diameter D.

5.12   *Depth of Focus.* Use Fig. 5.25 (cf. Fig. 5.19) to show that the diffraction-limited depth of focus is

$$\delta' = n\lambda/2(NA)^2 .$$

[Hint: When you are out of focus, all optical paths to the image point are not equal.] Compare to what you might get by using geometrical optics to calculate the depth of focus.

This result is generally valid for any diffraction-limited optical system, such as a microscope objective, when the image lies in a medium whose index of refraction is n.

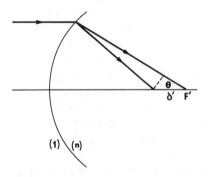

Fig. 5.25. Depth of focus

5.13   The optical-path difference between the mirrors of a particular interferometer is d. If the interferometer is illuminated with a beam whose wavelength is λ, $\cos^2$ fringes result, with

$$I_\lambda(d) = 4I_0 \cos^2 (\pi d/\lambda) ,$$

where I_0 is incident intensity. Suppose that the instrument is illuminated with a range of angular frequencies such that $I = I_0$ when $\lambda_0 - \Delta\lambda/2 < \lambda < \lambda_0 + \Delta\lambda/2$, and $I = 0$ otherwise. Assuming incoherent superposition, integrate $I_\lambda(d)$ over wavelength and show that the interference virtually disappears for all values of d that exceed $c\pi/\Delta\omega$.

5.14 A *Michelson stellar interferometer* consists of two diagonal mirrors separated by a long baseline B. The mirrors direct the (collimated) light from a distant star to the point directly between them, where interference fringes are formed. If the distance to the star is known, its diameter can be determined by measuring the spatial coherence of light of a given wavelength. Explain and find a formula for the diameter of the star in terms of B. What is the minimum value of B? Data: Largest star's angular diameter $= 10^{-8}$ rad.

5.15 The stars twinkle when observed with the naked eye; the planets do not. We conjecture that an interference effect is responsible: The beam is, perhaps, split by atmospheric turbulence so that rays arriving together at the eye have traveled difference optical paths. Is this explanation possibly correct? Data: Largest angular diameter of star $= 10^{-8}$ rad; smallest angular diameter of planet visible to the naked eye $= 10^{-4}$ rad.

5.16 A power meter for a very high-power laser consists of a diffuser and a detector located a distance L from the diffuser. The diameter of the laser beam is D, and that of the detector is d. The laser is pulsed, so each shot causes a different speckle pattern to fall onto the detector.

(a) What is the average number N of bright speckles to fall onto the detector?

(b) Treat the speckles as "billiard balls". What is the standard deviation of N? How large must N be to achieve an accuracy of 1%? Do you expect a higher or lower standard deviation in the case of a real speckle pattern? (Some knowledge of statistics is necessary to solve this problem.)

5.17 You are looking through a sheer curtain at a streetlamp across the street. There seem to be some fringes around the lamp, but only when you look through the curtain (so you know that your eyes are not failing). Are the fringes due to some sort of weird refraction in the fibers of the curtain, or could interference be responsible? Using reasonable numbers for the size and distance of the lamp, estimate the diameter of the coherence patch on the curtain. Is it larger than the mesh of the curtain? Could you be seeing an interference pattern? How can you test whether or not it is an interference pattern?

5.18 To make a quasi-thermal source, we illuminate a rotating disk of ground glass near its periphery. The diameter of the disk is 5 cm, and the diameter of the illuminated spot is 2 mm. The wavelength of the light is 633 nm. The light scattered by the disk falls onto an object 1 cm away, and the object is examined with the eye. How fast (in revolutions per second) must the disk spin in order for the speckle pattern to be invisible to the eye? Does the result change if we view the object with magnification, rather than with the naked eye? Note that we are here talking about the objective speckle that falls onto the object, not the speckle pattern that forms on the retina.

5.19 The condensing lens of a certain microscope has a diameter δ and is located a distance L from the object. The numerical aperture of this condensing lens is $NA_c = \delta/2L$. (a) Calculate the diameter d_c of the coherence patch in the plane of the object and express it in terms of NA_c. (b) The light in the plane of the object may be considered spatially coherent only if d_c is greater than the resolution limit of the objective lens. Show that this condition translates to $NA_c < 0.13\, NA$, where NA is the numerical aperture of the objective lens. That is, the light in a microscope may be considered coherent if the numerical aperture of the condensing lens is less than, roughly, one-tenth that of the objective lens.

5.20 Suppose that the optical system in Prob. 5.12 is defocused by considerably more than the depth of focus. Show that the defocus error can be partly corrected by placing a Fresnel zone plate in contact with the lens.

6. Interferometry and Related Areas

In Chap. 3, we discussed optical instruments that relied mainly on ray optics. Here we treat instruments whose operation depends on wave phenomena. These include division-of-amplitude and division-of-wavefront techniques, the physics of which was the subject of Chap. 5.

6.1 Diffraction Grating

In Chap. 5, we found that a multiple-slit screen diffracted light primarily into specific directions given by the *grating equation*

$$m\lambda = d\sin\theta \ . \tag{6.1}$$

Such a screen is known as a *diffraction grating*. In a given order m, different wavelengths are diffracted into different angles θ. A grating is thus said to *disperse* light and may be used for spectral analysis of a light source. A device that employs a grating for this purpose is a *grating spectrometer*. Spectrometers can be calibrated with sources of known wavelengths and are used in chemical analysis, astronomy, plasma diagnostics, and many other areas.

Most modern grating instruments use *reflection gratings*, such as that shown in Fig. 6.1. For such instruments, the grating equation is easily generalized to

$$m\lambda = d(\sin i + \sin\theta) \ , \tag{6.2}$$

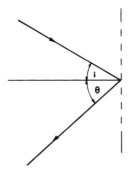

Fig. 6.1. Reflection grating

where i is the angle of incidence and i and θ have the same sign if they are on the same side of the normal.

Plane gratings are illuminated with collimated light. The source is focused on a narrow slit (Fig. 6.2). The light that passes the slit is collimated, frequently with a mirror. The relative aperture (F number) of the lens used to focus the source should be about equal to the relative aperture of the mirror. The grating disperses the light, which is then brought to a focus by a second mirror. Generally, the two mirrors are identical and, for best results, the slit width is held equal to the theoretical resolution limit of the mirror.

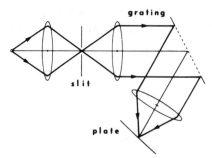

Fig. 6.2 Plane-grating spectrometer using lenses for collimating and focusing

In some instruments, wavelength is determined by placing a photographic plate in the focal plane of the second mirror. In others a slit is located at the focal point, and the grating is rotated in a known fashion, in order to direct different wavelengths through the slit. The latter device is usually called a *mono-chromator*. In either case, reference sources are needed for precise calibration.

In many instruments, the function of the grating and that of the mirrors is combined in a single, *concave reflection grating*, shown in Fig. 6.3. Such a grating disperses and focuses the light simultaneously. The entrance and exit slits are located on a circle, known as the *Rowland circle*, that is tangent to the grating.

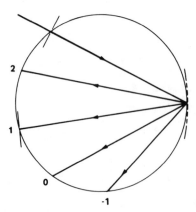

Fig. 6.3 Concave-grating spectrometer, showing various orders

6.1.1 Blazing

Reflection gratings are *ruled* approximately as shown in Fig. 6.4. Because of this, the spacing of the rulings is just equal to the width of each ruling. In Sect. 5.5.2, we found that important orders may be weak or *missing* in this case. The same would be true of the reflection grating, but for the fact that it is *blazed* for the wavelength of interest, usually in first order.

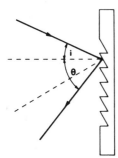

Fig. 6.4. Blazed grating

Blazing refers to the angling of the faces of the rulings as shown in Fig. 6.4. As with multiple-slit interference, the diffraction pattern of the individual ruling determines the relative strength of the light diffracted into a given order. With the reflection grating, that diffraction pattern is centered about the direction of the reflection from the faces of the individual rulings. By choosing the proper *blaze angle*, we can therefore locate the center of the single-slit diffraction pattern at any desired angle.

If we choose the blaze angle properly, we will not only find maximum intensity of the desired wavelength in the chosen order, but, in addition, we will find that the uninteresting zero-order diffraction image lies near a minimum of the single-slit diffraction pattern and is greatly suppressed. Energy that would otherwise be wasted in the zero-order image is thus conserved and increases the intensity in the chosen order.

High-quality gratings may also be made by *holography* (Chap. 7). Such gratings are highly efficient and scatter very little unwanted light.

6.1.2 Chromatic Resolving Power

We have found the shape of the diffraction pattern to be

$$I(\theta) = \frac{\sin^2\left(\dfrac{N\pi}{\lambda}\,d\sin\theta\right)}{\sin^2\left(\dfrac{\pi}{\lambda}\,d\sin\theta\right)}\,, \tag{6.3}$$

where N is the total number of rulings. A principal maximum occurs where both numerator and denominator are 0. When N is large, the numerator varies much more quickly than the denominator, so it alone determines the angular width of the maximum.

Just at a principal maximum, the numerator is 0. If we increase or decrease the argument slightly, the value of $I(0)$ falls sharply, and when the numerator reaches 0 for the next time, the denominator is no longer 0. Thus, the intensity of the diffracted light falls to 0 when θ is increased by a value of $\Delta\theta$ given by

$$(N\pi/\lambda)d\Delta(\sin\theta) = \pi, \quad \text{or} \tag{6.4}$$

$$\Delta\theta = \lambda/Nd\cos\theta. \tag{6.5}$$

This is also, approximately, the full angular width (at half maximum) of the diffracted beam.

We may use the Rayleigh criterion to estimate the smallest resolvable wavelength difference $\Delta\lambda$ between two sharp spectral lines. According to this criterion, the maximum of one wavelength coincides with the first zero of the other when the two spectral lines are just resolved. We relate $\Delta\lambda$ to $\Delta\theta$ by differentiating the grating equation to find that

$$\Delta\lambda = (d/m)\cos\theta\,\Delta\theta\,, \tag{6.6}$$

from which we find the *instrumental line width*,

$$\Delta\lambda = \lambda/mN\,. \tag{6.7}$$

This expression is commonly written as

$$\lambda/\Delta\lambda = mN\,. \tag{6.8}$$

$\lambda/\Delta\lambda$ is known as *chromatic resolving power*.

For resolution on the order of 1 nm or less, the product mN must be several thousand. Because m is generally 1 (sometimes 2 or 3), the total number of rulings on the grating must be sizeable. Typical gratings are made with 6000 rulings/cm or more, and a large, high-quality diffraction grating has tens of thousands of rulings.

6.2 Michelson Interferometer

This instrument is shown schematically in Fig. 6.5. A partially silvered mirror allows a single beam of light to fall on two mirrors, M_1 and M_2. The mirrors are adjusted to recombine the resulting two beams in a line with the observer's eye.

Looking into the beam splitter, we seem to see both M_1 and M_2 in approximately the locations of Fig. 6.6. The mirrors are apparently separated by

$$d = d_2 - d_1\,, \tag{6.9}$$

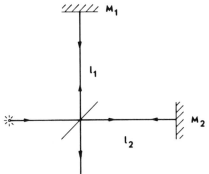

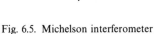

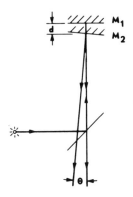

Fig. 6.5. Michelson interferometer

Fig. 6.6 Apparent positions of the mirrors in a Michelson interferometer

and interference maxima occur whenever

$$2d \cos \theta = m\lambda \, . \tag{6.10}$$

Because two beams are involved, the interference pattern is described by $\cos^2$ fringes. The interference pattern appears the same as that brought about by two nearly parallel surfaces with low reflectance. In the interferometer, however, the separation and orientation of the mirrors are adjustable. If the mirrors are tilted slightly with respect to one another, the fringes are nearly straight and parallel to the line of intersection of the mirror planes. If the mirrors are parallel, but separated somewhat, symmetry dictates circular fringes whose maxima occur at certain angles θ that depend on the value of d.

Michelson used the interferometer to measure the wavelength of light in terms of a material standard of length. Although the actual measurement is extremely difficult, it can be done conceptually in the following way.

First, illuminate the instrument with white light, whose coherence length is very nearly zero. At zero path difference only, interference can be observed as a single, black fringe. The fringe is black, not white, because only one beam undergoes a phase change of π on reflection from the half-silvered mirror.

Now, move (say) mirror M_1, maintaining it parallel to its initial position, by a distance equal to the length of the standard. Note the location of M_2 and move it until a black fringe is observed. The mirrors have now moved equal distances. Replace the white light with monochromatic light and count fringes as M_2 is slid back to its original position. If m fringes are counted, then $m\lambda/2$ is equal to the known length.

In fact the standard meter was longer than the coherence length of any available source, so Michelson had to develop several secondary material standards to complete the actual measurement.

The *visibility* V of fringes in a Michelson interferometer is defined as

$$V = \frac{I_{\max} - I_{\min}}{I_{\max} + I_{\min}} \tag{6.11}$$

and is used as a measure of the *degree of coherence* of the source. By measuring the relative mirror separation d at which the fringes disappear, we can determine the *coherence length* l_c of the source. If the source is known to consist of only one spectral line, l_c can be used to determine its *spectral width* $\Delta\omega$.

The Michelson interferometer is used today for *Fourier-transform spectroscopy* in the far infrared. It is possible to show that the visibility $V(d)$ of the fringes as a function of d is just equal to the Fourier transform of the spectral distribution $I(\omega)$ of the light. In this type of spectroscopy, $V(d)$ is measured over a wide range of d, and the transform is inverted on the computer to find $I(\omega)$.

6.2.1 Twyman-Green Interferometer

This instrument is closely related to the Michelson interferometer and resembles a Michelson interferometer illuminated with collimated light (Fig. 6.7). It is used to test flat optical windows and other optics whose transmission (as opposed to reflection) is important.

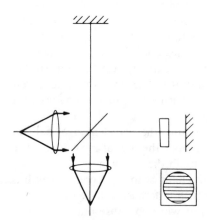

Fig. 6.7. Twyman-Green interferometer

The Twyman-Green interferometer is set up with collimated light; the mirrors are adjusted sufficiently parallel that a single fringe covers the entire field. A test piece, say an optical flat, is inserted in one arm. Any fringes that appear as a result of the flat's presence represent optical-path variations within the flat. For example, if the flat is slightly thicker on one side than on the other, it is said to have *wedge*. Looking through the flat makes the mirror appear slightly tilted and results in nearly straight fringes. Similarly, if the surfaces are not flat, but slightly spherical, the fringes will appear circular.

The Twyman-Green interferometer is also used for testing lenses. One mirror is replaced by a small, reflecting sphere, and the lens is positioned so that the center of the sphere coincides with the focal point of the lens. The beam returning through the lens is thus collimated if the lens is of high quality. Aberrations or other defects in the lens cause the returning wavefronts to deviate from planes and result in a fringe pattern that can be used to evaluate the performance of the lens.

6.2.2 Mach-Zehnder Interferometer

This instrument is shown in Fig. 6.8. Although not truly a relative of the Michelson interferometer, it is nevertheless a two-beam interferometer that works by division of amplitude. Because it is a single-pass interferometer, the Mach-Zehnder has only half the sensitivity of the Michelson or Twyman-Green interferometers. Like the latter, it may be used for testing optics for flatness, thickness variation and so on; it has also been used for measuring the index of refraction of gases.

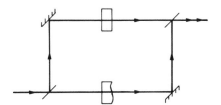

Fig. 6.8. Mach-Zehnder interferometer

The interferometer consists of two mirrors and two beam splitters. The first beam splitter divides the beam into two parts, whereas the second combines the parts after reflection from the two mirrors. In the figure, a test piece is located in the lower arm. The upper arm contains a compensator that is needed when the source has limited temporal coherence; the compensator has roughly the same optical thickness as the sample and ensures that the two arms have nearly equal optical path length. If the source is a laser, the compensator is superfluous.

An *interference microscope* is a system that allows transparent objects to be seen, provided that their index of refraction differs from the surroundings. The phase difference that results from the index difference causes interference and allows visualization of the objects. Most interference microscopes are based on the Mach-Zehnder interferometer; the highest-quality instruments use two matched objectives, one in each arm of the interferometer.

A *fiber interferometer* is an interferometer that uses an optical fiber in either or both of its arms. A common type of fiber interferometer is a Mach-Zehnder interferometer with single-mode fibers in both paths. The beam splitters may be replaced by *directional couplers* (Sect. 12.1); the mirrors are not necessary. Such an interferometer may be used as an element in a *fiber sensor*. This is an instrument that measures some quantity, such as temperature or strain (elongation), that changes the optical length of the fiber in one arm while the other arm is held constant. Usually the change of optical length is difficult to determine analytically, so such instruments are calibrated by comparison with known standards.

6.3 Fabry-Perot Interferometer

The *Fabry-Perot interferometer* consists of two highly reflecting mirrors maintained parallel to great precision. As we found before, a given wavelength is transmitted completely only when

$$m\lambda = 2d\cos\theta. \tag{6.12}$$

Like a diffraction grating, a Fabry-Perot interferometer is thus able to distinguish between wavelengths. It can be used in one of two ways. If the value of d is fixed and the interferometer illuminated with a slightly divergent beam, a given wavelength will be transmitted at several particular values of θ only. Another wavelength will be transmitted at other values of θ; the difference between the two is easily calculated. Often, only the difference is of interest; otherwise, a known wavelength must be introduced to calibrate the interferometer.

A Fabry-Perot interferometer is also used in a mode in which d is varied, generally with $\theta = 0$. Scanning may be accomplished by placing the entire instrument in a vacuum chamber and slowly lowering the pressure. This changes the optical thickness nd of the air between the plates, because n varies very nearly in proportion to the density or pressure of the air. Data collection by this method is relatively slow.

Most modern instruments have one mirror fixed to a piezoelectric crystal (whose thickness varies with applied voltage) or to a magnetic drive similar to a loud speaker. This mirror is rapidly driven back and forth through a few wavelengths, and the transmitted intensity is displayed on an oscilloscope.

6.3.1 Chromatic Resolving Power

As with the diffraction grating, we seek the smallest observable wavelength difference. The transmission profile of a single wavelength is, according to (5.52) and (5.53),

$$\frac{I_t}{I_0} = \frac{1}{1 + F\sin^2\dfrac{\phi}{2}}, \tag{6.13}$$

where we assume lossless mirrors for convenience and where $\phi = (2\pi/\lambda)2d\cos\theta$.

The maximum transmittance is 1 and occurs whenever $\phi = 2m\pi$. The transmittance is never quite 0, so we look instead for the width ε of the transmission function at half maximum, given implicitly by

$$\frac{I_t}{I_0} = \frac{1}{2} = \frac{1}{1 + F\sin^2\left(\dfrac{2m\pi + \varepsilon/2}{2}\right)} . \tag{6.14}$$

ε is the full width, so $\phi = 2m\pi + (\varepsilon/2)$ when the transmittance has decreased from 1 to 1/2. If F is large, the fringes are sharp and ε small. The sine term is expanded by the appropriate trigonometric formula and $\sin(\varepsilon/4)$ replaced by $\varepsilon/4$, so

$$\frac{1}{2} = \frac{1}{1 + F\left(\dfrac{\varepsilon}{4}\right)^2} . \tag{6.15}$$

Solving for ε, we find

$$\varepsilon = 4/\sqrt{F} . \tag{6.16}$$

It remains to relate ε with $\Delta\lambda$. We adopt the criterion that two wavelengths are resolved if their maxima are separated by ε. To find the change of ϕ that corresponds to a small change of λ, we differentiate the relation

$$\phi = (2\pi/\lambda)2d\cos\theta , \tag{6.17}$$

with respect to λ and find

$$\Delta\phi = (2\pi/\lambda^2)2d\cos\theta\,\Delta\lambda \tag{6.18}$$

(without regard to sign). If $\Delta\phi$ is set equal to ε, then $\Delta\lambda$ is equal to the minimum resolvable wavelength separation or *instrumental line width*.

We combine the last two equations and note that ϕ is very close to $2m\pi$, to find that $\Delta\lambda$ can be written

$$\Delta\lambda = \lambda/m\mathcal{N} \quad \text{(instrumental line width)} \qquad \text{or} \tag{6.19}$$

$$\lambda/\Delta\lambda = m\mathcal{N}, \qquad \text{where} \tag{6.20}$$

$$\mathcal{N} = 2\pi/\varepsilon \tag{6.21}$$

is known as the *finesse* of the instrument.

By analogy with the corresponding result for a diffraction grating, we may think of $\mathcal{N}$ as the effective number of transmitted beams. Using the relation between ε and F, we find

$$\mathcal{N} = \pi\sqrt{R}/(1 - R) . \tag{6.22}$$

$\mathcal{N}$ may be made as high as 30 or 50 before other factors, such as alignment of the mirrors, begin to influence resolution. The Fabry-Perot interferometer is nevertheless capable of very high spectral resolution . The mirrors may be separated by 1 cm or more, so the order m is typically several hundred thousand (for visible light). The product $m\mathcal{N}$ is therefore very large, and the interferometer is capable of much higher resolution than a grating spectrometer.

6.3.2 Free Spectral Range

Let a Fabry-Perot interferometer be operated in a scanning mode with $\cos\theta = 1$. It is illuminated with wavelengths λ and $\lambda + \Delta\lambda$. The approximate value of λ is known. To find $\Delta\lambda$, we could, for example, differentiate the relation $m\lambda = 2d$ to find

$$\Delta\lambda = 2\Delta d/m \ . \tag{6.23}$$

If one wavelength appears in order m at d and the other at $d + \Delta d$, then $\Delta\lambda$ is the difference between the wavelengths. The value of m is not known precisely, but $\Delta\lambda$ can be determined if the transmission maxima of the two wavelengths can be shown to have the same order. Usually this is true only if $\Delta\lambda$ is so small that the mth order of $\lambda + \Delta\lambda$ falls between the mth and $(m + 1)$th order of λ. Otherwise, the data are very difficult to interpret.

The limiting case occurs when the $(m + 1)$th order of λ coincides with the mth order of $\lambda + \Delta\lambda$. This value of $\Delta\lambda$ is the greatest wavelength range with which the instrument should be illuminated and is known as the *free spectral range* (FSR) or *range without overlap*. We find the free spectral range from the equations

$$(m + 1)\lambda = 2d \qquad \text{and} \tag{6.24}$$

$$m(\lambda + \Delta\lambda) = 2d \ . \tag{6.25}$$

Subtracting (6.24) from (6.25), we find

$$\Delta\lambda = \lambda^2/2d \quad \text{(free spectral range)}. \tag{6.26}$$

We can show further that the finesse $\mathcal{N}$ connects the instrumental line width or spectral resolution limit $(\Delta\lambda)_{\min}$ to the free spectral range $(\Delta\lambda)_{\text{FSR}}$ by the relation

$$(\Delta\lambda)_{\min} = (\Delta\lambda)_{\text{FSR}}/\mathcal{N} \ . \tag{6.27}$$

The free spectral range of a typical Fabry-Perot interferometer is small. Often light is predispersed with a grating whose spectral resolution is equal to the free spectral range of the interferometer. The interferometer thus offers an increase of 30 or more over the resolution of the grating.

6.3.3 Confocal Fabry-Perot Interferometer

A relatively recent variation on the Fabry-Perot interferometer is the *spherical, confocal Fabry-Perot interferometer*. This device uses two spherical mirrors, generally with equal radii, whose focal points coincide or nearly coincide. Because this geometry minimizes diffraction loss and the effects of mirror imperfections, the finesse $\mathcal{N}$ can be as great as 300. These instruments are almost always used to study the spectral purity of a laser whose entire spectral width is less than the free spectral range of the interferometer.

Relatively recently, Fabry-Perot interferometers have been designed with finesses in excess of 20 000. Although not confocal cavities, these use curved mirrors to reduce diffraction loss. To reduce scattering loss, the mirror coatings are deposited onto highly polished substrates that are often called *superpolished*, and the cavity is hermetically sealed. With a very short cavity, say, $d = 20\ \mu\mathrm{m}$, the instrument can display a spectrum that is 5 or 10 nm wide with good resolution.

6.4 Multilayer Mirrors and Interference Filters

6.4.1 Quarter-Wave Layer

When a light beam strikes the interface between media that have different indices of refraction, a fraction of the light is reflected. The reflectance R depends on the ratio μ of the indices of the two media. Near normal incidence,

$$R = \left(\frac{\mu - 1}{\mu + 1}\right)^2 . \tag{6.28}$$

In addition, the reflected wave changes phase by 180° when the incident light travels from the low- to the high-index material. The phase change of a wave traveling in the other direction is 0.

For an air-glass interface, μ is equal to the index n of the glass, and R is about 4%. It is possible to deposit a thin film of low-index material on the surface of the glass and thereby reduce the reflectance considerably. Suppose that the film has physical thickness d and index $n_f < n_g$. Both reflected waves have the same 180° phase change on reflection, which we therefore ignore.

The two reflected waves interfere destructively and there is a reflectance minimum when

$$2n_f d = \lambda/2 , \tag{6.29}$$

where $n_f d$ is the optical thickness of the film. Thus,

$$n_f d = \lambda/4 . \tag{6.30}$$

The layer is known as an *antireflection coating*. Any film whose optical thickness is $\lambda/4$ is known as a *quarter-wave layer*.

The waves will cancel completely only when the reflectances from the two interfaces are equal. This happens when the film index n_f is chosen so that the values of μ at the air-film interface equals that at the film-glass interface. That is,

$$n_f/1 = n_g/n_f . \tag{6.31}$$

The film should thus have index

$$n_f = n_g^{1/2} . \tag{6.32}$$

In practice, the most common material used for the film is MgF_2, whose index is about 1.38. Since the index of most optical glass is about 1.5, the match is not perfect, but R is reduced to about 1%. Special, three-layer coatings allow a greater reduction.

It is also possible to deposit a coating with index $n_f > n_g$. In this case, the wave reflected from the second interface does not undergo a phase change and the reflectance is made larger than 4%. Nonmetallic, and therefore nonabsorbing, beam splitters are often made in this way. For example, a quarter-wave layer of TiO_2 is an adequate beam splitter at 45° incidence.

6.4.2 Multilayer Mirrors

An efficient reflector can be made by using the *quarter-wave stack*, in which alternating high- and low-index layers, n_h and n_l, are deposited on a glass substrate. Each layer has an optical thickness of $\lambda/4$. Because of the phase change that occurs at alternate interfaces, the reflected waves add constructively. Fifteen or more layers can be deposited, and reflectance over 99.9% is possible when great care is taken to eliminate sources of light scattering within the films or on the surface of the substrate.

Quarter-wave stacks with relatively few layers are also used where only partial reflection is needed. These mirrors are efficient in the sense that very little is lost by absorption. Multilayer mirrors are also durable, and visible and near-ir lasers use them almost exclusively in resonators, interferometers, and other components.

6.4.3 Interference Filters

The most common *interference filters* are designed to transmit a narrow wavelength band and block all others. These are thin-film Fabry-Perot interferometers made of two thin, metal films separated by a thin, dielectric layer. (In addition to these layers, the outer surfaces of the metal layers are often antireflection coated with a stack that is relatively complicated because of the phase change on reflection from the metal.) These filters are sometimes called metal-dielectric-metal or *MDM filters*.

The two layers of metal are very thin and partially transparent. They serve as the Fabry-Perot mirrors, and their spacing is determined by the layer of dielectric. The optical thickness of this layer is determined by the relation $m\lambda = 2nd$, where m is almost always 1 or 2. Unwanted transmittance maxima occur at wavelengths greatly different from the design wavelength and are *blocked*, for example, with a colored-glass substrate.

Interference filters may be designed to pass very narrow bands, but usually sacrifice peak transmittance in that case. A typical interference filter for visible light may have a 7 nm *passband* with 70% peak transmittance or a narrower passband and lower transmittance. For example, for a 1 nm bandwidth, the peak transmittance may drop to 30% or so.

Multilayers are also designed to transmit or reflect relatively broad bands or to reflect infrared while transmitting visible light. The makeup of the layer is considerably more complicated than the quarter-wave stacks and MDM filters described here.

Problems

6.1 A diffraction grating is blazed for wavelength λ_1 in first order ($m = 1$). The slit width b is equal to the spacing d. (a) Show that the orders $m = 0$ and $m = 2$ are missing. (Just sketch the single-slit diffraction pattern (the envelope) and show that the orders 0 and 2 correspond to minima of the envelope. Remember that the center of the envelope depends only on the blaze angle and not on the wavelength.) (b) Consider the wavelength $\lambda_2 = 2\lambda_1$. Show that this wavelength appears in first order where $\sin\theta = 2\lambda_1/d$. Is this wavelength missing in first order? Is it missing in zero order?

6.2 A transmission diffraction grating has 714 grooves/mm. It is illuminated at normal incidence with a beam of visible light, whose wavelength varies from 400 to 700 nm. Show that there are two complete spectra, one between about $17°$ and $30°$, and one between about $35°$ and $90°$ (that is, two spectra on each side of the normal to the grating). (b) In (a), there is a range of about $5°$ between the first- and second-order spectra. At what number of grooves per millimeter would these spectra overlap? Explain why a coarser grating might give incorrect results if it is used with the wavelength range 400–700 nm.

6.3 If a diffraction grating has many orders, spectra in two different orders m and $m + 1$ may overlap one another. If the wavelength range of the light source is restricted such overlaps will not occur. Find the greatest wavelength range that can be employed with a diffraction grating in order m, without encountering overlapping of orders. This is known as the free spectral range of the grating.

6.4 A well corrected lens focuses the light transmitted by a plane-parallel Fabry-Perot interferometer onto a photographic plate. For one particular wavelength, $m\lambda = 2d$ at the center of the ring pattern. Show that other orders of wavelength λ emerge as circles whose radii are given by $f'(p\lambda/d)^{1/2}$, where p is an integer and f' is the focal length of the lens. (Note that order number *decreases* away from the center of the pattern.)

6.5 To estimate the effect of surface quality on a Fabry-Perot interferometer, consider an interferometer that has an accidental step in one mirror. The height of the step is h. Show that the maximum acceptable step height is approximately $\lambda/2\mathcal{N}$, where $\mathcal{N}$ is the finesse of the interferometer.

6.6 A perfect Fabry-Perot interferometer has diameter D. We estimate the divergence of any emerging ring or beam to be λ/D. Use this estimate to deduce the effect of diffraction on the limit of resolution of the interferometer. [Note: To solve this problem, you do not need much more than $m\lambda = 2d \cos \theta$ and the definition of $\mathcal{N}$. Answer: $\Delta\lambda = d(\lambda/D)^2$ with scanning operation ($\theta = 0$), or $\lambda^2\theta/D$ with photographic operation.]

6.7 An MDM filter exhibits peak wavelength λ_0 and passband $\Delta\lambda$. Roughly how much can you tilt the filter with respect to the incident beam before the transmitted wavelength changes by more than $\Delta\lambda$? This result determines the degree to which a beam must be collimated before it passes through the filter. [Answer: $(2\Delta\lambda/\lambda_0)^{1/2}$. In reality, the filter may be somewhat more sensitive to tilt because of coatings on the outer surfaces of the metallic layers.]

6.8 A spectrometer uses a grating with N lines in first order. The width of the output slit corresponds to the instrumental linewidth of the grating. To improve the resolution, we attach a Fabry-Perot interferometer to the output of the grating. Find the best mirror spacing d in terms of N and λ.

7. Holography and Image Processing

We begin this chapter by using simple arguments to describe holography and to treat such important aspects of holography as image position, resolution and change of wavelength. Under the general heading of Optical Processing, we include the Abbe theory of the microscope, spatial filtering, phase-contrast microscopy, and matched filtering; in short, what is often called Fourier-transform optics.

7.1 Holography

Holography or *wavefront reconstruction* was invented before 1948, about twenty years before the development of the laser. Its real success came only after the existence of highly coherent sources suggested the possibility of separating the reference and object beams to allow high-quality reconstructions of any object.

In Gabor's original holography, light was filtered and passed through a pinhole to bring about the necessary coherence. The source illuminated a small, semi-transparent object O that allowed most of the light to fall undisturbed on a photographic plate H, as in Fig. 7.1. In addition, light scattered or

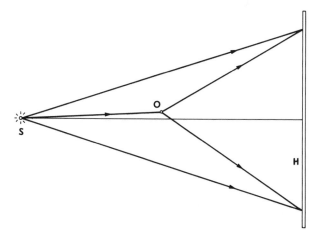

Fig. 7.1. Exposure of a Gabor hologram

diffracted by the object also falls on the plate, where it interferes with the direct beam or *coherent background*. The resulting interference pattern may be recorded on the plate and contains enough information to provide a complete *reconstruction* of the object.

To find the intensity at H we may write the field arriving at H as

$$E = E_i + E_0 , \tag{7.1}$$

where E_i is the field due to the coherent background and E_0, the field scattered from the object. The scattered field E_0 falling on H is not simple; both amplitude and phase vary greatly with position. We therefore write

$$E_0 = A_0 e^{i\psi_0} , \tag{7.2}$$

where A_0 and ψ_0 are implictly functions of position. We write a similar expression for E_i, even though E_i is usually just a spherical wave with nearly constant amplitude A_i. The field falling on the plate may then be expressed as

$$E = e^{i\psi_i}[A_i + A_0 e^{i(\psi_0 - \psi_i)}] , \tag{7.3}$$

and the intensity,

$$I = A_i^2 + A_0^2 + A_0 A_i e^{i(\psi_0 - \psi_i)} + A_0 A_i e^{-i(\psi_0 - \psi_i)} . \tag{7.4}$$

For our purpose it is convenient to characterize the photographic plate H by a curve of amplitude transmittance t_a vs exposure $\mathscr{E}$ rather than by the D vs log $\mathscr{E}$ curve of conventional photography. The t_a vs $\mathscr{E}$ curve is shown in Fig. 7.2. The curve is nearly linear over a short region; we call the slope there β.

In that region, the equation for the t_a vs $\mathscr{E}$ curve can be written as

$$t_a = t_0 - \beta\mathscr{E} \tag{7.5}$$

for the (negative) emulsion.

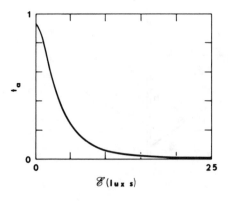

$\mathscr{E}$ (lux s)

Fig. 7.2. Film characteristic of Agfa-Gevaert 10E70 emulsion. [After F. H. Kittredge: *Diffraction Efficiency and Image Quality of Holograms Exposed on Agfa-Gevaert 10E75 Plates*. MSc Thesis, University of Waterloo, Waterloo, Ontario, Canada (1969)]

If we take the exposure time to be t, we find the amplitude transmittance of the developed plate to be

$$t_{\mathrm{a}} = t_0 - \beta t [A_i^2 + A_0^2 + A_0 A_i e^{i(\psi_0 - \psi_i)} + A_0 A_i e^{-i(\psi_0 - \psi_i)}] \,. \tag{7.6}$$

The term A_0^2 contributes to noise in the reconstruction, but we drop it here because of our assumption that the scattered field is small compared with the coherent background.

We may now remove the object and illuminate the developed plate with the original reference beam E_i. The developed plate is known as the *hologram*. The transmitted field E_t just beyond the hologram is

$$E_{\mathrm{t}} = A_i e^{i\psi_i} t_{\mathrm{a}} \,, \tag{7.7}$$

and the interesting part is

$$- \beta t A_i e^{i\psi_i} [A_i^2 + A_0 A_i e^{i(\psi_0 - \psi_i)} + A_0 A_i e^{-i(\psi_0 - \psi_i)}] \,. \tag{7.8}$$

We may factor A_i from the square bracket. Apart from real constants, the result is

$$e^{i\psi_i} [A_i + A_0 e^{i(\psi_0 - \psi_i)} + A_0 e^{-i(\psi_0 - \psi_i)}] \,. \tag{7.9}$$

The first two terms here are identical with the field that exposed the plate; the first term corresponds to the coherent background and the second to the wave scattered by the object. An observer looking through the plate would therefore seem to see the object located in its original position. Except for its intensity, this *reconstruction* is theoretically identical to the object.

The third term is identical with the second term, apart from the sign of the phase term $(\psi_0 - \psi_i)$. This term corresponds to a second reconstruction, located on the opposite side of the plate. This *conjugate reconstruction* is always present and is out of focus when we focus on the *primary reconstruction*. Its presence

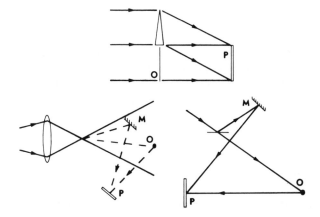

Fig. 7.3. Geometries for separating object and reference waves

therefore degrades the primary reconstruction and is a major obstacle to the production of high-quality holograms. For practical purposes, this obstacle was removed when the laser provided sufficient coherence to separate the object wave from the coherent background with a prism or beam splitter. Figure 7.3 shows several schemes for separating the two waves. The primary and conjugate reconstructions are spatially separated so that they do not interfere with one another.

7.1.1 Off-Axis Holography

The general theory of holography is too cumbersome to pursue further. Fortunately, a number of simpler arguments allow a good, working knowledge of holography.

We begin with Fig. 7.4. The *reference beam* (which replaces the coherent background) is for convenience taken normal to the plate H. We focus our attention on a small object, which may be regarded as a small portion of a larger object. The object is illuminated with a beam coherent with the reference beam; that is, the reference and object waves are derived from the same laser.

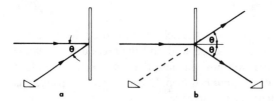

Fig. 7.4. (a) Exposure and (b) reconstruction of off-axis hologram

At a given location on the plate, the angle between object and reference beams is θ. The interference between the beams gives rise to a two-beam interference pattern. The separation between adjacent maxima is thus approximately

$$d = \lambda/\sin\theta \qquad (7.10)$$

at this location on the plate.

The developed plate or hologram therefore resembles a diffraction grating (or, for a complicated object, a superposition of diffraction gratings) whose spacing is d. When the reference beam illuminates the hologram, light is diffracted into several orders according to the grating equation

$$m\lambda = d\sin\theta' , \qquad (7.11)$$

where θ' is the direction of the diffracted waves. If the t_a vs $\mathscr{E}$ curve of the plate is approximately linear, the amplitude transmittance of the recorded fringe pattern is sinusoidal. It is possible to show that, for this case, only the orders $m = \pm 1$ exist, apart from the undiffracted, zero-order light. The hologram therefore diffracts light into the directions

$$\theta' = \pm \, \theta \; . \tag{7.12}$$

As shown in Fig. 7.4, the diffracted wave corresponding to the direction of the original object wave gives rise to a virtual reconstruction coincident with the object location. This is the primary reconstruction. The conjugate reconstruction is real (in this case) and comes about because of the second diffracted wave. What is important is that it is spatially separated from the primary reconstruction.

This diffraction-grating interpretation of holography has an important consequence. Suppose that a given film has resolving power RP. Then it cannot resolve a fringe pattern finer than $1/RP$. That is, the fringe spacing d cannot be finer than $1/RP$ if the hologram is to be recorded at all. This factor determines the greatest permissible angle θ_{max} between reference and object waves to be

$$\sin \theta_{max} = \lambda \, RP \; . \tag{7.13}$$

Provided that θ is less than θ_{max}, the hologram will be recorded and will, in principle, reconstruct the object wave precisely.

Looking at the primary reconstruction may be likened to looking through a window at the object. Resolution is determined by the relative aperture of the viewing optics, rather than by the resolving power of the photographic plate. The resolution limit is approximately equal to the radius of the Airy disk of the optical system.

The plate's resolving power does determine the highest-aperture hologram that can be recorded, however, and provides the ultimate limitation on the resolution of the reconstruction (Problem 7.2).

Problem. **Minimum Reference-Beam Angle.** An object subtends angle 2ϕ at the center of a holographic plate. (a) Find the highest spatial frequency that is brought about by interference between rays originating from different object points. Assume, for convenience, that the center of the object lies on a line perpendicular to the plate.

(b) The plate is exposed with a reference-beam angle θ and developed. The hologram is reconstructed with the same reference or reconstructing beam. Show that the reconstructing beam is dispersed into a range of angles by the self interference of the object beam and that θ must therefore exceed $\sin^{-1}(3 \sin \phi) \sim 3\phi$ to prevent the diffracted light, or *flare light*, from overlapping the desired reconstruction. (If the reference beam is much brighter than the object beam, the flare light may be weak, and θ may be reduced to ϕ.)

7.1.2 Zone-Plate Interpretation

If we look at the hologram as a whole, we find that the spacing of the interference maxima varies with position across the plate. The position and spacing of these maxima can be found by applying the relation that the OPD between the object and reference waves is equal to $m\lambda$. Clearly such a calculation would be carried out using *Fresnel's construction* for each object point and would merely repeat the derivation of the *zone plate*, whether or not the reference beam originated at ∞.

We therefore conclude that the hologram resembles a superposition of zone plates. Each zone plate is made by the interference of two beams and ideally has a sinusoidal amplitude transmittance, just as did the diffraction gratings in the preceding section. Indeed, these diffraction gratings can be regarded as portions of zone plates sufficiently small that the spacing of the maxima is substantially constant.

The sinusoidal diffraction grating gave rise to orders $+1$ and -1 only. Similarly, the sinusoidal zone plate has only one positive and one negative focal length. When the reference beam is collimated, these focal lengths are equal to the perpendicular distance of the object from the plate. When the reference beam originates at a point at a finite distance from the plate, the focal lengths f and f' are not in general equal but must be determined by application of the lens equation. The center or principal point of the zone plate lies on the line that joins the source of the reference beam with the object point (or its reflection in the plane of the plate if object and source lie on the same side of the plate).

The zone-plate interpretation of the hologram shows immediately not only why two reconstructions exist, but shows their locations as well. It is now clear why we earlier assigned one diffracted beam to the primary reconstruction and the other to the conjugate reconstruction.

Knowing the focal length of the elementary zone plates is important for other reasons. We may wish to reconstruct the hologram with a beam that differs from the original reference beam, either in location or wavelength. Although this will, in general, introduce aberrations, it may be desirable, for example, to achieve magnification. The magnification brought about by, say, a wavelength change is not obvious and is not, in general, equal to the ratio of the respective distances of the reconstructions from the plate. The magnification may be determined for a given case by finding the reconstructions of two object points and comparing their spacing with the original.

We leave as a problem to show that simple wavelength change does not result in a scale change when the reference and reconstructing beams are collimated.

7.1.3 Amplitude and Phase Holograms

A hologram recorded on film normally has a complicated amplitude-transmittance function and is often called an *amplitude hologram*. The *diffraction efficiency* of a hologram is the percentage of the reference beam that is diffracted into the primary reconstruction. The diffraction efficiency of an amplitude hologram depends on the constrast of the recorded fringes and is therefore greatest when the reference and object beams fall with equal strength on the plate. Even in that case, the maximum theoretical efficiency is about 5%, but plate nonlinearity dictates that the reference beam be three or more times stronger than the object beam. The diffraction efficiency of amplitude holograms is thus on the order of 2 or 3%.

Chemical processes known as *bleaching* convert the silver image into a silver salt that is transparent. The salt has a slightly different refractive index from the gelatin. In addition, the presence of developed silver produces a relief effect on the surface of the emulsion. These factors allow us to convert the amplitude hologram to a nearly transparent *phase hologram*. Even though the hologram has no variations of amplitude transmittance, it acts as a diffraction screen, in much the same way as a ruled reflection grating. Such a grating has only phase variations due to the ruling, but behaves in generally the same way as a transmission grating.

A phase hologram thus provides a reconstruction, just as an amplitude hologram does. Because the plate is not absorbing, though, the efficiency can be made over 10 times as great. Figure 7.5 shows diffraction efficiency as a function of optical density for bleached and unbleached holograms. The *balance ratio*, or the reference-beam intensity divided by that of the object beam, was in this case 7 to 1. This value was chosen because lower balance ratios cause higher-contrast fringes; phase holograms become nonlinear and display a great deal of scattered light or veiling glare. The balance ratio of 7 to 1 gives relatively good diffraction efficiency and relatively little veiling glare, whereas higher balance ratios result in lower diffraction efficiencies. Still, phase holograms are subject to more veiling glare than amplitude holograms. In addition, because the diffraction efficiency of the phase hologram is greatest when the optical density is 2 or more, phase holograms require substantially more exposure than amplitude holograms. They also may darken with age, especially when exposed to ambient light. These factors are usually outweighed by their high efficiency and relative insensitivity to errors in exposure.

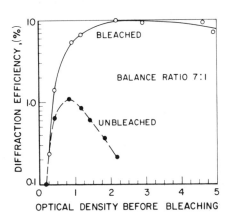

Fig. 7.5. Diffraction efficiency of amplitude-transmission and phase holograms exposed on 10E75 plates. [After M. Young, F. H. Kittredge: Appl. Opt. **8**, 2353 (1969)]

7.1.4 Thick Holograms

The photographic emulsion may be 10 μm or more thick; the interference fringes will therefore penetrate the emulsion. In this case, the hologram may resemble a pile of diffraction gratings or a venetian blind, as in Fig. 7.6a. Roughly speaking, emulsion thickness becomes a factor when the diffracted ray shown

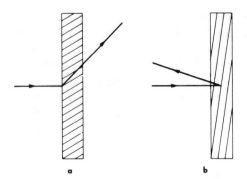

Fig. 7.6. (a) Thick transmission hologram, (b) thick reflection hologram

a   b

cannot "escape" the emulsion without interacting with another interference fringe on its way out. This depends on the emulsion thickness and fringe spacing, and on the angles involved. For most transmission holograms, the emulsion thickness seems to determine only that the diffraction efficiency is greatest when the original angles are maintained (Problem 7.3).

There is another kind of hologram, however, known as a *reflection hologram*, which is illustrated in Fig. 7.6b. A reflection hologram is exposed with the reference and object beams incident from opposite sides of the emulsion. The fringes, therefore, run roughly parallel to the emulsion and are always spaced about $\lambda/2n$ apart.

Any hologram made with the fringes parallel to the emulsion is reconstructed in reflection, rather than transmission. The process is very similar to Bragg reflection from planes in a crystal. Each fringe reflects a fraction of the reconstructing beam, and there is constructive interference only at the appropriate angle. The object is thereby reconstructed. In addition, the Bragg condition applies at only one angle of reflection, so the conjugate reconstruction does not exist.

Reflection holograms are always bleached, and the diffraction efficiency can theoretically approach 100%. Further, because Bragg reflection is a result of multiple-beam interference, it is highly wavelength selective. For this reason a reflection hologram can be reconstructed with a white-light source. The reconstruction appears colored because the hologram selects the proper wavelength. The process bears considerable resemblance to the Lippmann process of color photography (which dates to around 1900), and, as a result, reflection holograms are also called *Lippmann-Bragg holograms*.

A Lippmann-Bragg hologram is sometimes called a *white-light hologram* because it can be reconstructed in white light. Thin holograms may also be reconstructed in white light, provided that the reconstruction is very near to the plane of the hologram itself. To see this, consider a single zone plate, which is the hologram of a single point in the object. No matter what the wavelength of the laser light that exposed the hologram, the focal length of the resulting zone plate depends on the wavelength of the light used to reconstruct the hologram. If that focal length is very long, wavelengths at opposite ends of the visible spectrum will bring about reconstructions at very different points; the reconstruction will disappear into a blur of colors.

If, on the other hand, the focal length of the zone plate is short, close to 0, reconstructions with different wavelengths will appear in nearly the same planes, though there may be some colored fringes. A thin hologram may therefore be reconstructed in white light if its reconstruction lies close enough to the hologram plane.

A hologram with the required properties may be exposed by projecting an image with limited depth into the vicinity of the photographic plate and using that image as the object beam for the hologram. A hologram exposed in this way is called an *image-plane hologram*. To avoid distortion caused by the longitudinal magnification of a lens, the object beam of the image-plane hologram may itself be derived from a conventional hologram. If the image-plane hologram is backed with a mirror, its reconstruction may be viewed in reflection. Such holograms may be seen nowadays in many displays and are used on credit cards as a device to prevent counterfeiting.

7.2 Optical Processing

Optical processing and related areas are important branches of modern optics and cannot easily be done justice in a few pages. Here we offer an introduction and concentrate as usual on a physical understanding of what can sometimes require enormously complicated mathematical treatments.

We begin by considering the standard *optical processor* of Fig. 7.7. The object is illuminated by a coherent plane wave. Two identical lenses are placed in the locations shown in the figure. A simple, paraxial-ray trace will show that the lenses project an inverted image into the focal plane of the second lens.

The principal planes of the processor do not exist (or, rather, they are located at ∞). Still, we may use the Lagrange condition, *hnu* = constant, combined with the symmetry of the system, to argue that the magnification is numerically 1. Thus, the processor projects a real, inverted image at unit magnification.

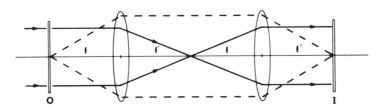

Fig. 7.7. Coherent optical processor. Solid lines show the path of the undiffracted light; dashed lines show how the image is projected by the lenses

7.2.1 Abbe Theory

We now approach the processor from the point of view of wave optics. The treatment that follows is identical to parts of the Abbe theory of microscopy, and we use the term optical processing to include many of the techniques of microscopy.

To simplify the explanation, let the object be a grating located in the *input plane* and centered about the axis of the system, as shown in Fig. 7.8. The grating spacing is d and the width of the slits is b. For convenience, let us also take $b = d/2$. The focal plane of the first lens is known as the *frequency plane* for reasons that will become clear later. The far-field diffraction pattern of the grating is displayed in the frequency plane. The electric-field amplitude may therefore be described by (5.41) multiplied by the diffraction pattern (5.57) of a single slit; that is,

$$E(\theta) = \exp[-i(N-1)\phi/2] \frac{\sin N\phi/2}{\sin \phi/2} \frac{\sin \beta}{\beta}, \tag{7.14}$$

where $\phi = (\pi/\lambda)d \sin \theta$ and $\beta = (\pi/\lambda)b \sin \theta$. The complex-exponential term is 1 if N is odd, so, again for convenience, we assume that N is odd and ignore the term.

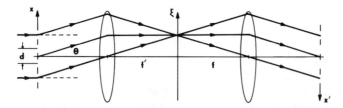

Fig. 7.8. Abbe theory of the microscope

Following our discussion of the diffraction grating, we find that the diffraction pattern consists of interference maxima located where

$$m\lambda = d \sin \theta, \quad m = 0, \pm 1, \pm 2, \dots. \tag{7.15}$$

In paraxial approximation $\sin \theta = \theta$, and principal maxima are located in the frequency plane at positions

$$\xi = m\lambda f'/d, \tag{7.16}$$

where ξ is the linear dimension in the frequency plane. Except for $m = 0$, the maxima may be grouped into pairs with orders $+m$ and $-m$. The amplitudes of the two maxima that have order $\pm m$ are proportional to the appropriate value of $\sin \beta/\beta$ in (7.14).

Figure 7.8 shows how a beam with order m is focused approximately to a point by the first lens. Beyond the focal plane, the beam diverges and is recollimated by the second lens. Each pair of orders gives rise to a pair of collimated beams that fall onto the image plane. Because of the symmetry of the system, the beams arrive at that plane with the diffraction angle $m\lambda/d$ at which they left the plane of the grating. Therefore, those two beams give rise to an interference pattern,

$$E^{(m)}(x') = \cos(2m\pi x'/d)(\sin \beta/\beta) , \tag{7.17}$$

where x' is the dimension in the image plane. The term $\sin \beta/\beta$ results from the finite width of the slits. We are interested in its value only at the angles θ at which we find the amplitude maxima in the frequency plane; that is, when $m\lambda = d \sin \theta$. For the case where $b = d/2$, we find that $\beta = \pi/2m$ and, therefore, that $\sin \beta/\beta = 2/m\pi$ when m is odd and 0 when m is even. (The even-numbered orders are missing for the special case where $b = d/2$ only.)

The total pattern in the image plane of the system is the sum of the electric-field amplitudes that result from all the orders. There is only one term with order 0. Its relative amplitude is 1 according to (7.17), so it adds a uniform value of $1/2$ to the total amplitude. If there is a total of M orders, the total amplitude is

$$E(x') = \frac{1}{2} + \sum_{m=1}^{M} \cos(2m\pi x'/d)(2/m\pi) , \tag{7.18}$$

where the sum is taken over odd values of m only. This is identical with the *Fourier-series* expansion of a square wave.

The value of M is determined by the numerical aperture of the first lens: Light that is diffracted into a high order will miss the lens entirely and not contribute to the diffraction pattern in the frequency plane. If the grating is comparatively small, the largest angle of diffraction that will pass through the lens is, in paraxial approximation, equal to the numerical aperture of the lens. Therefore, we may write the grating equation in the form

$$M\lambda = dNA , \tag{7.19}$$

from which we may easily determine M.

The optical system will not detect the grating at all unless at least one diffracted order passes through the first lens. In that case, the amplitude in the x' plane will be

$$E(x') = \tfrac{1}{2} + (2/\pi)\cos(2\pi x'/d) . \tag{7.20}$$

The intensity is the square of the amplitude and contains an extra term proportional to $\cos(4\pi x'/d)$, because $\cos^2 a = (1 + \cos 2a)/2$. The grating's image is not a simple $\cos^2$ intensity distribution, but its period is still d. It will not be visible if only the zero-order light passes through the system. The limit of resolution of the system is the spacing d of the finest grating that can be detected – that is, the finest grating whose $+1$ and -1 orders pass through the system. We may find the limit of resolution by setting M equal to 1 in (7.19). The result is λ/NA and differs slightly from our earlier, two-point resolution criterion. (In a one-dimensional analysis, that criterion yields $\lambda/2NA$; imaging points is not exactly the same as imaging a grating.)

If we allow the other orders to pass through the frequency plane, the grating becomes sharper than the sinusoidal fringe pattern and therefore more faithfully reproduces the object grating. We shall shortly return to this point in more detail. For now, let it suffice to say that the grating in the input plane will be recorded in the object plane only if the aperture in the frequency plane is large enough to pass both the 0 and either the $+1$ or -1 diffraction orders. It will be recorded as a sine wave, rather than as a square wave, but an image with the correct periodicity will nevertheless be detectable in the output plane.

Until now, we have tacitly assumed that the object is an amplitude transmission grating, consisting of alternately clear and opaque strips. We may take a glimpse of the power of optical processing if we allow the object to be a *phase grating* – that is, one that is transparent, but whose optical thickness varies periodically with the dimension x of the input plane. Ordinarily, the object is nearly invisible, because it is wholly transparent. Nevertheless, it exhibits diffracted orders, and one way (though not the best way) to render it visible would be to locate a special screen or *spatial filter* in the frequency plane and permit only the orders $+1$ and 0 to pass through holes in the screen. Then, as before, the image would be easily visible as a sinusoidal fringe pattern with the proper spacing between maxima. Spatial filtering is thus able to transform objects that are substantially invisible into visible images. This is the principle of *phase microscopy*, which is treated in more detail subsequently.

7.2.2 Fourier Series

The amplitude distribution in the frequency plane is the Fraunhofer-diffraction pattern of the object in the input plane. If the object is complicated and is characterized by an amplitude-transmittance function $g(x)$, we must generalize our earlier Fraunhofer-diffraction integral to include a factor $g(x)$ in the integrand,

$$E(\theta) = \frac{Ai\,e^{ikr}}{\lambda}\frac{}{r}\int_{-b/2}^{b/2} g(x)\,e^{-(ik\sin\theta)x}\,dx \; , \tag{7.21}$$

or, in terms of the dimension ξ in the frequency plane,

$$E(\xi) = \int_{-b/2}^{b/2} g(x)\,e^{-i\frac{2\pi\xi x}{\lambda f'}}\,dx \; , \tag{7.22}$$

apart from the multiplicative constants. If we define a new variable f_x,

$$f_x = \xi/\lambda f' ,$$ (7.23)

we have

$$E(f_x) = \int_{-b/2}^{b/2} g(x) e^{-i2\pi f_x x} dx .$$ (7.24)

The term $\xi/\lambda f$ has units of inverse length and is called *spatial frequency*, by analogy with temporal frequency.

Let us now return to the square grating in the input plane. For optical processing, we can most conveniently calculate the diffraction pattern in the frequency plane by expressing the grating function as a *Fourier series*. For convenience, we center the grating on the optical axis. In this case, only the *Fourier cosine series* need be considered.

According to the theory of Fourier series, any periodic function $g(\phi)$ may be expressed in the form

$$g(\phi) = \tfrac{1}{2} a_0 + a_1 \cos \phi + a_2 \cos 2\phi + a_3 \cos 3\phi + \ldots$$
$$+ b_1 \sin \phi + b_2 \sin 2\phi + b_3 \sin 3\phi + \ldots .$$ (7.25)

The values of the coefficients a_n and b_n may be found by multiplying both sides of the equation successively by $\cos \phi, \cos 2\phi, \ldots, \sin \phi, \sin 2\phi, \ldots$, and integrating from $-\pi$ to $+\pi$ with respect to ϕ.

The results are

$$a_n = \frac{1}{\pi} \int_{-\pi}^{\pi} g(\phi) \cos n\phi \, d\phi$$ (7.26)

and

$$b_n = \frac{1}{\pi} \int_{-\pi}^{\pi} g(\phi) \sin n\phi \, d\phi .$$ (7.27)

The square-wave grating centered on the axis is described by an *even* function, and we can verify that $b_n = 0$ in this case. The a coefficients take the values, $a_0 = 1$, $a_1 = 2/\pi$, $a_2 = 0$, $a_3 = -2/3\pi$, $a_4 = 0, \ldots$. This leads to the result (7.18) we obtained before on the basis of physical optics.

The Fourier-series operation can be said to describe the grating in terms of its *harmonics*. The a_0 term is called the *dc term*, and the a_n or b_n terms are called the nth harmonics. Figure 7.9 shows the square wave, the dc term, the result of including one harmonic and, finally, the result of adding the third harmonic. The more harmonics we include, the more nearly will the series sum approximate the square wave.

Now let us examine the distribution of amplitude in the frequency plane. For a grating with many rulings, the amplitude is very nearly 0, except at the discrete points where the grating equation is satisfied. Thus, the amplitude is appreciable

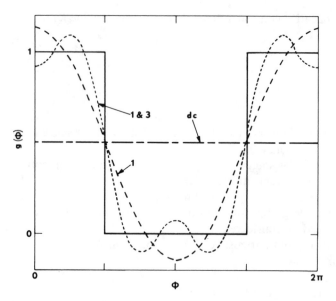

Fig. 7.9. Fourier synthesis of a square wave

only at those points in the frequency plane where

$$\xi = m\frac{\lambda f'}{d},$$ (7.28)

or, in terms of spatial frequency, where

$$f_x = m/d.$$ (7.29)

Each bright point in the frequency plane is one order of diffraction, and each pair of orders $\pm m$ corresponds to the mth harmonic of the square-wave grating.

Earlier, we arranged the optical processor to transmit only the ± 1 and the 0 orders of diffraction. The resulting image in the output plane was a sinusoidal grating. We now see that this grating is the first harmonic of the square wave. To project a good image of the square-wave grating, we must include many harmonics or, in other words, allow many diffraction orders to pass through the frequency plane. Figure 7.9, which shows the sum of the first few harmonics, also shows the effect in the output plane of restricting the frequency plane to only a few diffraction orders. The first lens of the processor is said to *analyze* the grating, the second to *synthesize* its image.

The importance of this lies in the fact that the synthesized image can be modified in a great number of ways by placing masks or phase plates in the frequency plane. For example, Fig. 7.10 shows the effect of excluding the first few orders and passing a few higher orders. The result is that continuous tones (low

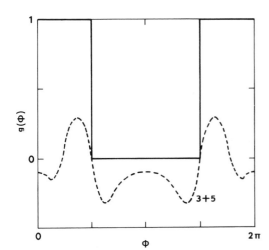

Fig. 7.10. Edge enhancement

spatial frequencies) are not recorded, whereas sharp edges are brought out because the amplitude far from the edges is small.

The lower spatial frequencies carry information about the overall distribution of light in the input plane, whereas the high spatial frequencies are required to reproduce sharp edges. This is generally true for all objects, although we have so far discussed only a grating object.

Finally, we have described the optical processor in terms of amplitudes only. This is because the functioning of the processor depends entirely on interference and diffraction. To obtain the intensity in the output plane (or anywhere else) it is necessary to calculate the absolute square of the amplitude. The image and object intensities, however, are not directly related to each other through the Fourier series. Since we observe intensity and not amplitude, this fact sometimes causes us to see artifacts in spatially filtered images.

7.2.3 Fourier-Transform Optics

In connection with Fourier series, we wrote the diffraction integral as

$$E(f_x) = \int_{-b/2}^{b/2} g(x)\exp(-2\pi i f_x x)\,dx\;, \tag{7.30}$$

where $g(x)$ is a mathematical function that describes the object in the input plane and $E(f_x)$ is a representation of the electric-field amplitude in the frequency plane.

$E(f_x)$ greatly resembles the *Fourier transform* of $g(x)$. To make the resemblance perfect, we have only to define $g(x)$ to be 0 outside the range $-b/2 < x < b/2$ and extend the range of integration from $-\infty$ to $+\infty$. The amplitude distribution in the frequency plane is then proportional to the

Fourier transform $G(f_x)$ of $g(x)$,

$$G(f_x) = \int_{-\infty}^{\infty} g(x)\exp(-2\pi i f_x x)\, dx \ . \tag{7.31}$$

We know from geometric optics that the processor casts an inverted image into the output plane. Thus, we define the positive x' axis to have the opposite direction from the positive x axis. When the x' axis is defined in this way, the amplitude distribution $g'(x')$ in the output plane will be identical to that in the input plane.

The theory of the Fourier transform shows that the *inverse transform*

$$g'(x') = \int_{-\infty}^{\infty} G(f_x)\exp(2\pi i x' f_x)\, df_x \tag{7.32}$$

is also equal to $g(x)$. We thus conclude that the second lens performs the inverse transform, provided only that the x' axis be defined, as above, to take into account the fact that the system projects an inverted image. Needless to say, the fact can also be derived by rigorous mathematics.

7.2.4 Spatial Filtering

This term is usually used to describe manipulation of an image with masks in the frequency plane. We have already encountered some examples in connection with the Abbe theory and the Fourier series.

The simplest kind of spatial filter is a pinhole located in the focal plane of a lens. It acts as a *low-pass filter* and is commonly used to improve the appearance of gas-laser beams.

A gas-laser beam is typically highly coherent. The presence of small imperfections in a microscope objective, for example, results in a certain amount of scattered light. In an incoherent optical system this is of minor importance. Unfortunately, the scattered light in a coherent system interferes with the unscattered light to produce unsightly ring patterns that greatly resemble Fresnel zone plates.

Fortunately, the rings have relatively high spatial frequencies, so these frequencies can be blocked by focusing the beam through a hole that transmits nearly the entire beam. The hole should be a few times the diameter of the Airy disk, so that very little other than the scattered light is lost.

Another important spatial filter is the *high-pass filter*. This consists of a small opaque spot in the center of the frequency plane. The spot blocks the low-frequency components of the object's spatial-frequency spectrum and allows the high-frequency components to pass. We have already seen the result in Fig. 7.10: continuous tones are not recorded and edges are greatly enhanced. High-pass filtering or *edge enhancement* can be used to sharpen photographs or to aid in examining fine detail.

Figure 7.11a shows a three-bar target with 10-μm wide lines. The picture was taken with a 40 × microscope objective. Figure 7.12 shows the intensity in the transform plane of the optical processor; the small square is located in the center of the plane to prevent the very bright zero-order light from saturating the photographic plate. The target has been processed with a high-pass filter that occupied approximately one-quarter of the frequency plane; that is, the filter blocked spatial frequencies up to one-quarter of the highest spatial frequency passed by the microscope objective. The intensity of such an image is low because most of the light is blocked by the spatial filter, but it can be seen readily with the aid of a video microscope (Sects. 3.5, 7.4).

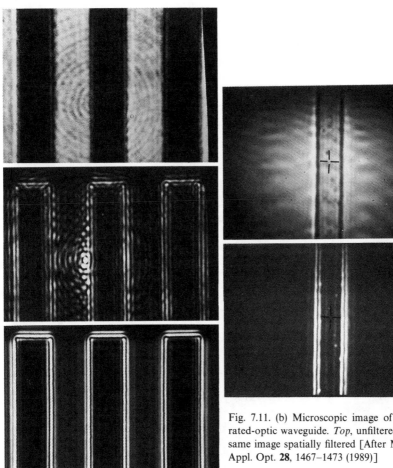

Fig. 7.11. (b) Microscopic image of an integrated-optic waveguide. *Top*, unfiltered. *Bottom*, same image spatially filtered [After M. Young, Appl. Opt. **28**, 1467–1473 (1989)]

Fig. 7.11. (a) Microscopic images of 10-μm lines. *Top*, coherent image. *Center*, spatially filtered image. *Bottom*, spatially filtered with quasi-thermal source.

Fig. 7.12. Diffraction pattern of the lines in Fig. 7.11

The uppermost image is not spatially filtered. The central image shows the three bars spatially filtered with a helium-neon laser source (Sect. 8.4.4); the unsightly ring pattern is the diffraction pattern of the aperture stop of the microscope objective. To eliminate these rings, the bottommost image was exposed with a small quasi-thermal source (Sect. 5.7.3) whose coherence area was smaller than the diameter of the aperture stop.

The edges of the lines are characterized by a minimum of intensity enveloped by a halo; the minimum of intensity is related to the zero of amplitude in Fig. 7.10. It can be used as an aid to making dimensional measurements, but the halo is an unwanted artifact that sometimes makes spatial filtering undesirable for microscopic imaging of, say, biological specimens.

Figure 7.11b is a transparent object, an integrated-optical waveguide, filtered like the lines in Fig. 7.11a. The upper, unfiltered image is not especially useful, but the lower image shows defects in the image and can be used to measure the width of the waveguide very precisely.

Spatial filtering may also be used to remove unwanted detail from a photograph or to identify a defect in a photograph. For example, Figure 7.13 (left) is

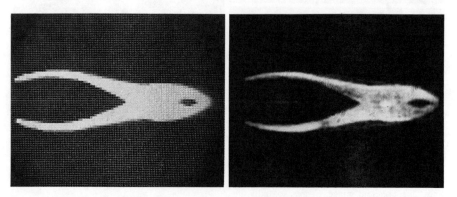

Fig. 7.13. *Left*, coherent image of a pair of pliers displayed on a liquid-crystal television monitor. *Right*, the same image spatially filtered to remove the grid. [Courtesy of Matthew Weppner]

a photograph of an image on a liquid-crystal television receiver, an example of a *spatial light modulator*, that has been illuminated in transmission. The light source, a He–Ne laser, is coherent. The picture is composed of approximately 150 × 100 picture elements separated by the fine wires used to apply a voltage to the liquid crystals. The spacing of the wires determines the highest spatial frequency in the photograph. If we place the photograph in the input plane of the processor, we will see very strong diffraction orders in the frequency plane. These orders correspond to the harmonics of the grating formed by the horizontal lines.

To eliminate the wires from the image, we carefully insert two knife edges in the frequency plane (Fig. 7.14). We locate the knife edges so that they cut off the + 1 and − 1 diffraction orders of the grating but pass all lower spatial frequencies. The result is that the picture is passed virtually unchanged, but the lines are eliminated completely. The picture has not been blurred, and the finest details are still visible in the output plane. Only the lines are absent (Fig. 7.13, right).

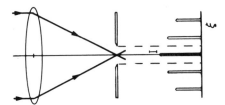

Fig. 7.14. Low-pass filter

We can make more complicated and more useful frequency-plane masks by photography. For example, suppose we needed to compare two objects. The objects could be a set of alphanumeric characters, healthy and diseased cells, or transparencies suspected of having small defects.

To begin, we place an object $h(x)$ in the input plane. Its transform is $H(f_x)$. Suppose $h(x)$ is a master of some sort, and we wish to examine a set of reproductions $g(x)$ for defects. Since photographic film responds to intensity, we record the square of $H(f_x)$ as a photographic negative and position the negative carefully in the frequency plane. A mask made in this way attenuates $H(f_x)$ severely and transmits very little to the output plane.

On the other hand, suppose $g(x)$ is slightly different from $h(x)$. Its transform $G(f_x)$ will not be precisely identical with H, and some light will pass through the frequency plane to the output plane. This light forms an image of the area in which there is a defect. In a complicated integrated-circuit mask, it is nearly impossible to find small defects visually; spatial filtering renders them easily visible.

We can most easily see how this technique works by discussing an object made up of several gratings, each having a different spatial frequency. When we record the transform of the object, the negative will display several sets of dark spots corresponding to the orders of the different gratings.

We now test a number of similar gratings. Suppose one of them contains a strip with the wrong spatial frequency. The light from that strip will pass through the frequency plane unobstructed. Geometric optics shows that an image of that strip will appear in the output plane. Similarly, if there is a defect in one of the gratings, light will be diffracted by the defect and will, in part, pass around the opaque areas in the frequency plane. The defect will thus be seen in the output plane.

This technique works best on objects that have a relatively limited number of spatial frequencies and does not work well on objects that have a continuous range of frequencies. It has been applied successfully, for example, to the detection of defects in arrays of integrated circuits and photographically reduced masks for manufacture of integrated circuits. For a more powerful method based on holography, see Sect. 7.2.6.

7.2.5 Phase Contrast

Many objects in microscopy are virtually transparent and, hence, nearly invisible. They are often characterized by refractive-index variations that influence the phase of a transmitted light wave. Zernike's *phase-contrast microscopy* provides a way of viewing such invisible *phase objects*.

As before, we begin for simplicity by considering a grating in the input plane of an optical processor. This time, however, we assume the object to be transparent but to have refractive-index or thickness variations such as the grating in Fig. 7.15. Such an object is described by an amplitude-transmittance function

$$g(x) = e^{i\phi(x)} , \tag{7.33}$$

where $\phi(x)$ is the phase variation across the object.

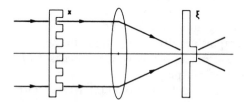

Fig. 7.15. Zernike's phase-contrast microscopy

If the phase variation ϕ is small, we may expand the exponential in a Taylor series and drop all but the first two terms,

$$g(x) \cong 1 + i\phi(x) . \tag{7.34}$$

If the amplitude distribution in the frequency plane is not manipulated with a spatial filter, the image in the output plane will be identical to that in the input plane; that is,

$$g'(x') = 1 + i\phi(x') \,. \tag{7.35}$$

The irradiance in the output plane is just

$$|g'(x')|^2 = 1 + \phi^2(x') \,. \tag{7.36}$$

To first order in ϕ, this is just 1, as we would expect.

Let us examine the electric-field amplitude distribution $G(f_x)$ in the frequency plane,

$$G(f_x) = \int_{-b/2}^{b/2} [1 + i\phi(x)] e^{-i2\pi f_x x} dx \,. \tag{7.37}$$

We may divide the integral into two terms,

$$G(f_x) = G_1(f_x) + iG_2(f_x) \,. \tag{7.38}$$

The first term

$$G_1(f_x) = \int_{-b/2}^{b/2} e^{-i2\pi f_x x} dx \tag{7.39}$$

is just the diffraction pattern of the aperture located in the input plane. We may think of it as zero-order diffraction. G_1 therefore describes the Airy disk, a sharp, bright point of light in the neighborhood of the point $f_x = 0$. G_1 is nearly 0 at all other points in the frequency plane.

The second term,

$$G_2(x) = \int_{-b/2}^{b/2} \phi(x) e^{-i2\pi f_x x} dx \,, \tag{7.40}$$

is the diffraction pattern of the phase object and corresponds to an amplitude distribution in the frequency plane. It is identical with the diffraction pattern of an object whose amplitude transmittance is $\phi(x)$. Because $\phi(x) \ll 1$, the diffraction orders are much weaker than the central order.

Further, the term $G_2(x)$ is preceded by a factor i. Because

$$e^{i\pi/2} = i \,, \tag{7.41}$$

this factor corresponds to a phase difference of $\pi/2$ between the zero order and the higher orders.

We could retrieve the information about the phase object by placing a high-pass filter in the frequency plane. The second lens would synthesize an amplitude pattern from the higher orders. Unfortunately, this technique would be inefficient because most of the energy passes through the central spot, and it would also exclude low-frequency information.

Instead, we locate a *phase plate* in the frequency plane. Such a plate is usually a good optical flat with a layer of transparent material deposited in the center. We choose the thickness of the transparent layer so that it changes the phase of the G_1 term by $\pi/2$, while leaving the G_2 term as a whole intact. (The layer must be deposited over an area roughly equal to that of the Airy disk.)

The presence of the phase plate changes the amplitude distribution in the frequency plane from $G(f_x)$ to

$$G'(f_x) = i[G_1(f_x) + G_2(f_x)] \, . \tag{7.42}$$

The G_1 and G_2 terms are now in phase with one another.

The second (or synthesizing) lens performs an inverse transform on $G'(f_x)$, giving an image $g'(x')$. Because G_1 represents the entire input plane, its inverse transform is a constant (which we took to be 1). G_2 is the transform of the grating function $\phi(x)$, so the inverse transform of G_2 is an amplitude grating with the same spacing as the original phase grating.

Thus,

$$g'(x') = i[1 + \phi(x')] \, . \tag{7.43}$$

The factor of i is common to both terms and is therefore inconsequential. It vanishes when we calculate the intensity,

$$|g'(x')|^2 \cong 1 + 2\phi(x') \, , \tag{7.44}$$

to first order in ϕ. Not only has the phase plate rendered the object visible, but it has also rendered the intensity variations proportional to the phase variations in the original object.

Although we used a grating for conceptual simplicity, the mathematics have been completely general, and phase-contrast microscopy can be used for any phase object for which $\phi(x) \ll 1$. When ϕ is large, the technique will allow phase objects to be visualized, but the intensity variations will no longer be proportional to the phase variations.

Phase-contrast is used primarily in microscopy to observe minute, transparent structures that have different refractive indices from the index of their surroundings. To increase contrast, phase contrast is often combined with high-pass filtering; that is, the phase-contrast filter includes a metal film with a transmittance of 10 or 20%, as well as the phase plate. Unfortunately, the phase-contrast image exhibits a halo like that in Fig. 7.11. To increase efficiency, commercial phase-contrast microscopes use an annular condensing lens and an annular filter, but the principle is the same.

7.2.6 Matched Filter

This is a kind of spatial filtering that can be described adequately only with Fourier-transform optics. It is closely related to holography in that a reference

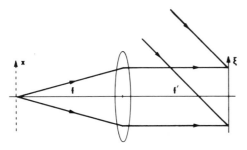

Fig. 7.16. Synthesis of matched filter

beam is introduced in the frequency plane. Figure 7.16 shows the first half of the processor, with a collimated reference beam falling on a photographic plate located in the frequency plane. The reference beam makes an angle θ with the normal to the plate. Thus, the (complex) amplitude on the plate is

$$\exp\left(-\frac{2\pi}{\lambda}i\xi\sin\theta\right) = \exp\left(-2\pi i\alpha f_x\right), \tag{7.45}$$

where $\alpha = f'\sin\theta$. If the input to the processor has amplitude $h(x)$, then the total amplitude falling on the plate is

$$H(f_x) + e^{-2\pi i\alpha f_x}, \tag{7.46}$$

and the film records intensity

$$I(f_x) = 1 + H^* e^{-2\pi i\alpha f_x} + H e^{2\pi i\alpha f_x} + |H|^2. \tag{7.47}$$

If the variation of exposure along the plate is small enough, the amplitude transmittance $t(f_x)$ of the developed plate will be roughly proportional to $I(f_x)$, so

$$t(f_x) = 1 + H^* e^{-2\pi i\alpha f_x} + H e^{2\pi i\alpha f_x} + |H|^2, \tag{7.48}$$

apart from a multiplicative constant.

Suppose, now, that we place the developed film in its original location in the frequency plane. We replace the object with another, whose amplitude transmittance is $g(x)$. The amplitude of the wave leaving the frequency plane is the product of $t(f_x)$ and $G(f_x)$. The important terms are

$$GH^* e^{-2\pi i\alpha f_x} + GH e^{2\pi i\alpha f_x}. \tag{7.49}$$

Suppose first that the inputs g and h had been identical. Then the first term would become

$$GH^* e^{-2\pi i\alpha f_x} = |H|^2 e^{-2\pi i\alpha f_x}. \tag{7.50}$$

Because $|H|^2$ is real, the first term describes an amplitude-modulated plane wave traveling in precisely the same direction as the original reference wave. The second lens focuses the wave to a sharp point.

In general, however, g and h are different and GH^* is not real. The wave is therefore only roughly a plane wave and is not focused to a sharp point. Rather, it becomes a more diffuse spot in the output plane.

This principle may be used in *pattern recognition* or *character recognition*, in which a filter is made and used to distinguish between h and a family of characters, $g_1, g_2, \ldots$. This type of filter is known as a *matched filter*. The outputplane distribution in the neighborhood of the focused beam is equivalent to the mathematical *cross-correlation* function between the characters g and h.

The term GH, incidentally, is almost never real, even if $g = h$, but results in a second distribution known as the *convolution* of g and h. Convolution and cross correlation are shown in Fig. 7.17.

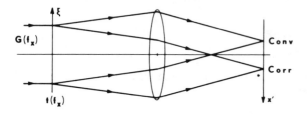

Fig. 7.17. Cross-correlation and convolution from a matched filter

Matched filtering is closely related to holography; the matched filter is no more than the hologram of the object's Fourier transform, that is, a *Fourier-transform hologram*. However, the hologram is worked in reverse, and the object wave is used to reconstruct the reference beam.

7.2.7 Converging-Beam Optical Processor

The optical processor shown in Fig. 7.7 is often chosen for precise applications because both the transform and the image lie on planes (provided that the lenses are properly designed for imaging and for Fourier transformation). When a flat field is not important, however, an optical processor may be set up with a single lens, as in Fig. 7.18. This is sometimes called a *converging-beam optical processor*.

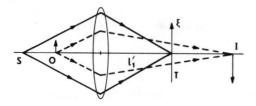

Fig. 7.18. Converging-beam optical processor

The lens may be anything from a microscope objecive to a telescope doublet or a high-quality enlarging lens, depending on the size of the object and the need for precision.

The lens projects the image of the source S into the transform plane T. The transform of the object O does not really lie in a plane but on a sphere. The projection of the transform onto the plane is equal to the transform multiplied by a phase factor with the form $\exp(iA\xi^2)$, where A is a constant. The argument of the exponential function is known as a *quadratic phase factor*; the quadratic form is the result of the sag formula (5.65). In many instances, such as spatial filtering, the phase factor has no importance, so the converging-beam processor may be adequate.

The transform plane lies a distance l_1' from the lens, not f', as before. This results in scaling the transform by the factor $l_1'/f' = (1 + m_1)$, where m_1 is the magnification of the source. That is, the factor f' in previous equations [such as (7.23)] must be replaced by l_1'.

The image I is located in the plane predicted by geometrical optics. In essence, the lens automatically performs the inverse transform. The image also lies on a sphere (that is generally true of lens systems), but again this may have no consequence. The magnification may be varied by translating the object and the image planes axially; this is one advantage of the converging-beam optical processor over the two-lens arrangement, whose magnification is always 1.

Most optical processing may be carried out with the converging-beam processor as with the conventional system, except for the scale changes of image and transform.

The converging-beam processor has an additional advantage in cases where only the transform (not the image) is desired; this may be the case in certain electronic applications where the Fourier transform of an electronic signal is to be generated optically.

Suppose that the object of Fig. 7.7 has fine detail near the edge of the field. To resolve that detail with the correct spatial frequency, the lens will be required to transmit both the $+1$ and -1 diffraction orders. Therefore the lens will have to be larger than the object. If we require diffraction-limited performance at the edge of the object, the lens diameter will have to be equal to $h + \lambda f'/d$, where h is the radius of the object and d is the required resolution limit. This can result in a lens diameter twice that required for the same resolution limit on the axis.

To see this, suppose we require a limit of resolution d; then the lens diameter D will be $2\lambda f'/d$. Because the incident light is collimated, the largest useable object height is $D/2$; however, near the edge of the object, no spatial frequency will pass both the $+1$ and -1 orders through the lens. To ensure that both orders pass through the lens, we would have to increase its diameter by $\lambda f'/d$, while leaving the object's diameter unchanged. This results in a diameter of $2D$, or twice the object size.

The resolution limit of the lens is now $\lambda f'/D$ at the edge of the object and $\lambda f'/2D$ at the center (if the lens is diffraction limited). Therefore, to achieve the required resolution limit away from the axis, we require a high-aperture lens that, on the axis, displays higher performance than we require.

The same effect need not occur with the converging-beam apparatus; the object may be placed in contact with the lens (and on either side of it). Because there is no separation between the object and the lens, beams diffracted from all positions on the object pass through the lens; the lens need be no larger than the object to pass diffraction orders from extreme points on the object.

7.3 Impulse Response and Transfer Function

7.3.1 Impulse Response

Consider, for the moment, an object that consists of a single point. The image of that object is not a point, but rather it is a small, diffuse spot. If the lens is diffraction limited, the image is the Airy disk; otherwise, the image is a spot whose properties are determined by the aberrations of the lens. In either case, the image of a single point is known as the *point-spread function* or the *impulse response* of the system.

In systems that employ coherent light, where we add amplitudes, the appropriate impulse response is the amplitude response to a point source. With diffraction-limited optics, the impulse response is the amplitude diffraction pattern of a point source; otherwise, in an aberration-limited system, it is the Fourier transform of the amplitude distribution across the exit pupil of the lens. This amplitude distribution is known as the *pupil function P*.

Similarly, in systems employing incoherent light, where we add intensities, the appropriate impulse response is the intensity response to a point source. The intensity impulse response is thus the absolute square of the amplitude impulse response.

When a system is diffraction limited, the impulse response is just the diffraction pattern of the aperture. In one dimension, this means either $\operatorname{sinc} \beta = (\sin \beta)/\beta$ or its square (Fig. 5.13), depending on whether the light is coherent or incoherent; see (5.59a) and (5.59b). In a two-dimensional system that has circular symmetry, the sinc function must be replaced by the *sombrero function*, $\operatorname{somb} \beta = 2J_1(\beta)/\beta$, where J_1 is a Bessel function. The sombrero function resembles a sinc function that has been rotated about the $\beta = 0$ axis; most conclusions about one-dimensional systems also apply qualitatively to two-dimensional systems.

Systems whose impulse response does not vary with the positions of object and image are known as *shift invariant*. (Sometimes they are called *aplanatic*, but this definition of aplanatic is not quite the same as the lens designer's definition.) In the following pages, we assume that the systems are shift invariant. (In reality, however, coherent-optical systems are never shift invariant. To see this, consider a small grating well off the axis of a lens. The lens may transmit a larger number of positive diffraction orders than negative; as a result, the appearance of the image will be different from that of an on-axis object. Fortunately, an optical system can usually be considered shift invariant over a small region around the optical axis. See Sect. 7.2.7.)

What is the image of an extended object when the impulse response of the lens is $r(x')$? Suppose that the object is illuminated incoherently and that the intensity in the object plane is $I(x)$. Because the image of a point is a spread function, each point x' in the image plane contributes to the intensity $I'(x_i')$ at a particular point x_i'. Therefore, the image intensity $I'(x_i')$ must be expressed as an integral over the entire x' plane, though often an integral over a few Airy-disk diameters will suffice.

Consider a point x in the object plane. It gives rise to a spread function surrounding the point x' in the image plane. If x' is located a distance ξ from x_i', then the contribution of the point x' to the intensity at x_i' is proportional to $r(\xi)$, the value of the impulse response at x_i' (Fig. 7.19).

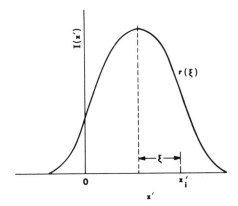

Fig. 7.19. Intensity at image point x_i' due to a neighboring image point located at x', a distance ξ away from x_i'

Further, we assume that the contribution of the point x' is proportional to the intensity of the geometric-optics image at x'; that is, we assume that bright object points contribute proportionately more than dim ones. The geometric-optics image intensity $I_g'(x')$ has the same functional form as the object intensity $I(x)$ but is scaled to take into account the magnification of the optical system.

The distribution of intensity in the image is therefore given by the integral

$$I'(x_i') = \int_{-\infty}^{\infty} I_g'(x') r(x_i' - x') dx' , \qquad (7.51)$$

where we have used the fact that $\xi = x_i' - x'$, as shown in Fig. 7.19. In the language of Fourier theory, the image intensity is the *convolution* of the object intensity and the impulse response of the lens.

Similarly, when the light is coherent, the distribution of *electric-field ampli-tude* in the image is the convolution

$$E'(x_i') = \int_{-\infty}^{\infty} E_g'(x')r_c(x_i' - x')dx' , \qquad (7.52)$$

where r_c represents the amplitude or coherent impulse response and E_g' repres-ents the image amplitude predicted by geometrical optics. The distribution of intensity in the image is the square of the convolution, not the convolution itself, when the light is coherent. This is analogous to the rule to add intensities when light is incoherent, but to add amplitudes, then square, when the light is coherent.

Convolution between two functions may be visualized as inverting one of the functions and sliding it across the other. A simple change of variables in (7.51) or (7.52) shows that either function may be slid across the other with the same result. To calculate the value of the convolution function at each position along the horizontal axis, multiply the two functions together and integrate the result; this is the value of the convolution at that point. Repeat the operation at every point to generate the complete convolution function. For more detail, see Sect. 7.4.3.

7.3.2 Edge Response

If the object is a sharp edge, its image is the convolution of a step function with the impulse response of the lens. This image is called the *edge response*; it is a function of amplitude when the light is coherent and a function of intensity when the light is incoherent. The edge response is not defined when the light is partially coherent.

For simplicity, consider the one-dimensional case with incoherent light. If we replace I_g' in (7.51) with a step function, $I_g'(x') = 1$ when $x' < 0$ and 0 otherwise, we find that the value of the edge response at any point x_1' is

$$I'(x_i') = \int_0^{\infty} r(x_i' - x')dx' . \qquad (7.53)$$

If we let $\xi = x_i' - x_i$, we find that

$$I'(x_i') = -\int_{x_i'}^{\infty} r(\xi)d\xi = \int_{-\infty}^{x_i'} r(\xi)d\xi , \qquad (7.54)$$

which is just the integral of the impulse response from $-\infty$ to x_1'. The integral is similar in two dimensions, except that integration is taken over the entire y axis as well. Figure 7.20, short dashes, shows the edge response of an incoherent, diffraction-limited system in two dimensions. The horizontal axis is expressed in units of the resolution limit $RL = 1.22\lambda f'/D$, (5.62). Because the

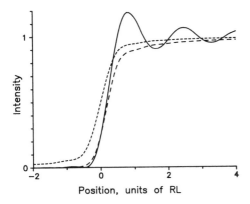

Fig. 7.20. Diffraction-limited images of an edge. *Solid curve*, coherent light. *Short dashes*, incoherent light. *Long dashes*, scanning confocal microscope. [Courtesy of Gregory Obarski]

impulse response is always positive, the integral (7.54) rises steadily, and the edge response approaches its asymptotes 0 and 1 only relatively slowly.

When the light is coherent and specular (Sect. 5.7), we must calculate the coherent edge response (in electric-field *amplitude*) and then square the result to find the intensity; the intensity function itself is not the edge response. In one dimension, the coherent impulse response of a diffraction-limited lens is proportional to $(\sin \beta)/\beta$. Away from the principal maximum, this function is oscillatory; this oscillation in turn gives rise to oscillation, or *edge ringing*, in the edge response and therefore in the image of an edge. The solid curve in Fig. 7.20 shows the image of an edge calculated for a diffraction-limited system in two dimensions. Very near the edge, there is an overshoot of nearly 20%. This overshoot gives rise to artifacts in images that contain many sharp edges. Incoherent images are free of these artifacts because the edge response in incoherent light does not display edge ringing.

Finally, because the impulse response of a rotationally symmetric lens is rotationally symmetric, the geometrical image of the edge may be located, when the light is incoherent, by finding the position at which the value of the intensity is 1/2 the asymptotic value. When the illumination is coherent, the edge may be found by locating the position at which the *amplitude* is 1/2 the asymptotic value; squaring to get the intensity, we find that the geometrical image of the edge is located at the point where the intensity is 1/4 the asymptotic value. We call these values the respective *edge-finding levels*.

7.3.3 Impulse Response of the Scanning Confocal Microscope

Suppose that a scanning confocal microscope (Sect. 3.6) is illuminated with a laser, so the light may be considered highly coherent. The impulse response of this microscope is not obvious but may be deduced from a simple argument.

The microscope in Fig. 7.21 is shown for clarity as a transmission microscope, but the argument applies equally to reflection microscopes. Assume that

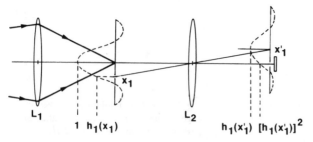

Fig. 7.21. Scanning confocal microscope

the objectives are identical and that the object is scanned across their common focal plane. For convenience, take the magnification of the second objective to be 1.

The impulse response of any coherently illuminated optical system is by definition the amplitude image of a point source. In a scanning microscope, however, the image is generated by scanning an object across the object plane, that is, the plane that contains the image of the source. Therefore, the impulse response we seek is the image obtained by scanning a tiny pinhole – a mathematical point – across the object plane; that is, it is the amplitude that falls on the detector in the image plane as a function of the position of the pinhole in the object plane. This pinhole is a mathematical construction and should not be confused with a real pinhole such as that in front of the detector.

Figure 7.21 shows the impulse response $h_1(x)$ of the first lens L_1. The fictitious pinhole is located a distance x_1 from the axis, that is, from the center of the impulse response of L_1. The electric field amplitude transmitted by the pinhole is $h_1(x_1)$. Because the pinhole is infinitesimal, its far-field diffraction pattern is a uniform wave; the amplitude that falls onto the second lens L_2 is proportional to $h_1(x_1)$. Because the wave that falls onto L_2 has constant amplitude, the image that L_2 projects onto the plane of the detector is just its impulse response $h_1(x')$, where we assume that $h_2 = h_1$. Because the pinhole is off-axis by x_1, however, the center of that impulse response is located at x_1', the position of the geometrical image of the pinhole. The distribution of amplitude in the detector plane is therefore proportional to $h_1(x_1) \cdot h_1(x_1' - x')$.

Consider now the case where an infinitesimal pinhole lies in front of the detector, where $x' = 0$. The net amplitude incident on that pinhole is $h_1(x_1) \cdot h_1(x_1')$. Because the magnification is 1, $x_1' = x_1$, and the amplitude incident on the pinhole is $[h_1(x_1)]^2$. That is, the impulse response of the scanning confocal microscope is the square of the usual impulse response for coherent light – even though the light is coherent. If the lenses are not identical or if the magnification of L_2 is not 1, the impulse response of the microscope is just the product of the impulse response of L_1 and that of L_2 as seen in the object plane; that is, the diffraction pattern in the detector plane is projected through L_2 into the object plane and demagnified accordingly to find the effective impulse response in the object plane.

This result – that the impulse response of the scanning confocal microscope is the square of the ordinary coherent impulse response – has an important consequence. If a lens is diffraction-limited, its coherent impulse response displays strong oscillation outside the principal maximum (see also Sect. 5.7.2). The image of a sharp edge as a result displays a significant overshoot. In the scanning confocal microscope, by contrast, the impulse response has the same functional form as that of an incoherent system and therefore displays a secondary maximum of only 4% of the principal maximum and no undershoot. Even though the light is coherent, therefore, images in the scanning confocal microscope do not suffer from the edge ringing that coherent systems usually display.

Further, the image of an isolated point is the square of the sombrero function in a conventional microscope, whether coherent or incoherent (though the edge responses are different). In a scanning confocal microscope, however, the image of a point is the square of its impulse response, or the *fourth* power of the sombrero function. Figure 7.22 compares the image of an isolated point in a conventional microscope (solid curve) and a scanning confocal microscope (dashed curve). The dashed curve is the square of the solid curve. Besides being narrower by about 30%, the dashed curve displays no visible secondary maximum and is therefore said to be *apodized* (Problem 7.8).

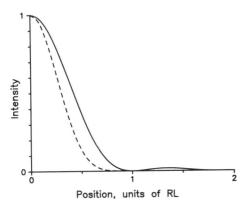

Fig. 7.22. Images of an isolated point. *Solid curve,* conventional microscope, either coherent or incoherent light. *Dashed curve,* scanning confocal microscope. [Courtesy of Gregory Obarski]

Finally, the detected image of an edge in a scanning confocal microscope is the square of the coherent edge response. It is shown as the long dashes in Fig. 7.20 and is slightly sharper than the image in a conventional, incoherent microscope. The scanning confocal microscope can be used to measure linewidth with precision less than 0.1 μm and, because of the coherence of the light, may present fewer problems of analysis than conventional microscopes (Sect. 7.4.5).

7.3.4 Image Restoration

This is a very powerful technique whereby an image (usually a recorded image) that has been blurred in a known way may be sharpened or *deblurred*. The simplest case to analyze is that of blurring by steady, one-dimensional motion. For example, suppose that a camera has moved a distance x_1 during the exposure of a transparency. Then the image of a single point will be a line whose length is x_1. That is, the impulse response is

$$r(x) = 1, \quad -x_1/2 < x < x_1/2, \quad \text{and} \tag{7.55a}$$

$$r(x) = 0, \quad \text{otherwise} . \tag{7.55b}$$

We place the transparency in the input plane of an optical processor and illuminate it coherently. The amplitude in the frequency plane is the Fourier transform of the impulse response,

$$E(f_x) = 2ix_1 \, \text{sinc}(\pi f_x x_1) , \tag{7.56}$$

where $\text{sinc}\,\alpha = (\sin \alpha)/\alpha$ (Problem 7.10).

The Fourier transform of a single point is a constant; this is the frequency-plane amplitude we would like to have had. Therefore, we devise a filter whose amplitude transmittance is

$$t(f_x) \propto 1/\text{sinc}(\pi f_x x_1) . \tag{7.57}$$

The filter can only be approximated, because there are zeroes in the denominator. The filter must also include the phase shift of π at those values of f_x where the sine changes sign or, equivalently, $t(f_x)$ undergoes a phase shift of π. Such filters are complicated to fabricate and are best done using techniques of matched filtering or holography. The effect of such a filter is to correct the amplitude distribution in the frequency plane by making it more nearly uniform. The second lens of the processor is then exposed to a nearly uniform plane wave, which it focuses to a point. This is the deblurred image we seek.

This treatment has tacitly assumed an object that consists of a single point. In reality, the Fourier transform of a complicated object is the superposition of the transforms of all the blurred points in the transparency. We may see this by considering the transparency to be a screen that consists of many identical apertures; each aperture represents the blurred image of a point. According to Fraunhofer-diffraction theory, the diffraction patterns of these apertures are centered on the axis of the system and are identical except for phase. (This result may be expressed mathematically by the *convolution theorem*.) Thus, the transform of the blurred transparency will be very similar to the transform of the impulse response; it is possible to show rigorously that the optimum deblurring filter remains the reciprocal (7.57) of the sinc function in the general case.

A precisely similar analysis applies to the two-dimensional case in which the image is blurred by defocusing. If the exit pupil of the optical system is circular, the impulse response is a uniformly illuminated circle, and the Fourier transform has radial symmetry. The deblurring filter is the reciprocal of the Fourier transform.

7.3.5 Optical Transfer Function

Let the object intensity vary sinusoidally with dimension x in the object plane; then the geometric-optics image in the image plane varies sinusoidally with x'. For convenience, we use complex-exponential notation and describe the geometric-optics image by the expression

$$I'_g(x') = e^{2\pi i f_x x'} . \tag{7.58}$$

If the impulse response of the optical system is $r(x')$, then the image intensity $I'(x'_i)$ is

$$I'(x'_i) = \int_{-\infty}^{\infty} e^{2\pi i f_x x'} r(x'_i - x') dx' . \tag{7.59}$$

If we change variables, letting $\xi = x'_i - x'$, we find that

$$I'(x'_i) = e^{2\pi i f_x x'} \int_{-\infty}^{\infty} r(\xi) e^{-2\pi i f_x \xi} d\xi . \tag{7.60}$$

The image intensity is equal to the object intensity (or the geometric-optics image intensity) multiplied by a factor known as the *optical transfer function* $T(f_x)$, where

$$T(f_x) = \int_{-\infty}^{\infty} r(x') e^{-2\pi i f_x x'} dx' . \tag{7.61}$$

(We replace the dummy variable ξ with the variable x'.) The optical transfer function is equal to the Fourier transform of the impulse response. It is a measure of the properties of the image of a sinusoidal grating of specified spatial frequency. (f_x is the spatial frequency of the image; the spatial frequency of the object is f_x divided by the magnification of the optical system.)

The magnitude of the optical transfer function is called the *modulation transfer function*, frequently abbreviated MTF. The MTF is a measure of the contrast of the optical image at a specified spatial frequency. At high spatial frequencies, the MTF becomes very small or 0; image detail cannot be detected at these spatial frequencies, because $1/f_x$ is less than the resolution limit of the optical system. Unlike resolution limit, however, the MTF gives information regarding the character of the image at all spatial frequencies.

The phase of the optical transfer function, or OTF, is not in general 0. A nonzero phase term indicates a shift of the pattern from the position predicted by geometric optics. In particular, the phase of the OTF is sometimes equal to π; then, the image of a sinusoidal grating is shifted from the geometrical-optics image by precisely one-half period. The phenomenon occurs, for example, when detail is below the limit of resolution of a defocused optical system. It is known as *spurious resolution*; Fig. 7.23 shows how spurious resolution can occur as

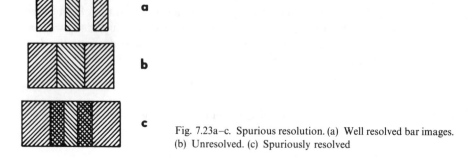

Fig. 7.23a–c. Spurious resolution. (a) Well resolved bar images. (b) Unresolved. (c) Spuriously resolved

a result of the overlapping of defocused bar images. In (c), the three-bar pattern appears as two bars with low contrast and π out of phase with the well resolved image.

7.3.6 Coherent Transfer Function

Using precisely the same reasoning as above, we can show, for the coherent case, that the electric-field amplitude $E'(x_i')$ in the image plane is given by the relation

$$E'(x_i') = \int_{-\infty}^{\infty} E_g'(x')h(x_i' - x')dx' , \tag{7.62}$$

where $E_g'(x')$ is the amplitude image predicted by geometric optics and h is the amplitude impulse response. Similarly, the *coherent transfer function* $H(f_x)$ is the Fourier transform of the coherent impulse response,

$$H(f_x) = \int_{-\infty}^{\infty} h(x')e^{-2\pi i f_x x'}dx' . \tag{7.63}$$

The coherent and incoherent impulse responses are connected by the relation,

$$r(x') = |h(x')|^2 . \tag{7.64}$$

7.3.7 Diffraction-Limited Transfer Functions

A simple, geometrical argument may be used to calculate the coherent transfer function of a lens.

First, consider a sinusoidal grating. We may visualize the grating as resulting from the interference of two plane waves. If the period of the grating is d, then its spatial frequency is

$$f_x = 1/d . \tag{7.65}$$

From the grating equation, $m\lambda = d \sin \theta$, we find that the angle between either wave vector and the normal to the plane of the grating is related to the spatial frequency of the grating by the equation

$$f_x = \theta/\lambda , \qquad (7.66)$$

where we assume that θ is sufficiently small that $\sin \theta \cong \theta$.

We may write any object distribution alternatively in terms of the angular distribution of plane waves it diffracts; such a representation is known as the *angular spectrum*. As in calculating the coherent transfer function, it is sometimes convenient to trace plane waves through an optical system, rather than to consider individual object points.

Suppose that the sinusoidal grating is placed in the object plane of an optical system as in Fig. 7.24. The grating is assumed to be much larger than the entrance pupil of the optical system. Pairs of waves composing the angular spectrum of the grating pass through the lens but are vignetted by the finite aperture. Each plane wave therefore illuminates only a finite area in the image plane. If the two *truncated waves* overlap, we see a sinusoidal interference pattern that we associate with the image of the grating. That is, the grating is visible in the image plane only where the two waves overlap. Where the waves do not overlap, the interference pattern is not seen, and the grating image is not transmitted by the lens.

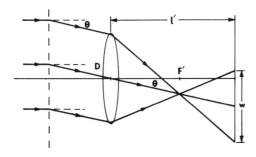

Fig. 7.24. Truncated-wave analysis of the coherent transfer function

On the axis of the lens, the waves with $m = \pm 1$ will overlap until the angle θ becomes greater than a certain value. This occurs when both truncated waves barely touch the axis. (For simplicity, only one of the waves is shown in Fig. 7.24.) The size of the illuminated area w produced by one of the waves is given by

$$w/(l' - f') = D/f' , \qquad (7.67)$$

provided that θ is small. Because $l' = f'(1 - m)$, we find that $w = mD$, where m is the magnitude of the magnification.

At *cutoff*, one of the extreme rays from each bundle intersects the axis in the plane of the image. Then

$$\theta_c = w/2l' , \qquad (7.68)$$

where θ_c is the value of θ at cutoff. Using the relation between w and D, and the definition $m = l'/l$, we find that

$$\theta_c = D/2l , \qquad (7.69)$$

or, in terms of spatial frequency,

$$f_c = D/2\lambda l . \qquad (7.70)$$

f_c is known as the *cutoff frequency*; objects with spatial frequencies higher than f_c are not recorded by the lens. Further, images that are recorded are recorded with high contrast because both plane waves have the same amplitude and are coherent with one another. The coherent transfer function is 1 out to the cutoff frequency and falls abruptly to 0 thereafter, as shown in Fig. 7.25.

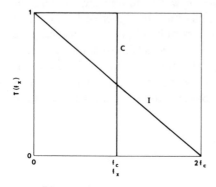

Fig. 7.25. Coherent (C) and incoherent (I) transfer functions of a diffraction-limited optical system

The cutoff frequency f'_c observed in image space, rather than object space, is

$$f'_c = f_c/m = D/2\lambda l' . \qquad (7.71)$$

The (incoherent) optical transfer function is difficult to calculate by this approach. A spatially incoherent source may be regarded as a spatial distribution of mutually incoherent point sources. Each such source illuminates the object from a different angle and therefore gives rise to an interference pattern somewhat displaced from the patterns of other sources. These interference patterns must be added incoherently.

One result of using an incoherent source is that detail can be seen up to the spatial frequency $2f_c$. This is so because part of the source is located off the axis of the system.

Further details of calculating the optical transfer function are left to the references. The result is that the MTF of a diffraction-limited lens falls off linearly with increasing spatial frequency, as shown in Fig. 7.25. The MTF falls to 0 at twice the cutoff frequency $2f_c$; it remains 0 at higher frequencies.

It may be tempting to infer that an incoherent image has twice the resolving power of a coherent image. We have seen, however, in our discussion of speckle and limit of resolution that coherent and incoherent imaging are different in kind; therefore no such inference is justified.

The transfer function of a lens that has aberrations decreases rapidly with increasing spatial frequency. The MTF may fall to 0 at a spatial frequency nearly an order of magnitude less than $2f_c$; it remains small thereafter and is identically 0 beyond $2f_c$. The resolution limit of such a lens is poor, and certain spatial frequencies are completely absent from the image.

7.3.8 MTF of Photographic Films

The concept of MTF is not restricted to lens systems but may be extended to other imaging devices such as television cameras and screens or to a medium such as photographic film. The MTF of the photographic emulsion may be defined as the contrast of the recorded image when the object is a sinusoidal grating with unit contrast. The MTF of a particular emulsion is 1 (or even slightly more than 1) for low spatial-frequency gratings. At higher spatial frequencies, it falls off rapidly because of scattering within the emulsion. The resolving power of the film is equal to the spatial frequency at which the MTF falls to 0.1 or 0.2.

The MTF of a film-lens combination is equal to the product of the individual MTFs. Unfortunately, the MTF of a combination of lenses is not the product of the MTFs of the individual lenses but must be calculated or measured for the complete system. This is so because one lens could correct (or worsen) the aberrations due to another lens, whereas film records the intensity distribution in the image plane of a lens, modified by its own MTF.

7.4 Digital Image Processing

7.4.1 Video Camera

A modern video camera consists, in essence, of an array of about 780×480 semiconductor detectors (Sect. 4.3), most commonly *charge-coupled devices*, or CCDs. The physics of the CCD array is not important here, but it is relevant that the array consists of silicon detector elements and is sensitive to wavelengths from about 400 nm (blue) to 1.1 μm (near ir). Most cameras are therefore equipped with an *ir-cutting filter* or window directly over the array. The filter is important in part because most lenses display unacceptable chromatic aberration unless the illumination is somehow limited to the visible spectrum. For work with nearly monochromatic sources at 850 nm or 1.06 μm, however, the filter must be removed or replaced by a clear-glass window.

The CCD array is rectangular and typically has dimensions of 8.8×6.6 mm, horizontal by vertical. Each detector receives a portion of the image known as a *picture element*, or *pixel*. Unfortunately, it is not practical to transmit the output (usually a voltage) of each pixel individually. The array is therefore *scanned* horizontally, and the information about each row is transmitted as an analog voltage as a function of time.

Rows are not transmitted sequentially, but rather odd rows are transmitted first and even rows next. A picture transmitted this way is *interlaced*. A set of odd rows or even rows is called a *field*; two consecutive fields are a *frame*. Usually, one complete frame is transmitted (and displayed on a monitor) about every 30 ms.

The signal transmitted by even a black-and-white video camera is complex. The image itself is a positive voltage waveform less than 1 V. Preceding each line is a negative-voltage pulse called the *horizontal-trigger pulse*; each field is preceded by a *vertical-trigger pulse*. The pulses collectively are called *composite synch* (short for synchronization), and the signal and the synch together are called *composite video*. Color video is similar but more complex, and we will not discuss it here.

A *frame digitizer*, or frame grabber, includes analog-to-digital circuitry that can digitize and store the signal portion of the camera's output. A computer can then perform complex operations such as matched filtering, Fourier transformation, and high-pass filtering (Sect. 7.2), as well as store sequential frames for later analysis. These operations can be time consuming when performed digitally, but the computer can perform them on images formed with incoherent light; this is often a big advantage over optical processing. Further, the computer can increase the contrast of a weak image, suppress noise in various ways, or perform two-dimensional convolutions that have effects similar to high- or low-pass filtering. In addition, computer processing can be used in conjunction with a coherent-optical processor, for example, to locate the output of a coherent-optical matched filter or to analyze an optical Fourier transform (Sect. 7.2).

7.4.2 Single-Pixel Operations

Here is an example of the power of computer processing: Consider an image that is, for some reason, covered with noise that has the form of isolated pixels with very high apparent brightness (Fig. 7.26, upper left). This might happen because of interference during the electronic transmission of an image, for example. A digital processor can be programmed to examine the neighborhood of each pixel and compute the average of the four or eight nearest neighbors. If the pixel at hand has an apparent intensity that exceeds that average by more than a fixed factor, say, 1.5 or 2, then that pixel's intensity is assumed to be spurious and is replaced with the average intensity of the neighbors. As long as the image does not display very sharp edges or other high-contrast features, the noise is eliminated (Fig. 7.26, upper right) with little or no degradation of other areas of the image.

Noise reduction by this method is an example of *nonlinear filtering* and cannot be performed by a coherent-optical processor. Another operation that can be performed by digital processing is *contrast enhancement*. Imagine, for

Fig. 7.26. Digitally processed images. *Upper left*, image with isolated points of noise. *Upper right*, noise filtered from same image. *Lower left*, image subjected to a low-pass filter. *Lower right*, image subjected to a high-pass filter. [Processed images courtesy of Matthew Weppner. Original photograph copyright 1990, The Art Institute of Chicago. All rights reserved. Grant Wood, American, 1892–1942, American Gothic, oil on beaverboard, 1930, 76 × 63.3 cm, Friends of American Art, 1930.934]

example, that the video image is digitized so that black is represented by 0 and white by 255; intermediate intensities are denoted by digits between 0 and 255. Suppose, however, that a certain image has very low contrast and its intensity when digitized falls between two values L and U, where $U > L$, but $U - L$ is less than 255. We may enhance the contrast of the image by changing the intensity I at each pixel to

$$I' = 255(I - L)/(U - L) . \tag{7.72}$$

This operation will ensure that the lowest intensity in the image is 0 and the

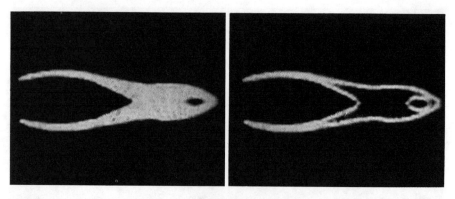

Fig. 7.27. Pliers of Fig. 7.13 (right). *Left*, hard-clipped. *Right*, edge-enhanced. [Courtesy of Matthew Weppner]

highest, 255. Figure 7.27 (left) shows the image of Fig. 7.13, but here the contrast has been adjusted so that intensities above a certain value are set equal to 255, whereas those below a certain value are set equal to 0. This is sometimes called *hard clipping*.

Similarly, we could display only those parts of the image that have intensity between two values U and L using the rule that the intensity of each pixel is changed to 0 if its original intensity lies outside the range $L < I < U$. Then, we could, say, use (7.72) to enhance the contrast and examine very low-contrast structure.

Such operations are sometimes called *single-pixel operations* because they alter the value of the intensity at single pixels with no regard for the intensity at neighboring pixels.

7.4.3 Cross-Correlation

This is defined by the sum

$$C(x_i) = \sum_{k=-K}^{K} h(x_{i+k})g(x_k) \qquad (7.73)$$

and is closely related to the convolution integral (7.51). Figure 7.28 sketches how to calculate one term in the cross-correlation between a discrete function $h(x_i)$, where i is an integer, and a *convolution kernel* $g(x_k)$, where k is an integer. This term is calculated by multiplying corresponding nonzero terms of the two functions and adding the products. The next term is calculated by sliding $g(x_k)$ to the right one position and performing a similar calculation; the preceding term is calculated by sliding $g(x_k)$ to the left. The aggregate of all the terms is a function of x_i and is called the cross-correlation function $C(x_i)$. Convolution is the cross-correlation between $h(x_i)$ and the mirror image of $g(x_i)$ and is import-

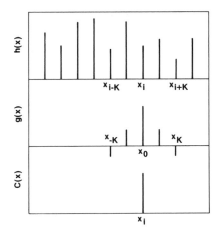

Fig. 7.28. Calculation of the cross-correlation function $C(x)$ between a function $h(x)$ and a convolution kernel $g(x)$

ant in Fourier theory; the distinction is unimportant for our purposes here, however.

Now suppose that we have stored an image digitally. We may want to calculate the cross-correlation of that image with an arbitrary kernel. For example, suppose that the image is a line or stripe represented by the one-dimensional matrix

$$\ldots 0\ 0\ 0\ 0\ 1\ 1\ 1 \ldots 1\ 1\ 1\ 0\ 0\ 0\ 0 \ldots \qquad (7.74)$$

Here, for illustration, we normalize the peak intensity to 1; in fact, it could represent any voltage between 0 and 1 V or any digital level between 0 and 255 (the larger number determines the number of gray levels between black and white). If we cross-correlate this function with a kernel such as $-1\ 0\ 1$, which has zero net area, then we will calculate values of 0 everywhere except in the vicinity of the edge. Near the left edge, however, we calculate $\ldots 0\ 0\ 1\ 1\ 0\ 0 \ldots$, whereas near the right edge, where the intensity drops to 0, we get $\ldots 0\ 0\ -1\ -1\ 0\ 0 \ldots$. The kernel behaves as a *digital filter* that differentiates the image and locates sharp edges. That is, the cross-correlated or *filtered* image is approximately 0 everywhere except near an edge. Since intensity can never be less than 0, it might be necessary to take the absolute value of the filtered image in order to display it, but, to a digital system, the negative intensity conveys whether the edge rises or falls with increasing x.

A two-dimensional analog of the kernel $-1\ 0\ 1$ is a matrix such as

$$
\begin{array}{ccc}
0 & -1 & 0 \\
-1 & 0 & 1 \\
0 & 1 & 0 .
\end{array}
\qquad (7.75)
$$

This kernel differentiates a two-dimensional image and locates edges, regardless

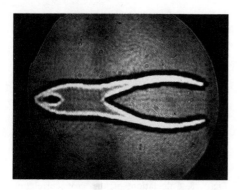

Fig. 7.29. Differentiated image of the pliers of Fig. 7.27 (left). [Courtesy of Matthew Weppner]

of their orientations. Figure 7.29 shows the result of edge-enhancing the image of Fig. 7.27 (left) with a kernel like (7.75). A constant value has been added to the result so that the negative values could be displayed. In Fig. 7.27 (right), however, only the inside edge is displayed; that is, values less than 0 are ignored. This image was used instead of the original image in a matched-filtering experiment (Sect. 7.2.6) because the low spatial-frequency information or continuous tones convey little information to distinguish between objects that have similar sizes or shapes.

Other kernels can perform other operations; for example, one-dimensional kernels like 1 1 1 or 1/2 1 1/2 are equivalent to low-pass filters and will smooth or taper sharp edges. Such kernels, because they average neighboring pixels, can also smooth or reduce noise, but only at the expense of blurring the picture. The bottom row in Fig. 7.26 shows an image processed with both low- and high-pass filters.

Finally, cross-correlation with a large kernel will often cause the calculated intensities to grow so large that the image cannot be displayed on a video monitor. For this reason, practical kernels are usually normalized to 1 in the sense that the sum of all the elements is set equal to 1. For example, if we were to cross-correlate with the kernel 1 1 1, we would then divide each of the elements in the resulting matrix by 3. Other kernels, however, have a sum of 0, like -1 0 1. Most such kernels are antisymmetric, and each half of the kernel must be normalized independently if it is necessary to keep the result within limits.

7.4.4 Video Microscope

The simplest video microscope is a microscope objective with a video camera located at the proper tube length (Sect. 3.5). The ir-cutting filter must be present if the source has a broad spectrum, because of chromatic aberration. Ordinary objectives may be used, however, with spectrally narrow sources at almost any wavelength up to 2 μm, provided that an ir-sensitive video camera is used. Beyond 2 μm, the glass in the objectives becomes opaque.

A typical 40 × objective has a numerical aperture equal to 0.65. According to the Abbe theory (Sect. 7.2), this objective can resolve a grating whose period is λ/NA, or about 0.85 μm when $\lambda = 0.55\,\mu$m. The image of the grating has a period 40 times larger, or 34 μm. We may think of this image as a series of bright and dark bands one-half that distance, or 17 μm, apart. A CCD array will not detect this image unless its pixels are spaced by this distance or less; this statement is roughly equivalent to the *sampling theorem*, which states, in part, that a sinusoidal signal can be sampled and re-created exactly only if it is sampled at least twice per period.

Sampling at a lower spatial frequency than the *Nyquist frequency* or *Nyquist sampling rate*, incidentally, creates spatial frequencies that may not exist in the object; this effect is called *aliasing* and is illustrated in Fig. 7.30, where a square wave is sampled at one-third the correct spatial frequency. Because the sampling, which is indicated by the arrows, has a fixed phase relationship with the sampled wave, alternate samples detect 0 and 1; this causes a spurious spatial frequency at one-fourth the spatial frequency of the sampled wave. Aliasing is commonly seen on television as colored fringes that appear when someone wears a finely striped or checked suit.

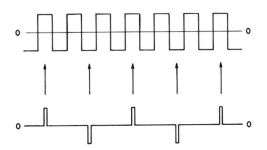

Fig. 7.30. Aliasing as a result of undersampling. *Top*, a periodic signal. *Center*, points at which samples are taken. *Bottom*, the result, a periodic signal at the wrong spatial frequency

Many CCD arrays are 8.8 mm wide and have 780 pixels per row; the distance between pixels is therefore slightly over 11 μm. Such an array can resolve the image of the microscope objective, with only slight empty magnification (Sect. 3.8). Here, however, empty magnification may be a benefit: Because the pixels on the CCD array are fixed and regular, fine structure that exactly matches the period of the pixels can be resolved only if it is roughly in phase with the distribution of pixels; if it is exactly out of phase, all pixels detect the same intensity, and the fine structure vanishes. We therefore say that imaging by a discrete array is *shift variant*, because a slight shift of the position of the image will cause a change of its appearance. Having pixels with a higher spatial frequency than the highest spatial frequency of the imaging optics reduces the effect of the shift variance of the array.

The field of view of the microscope is given by the dimensions of the CCD array divided by the magnification of the objective lens; for the example of the 40 × objective, the field of view is therefore 8.8/40 × 6.6/40 mm, or 220 × 165 μm.

If the objective magnification were $100 \times$ and the NA were 1.3, then the field of view would be reduced to 88×66 μm. We leave it as a problem to show that the spatial period of the detectors in the CCD array still about matches the highest spatial frequency the lens can resolve.

7.4.5 Dimensional Measurement

The CCD array is manufactured with great precision. In addition, even inexpensive microscope objectives may give nearly rectilinear imaging. If the magnification of the system can be determined accurately, it is possible to perform very precise dimensional measurements. Consider, for example, a sharp edge. According to (7.51), the image of that edge is the convolution of the edge function ... 0 0 1 1 ... with the impulse response of the objective lens.

For simplicity, consider the one-dimensional case, and let the light be wholly incoherent. Because the light is incoherent, we add intensities and therefore convolve the edge with an intensity impulse response to calculate the image of the edge (Sect. 7.3). This image is the edge response of the system.

As long as the impulse response is a symmetric function, convolution is the same as cross-correlation. Further, by symmetry, the convolution has the value 1/2 precisely at the location of the geometrical image of the edge. In addition, also by symmetry, the image displays an inflection point precisely at the location of the geometrical image. Either of these facts may be used to locate the edge precisely by digital image processing; if the image quality is good, the edge can usually be located by interpolation within about one-tenth pixel, or roughly one-fifth resolution limit. For a microscope objective with $NA = 0.65$ and $\lambda = 0.55$ μm, this is equivalent to a precision of 0.1 μm.

Unfortunately, the light in a microscope is almost never completely incoherent. To approximate incoherence, the condenser's numerical aperture would have to be at least twice that of the objective (Sect. 5.6.4), and this is impossible for objectives that have numerical apertures greater than 0.5. For this reason, the true position of the edge is almost always at an intensity somewhat less than 1/2. To see this, consider a microscope that uses highly coherent light. Then, by the same argument as above, the geometrical edge is located by finding position at which the *amplitude* falls to 1/2. Since we detect only intensity, or the square of the amplitude, we find that the edge is located where the intensity falls to 1/4. The edge in a partially coherent system usually lies somewhere between 1/2 and 1/4, but the precise value is difficult to predict mathematically.

In Fig. 7.31, the geometrical edge should be found by locating a point whose intensity is between 1/4 and 1/2, owing to partial coherence. If, however, we use the value 1/2 to measure the linewidth, as indicated by the abbreviation "meas.", bright lines will always appear slightly narrower than their true width, whereas dark lines on a bright background will always appear wider. Therefore, if it is necessary, as with semiconductor or integrated-optical measurements, to measure widths with accuracy less than about one-half resolution limit, it is usually necessary to acquire a *linewidth standard* that consists of calibrated lines

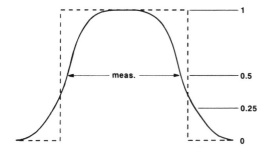

Fig. 7.31. Image of a sharp-edged line in partially coherent light. The linewidth as measured between the half-intensity points is in error

whose widths are known from other measurements. This standard usually consists of vacuum-deposited chromium lines on a glass substrate. The deposited layers are much thinner than 1 wavelength of visible light, so that the wave reflected from the glass substrate will not be out of phase with the wave reflected from the chromium layer.

Lines that are narrower than a few resolution limits present the additional difficulty that the edge response of one edge overlaps the other edge. Especially if the light is coherent or partially coherent, this factor makes it extremely difficult to measure linewidth accurately with an ordinary optical microscope. Lines or structures that are thicker than $\lambda/10$ or so cannot be measured accurately because of the phase difference between the wave reflected by the substrate and that reflected by the structure itself. Similarly, the phase shift that is inherent on reflection from any metallic surface can cause errors in measuring the width of even physically thin metallic lines. For these reasons, how to measure widths of narrow lines, or thick lines or stripes accurately is a subject of much current research.

Problems

7.1 Suppose that the exposure range of a hologram is so great that the t_a vs λ curve cannot be considered linear but may be described by the equation

$$t_a = t_0 - \beta \mathscr{E} - \beta' \mathscr{E}^2 \ .$$

Show that this nonlinearity results in two additional reconstructions.

7.2 If a hologram is large compared with the object distance, the spacing of the carrier fringes will vary with position. Use this fact to determine the largest off-axis hologram that can be recorded by a film whose resolution limit is RL. Show that the maximum theoretical resolution limit in the reconstruction is about equal to that of the film used to record the hologram.

7.3 Show that a hologram must be considered thick when the thickness of the emulsion exceeds $2nd^2/\lambda$, where n is the index of refraction of the emulsion and

d is the spacing of the carrier fringes. For convenience, let the reference and object waves fall with equal angles onto the plate. It is further helpful to assume that the recording medium has high contrast, so that sharp, narrow fringes are recorded. (Exact theory shows that a hologram is thick when the parameter $Q = 2\pi\lambda t/nd^2$ exceeds 10, where t is the emulsion thickness. For smaller values of this parameter, thickness effects are less evident.)

7.4 A point C on a three-dimensional object is located a distance l from the center O of a holographic plate. The optical path length from the beam splitter to O by way of C is equal to the distance traversed by the reference beam. If the coherence length of the laser is l_c, the depth of the reconstruction (along the line OC) will be equal to l_c. Ignoring the finite size of the plate, show that temporal coherence also limits the width of the reconstruction to the value $[l_c(l_c + 2l)]^{1/2}$, or, when l_c is small, $(2ll_c)^{1/2}$.

7.5 *Abbe Theory Applied to Resolution in Optical Systems.* (a) If the period d of the grating in Fig. 7.8 is small, only the zero-order or undiffracted light will enter a lens. Use this fact to show that the resolution limit of a lens is λ/NA, where NA is the numerical aperture of the lens.

(b) Suppose that the object is illuminated obliquely rather than normally. Show that the limit of resolution of a small, on-axis object may be reduced to $\lambda/(2NA)$ by directing the incident beam just to the edge of the lens.

(c) Show that objects that exhibit more than one order of diffraction can be seen even when the zero-order diffraction does not enter the lens aperture. Show further that the limit of resolution is $\lambda/(2NA)$ in this case also. (Use small-angle approximation in all cases.) Case (c) is the special case of *dark-field* or *dark-ground* illumination. Because the zero-order diffraction is suppressed, the background appears dark, and objects (whether amplitude or phase objects) appear as bright images against a dark background. A microscope that uses oblique illumination is often called an *ultramicroscope*.

7.6 A slit or other linear object is illuminated with coherent light and observed with a lens that has a circular entrance pupil. Explain why the limit of resolution is given by the one-dimensional formula $\lambda f'/D$, rather than by $1.22\lambda f'/D$.

7.7 Use integration by parts to show that the Fourier transform of $dg(x)/dx$ is proportional to $f_x G(f_x)$, where G is the Fourier transform of g. Explain why it is possible to display the derivative of a phase object by placing in the frequency plane a mask whose amplitude transmittance is $t_a = t_0 + bf_x$. What is the function of the constant term t_0?

7.8 *Apodization.* Suppose that the function $g(x)$ in (7.21) or (7.30) is given by $\cos(\pi x/b)$, where b is the width of the aperture. Show by direct integration that the Fraunhofer amplitude-diffraction pattern of the aperture is the sum of two displaced $(\sin u)/u$ functions. Sketch the functions carefully and show that each is displaced from the center by $\lambda f'/2b$. Show further that the secondary maxima, or sidelobes, of one function partially cancel those of the other. Show from the sketch that the width of the diffraction pattern is $1.5\lambda/b$. Show that the intensity of the first secondary maximum is about 0.006 times that of the principal maximum.

The intensity of the first sidelobe has been reduced substantially compared with that of the ordinary, clear aperture, at the expense of a slight broadening of the principal maximum. This is known as *apodization*. Apodization is accomplished with apertures whose transmittance falls gradually to 0 at the edges and is used when the intensity of the sidelobes is more important than resolution.

7.9 In phase-contrast microscopy, intensity variations are proportional to phase variations only when $\phi(x) < 1$. To estimate how severe this requirement is, carry out the expansion of $\exp(i\phi)$ to three terms. Find the Fourier transform in terms of three terms, G_1, G_2, and a new term G_3 [cf. (7.42)]. Insert the phase plate to find G'. Take the inverse transform to find the intensity in the output plane. There will be an additional term in (7.44). Suppose we require this term to be less than 10% of $2\phi(x')$. Find the maximum acceptable value of $\phi(x)$ and express the result as a fraction of the wavelength.

7.10 Calculate the optical transfer function for the case that $r(x')$ is a constant r_0 when $-d/2 < x' < d/2$ and is 0 elsewhere. [Hint: The integral you will have to evaluate should resemble a familiar diffraction integral.] This is approximately the case of defocused system. Relate the phase change of the OTF to the appearance of spurious resolution in Fig. 7.23.

7.11 Show that the cutoff frequency a distance $\delta/2$ from the lens axis is $f_c(1 - \delta/mD)$.

7.12 The modulation transfer function falls to 0 at the spatial frequency $D/\lambda l'$. What is the value of the MTF at the Rayleigh limit? Note: The MTF of a circular diffraction-limited lens may be approximated by

$$T(u) = 1 - 1.30u + 0.29u^3,$$

where $u = f_x/f_c$. This approximation is accurate to less than 1%.

7.13 Consider the image of an edge in coherent illumination (Fig. 7.20). Use simple, qualitative arguments to show that the location where the intensity is equal to 1/4 is exactly one resolution limit from the first maximum of intensity. For convenience, consider the one-dimensional case, and assume diffraction-limited imagery. [Hint: At what amplitude does the integral (7.54) stop increasing in value? The functional form of the impulse response is irrelevant.] This calculation shows that the location of the edge can be uncertain by a sizeable fraction of one resolution limit if the coherence of the source is uncertain. Why?

7.14 Consider a one-dimensional optical system whose impulse response can be approximated as an isosceles triangle. Assume that the illumination is incoherent. *Sketch* the edge response (the convolution of the impulse response with a unit step function) and show that the geometrical image of the edge is located where the intensity is 1/2. Suppose that the impulse response were a skew triangle, perhaps due to an off-axis aberration or because the lens was manufactured with one or more elements decentered with respect to others. Explain why the image of the edge would not then be located where the intensity is 1/2.

7.15 Repeat Problem 7.14 for coherent light. Sketch the intensity and show that the edge is located where the intensity is 1/4, as long as the impulse response is symmetric.

7.16 A scanning confocal microscope uses a $40\times$, $0.65\text{-}NA$ objective lens. A Nipkow disk is illuminated with a circle of incoherent light just large enough that the field of view of the microscope is a $250\text{-}\mu m$ (diameter) circle. (a) What is the optimum diameter of each hole in the Nipkow disk? (b) What fraction of the total radiant power in the illuminated patch would pass through the holes if they were made this diameter?

7.17 A video microscope uses a $100\times$, $0.65\text{-}NA$ objective lens with a 160-mm tube length. The image is focused directly onto a CCD array that consists of 780×480 detector elements and has an overall dimension of 8.8×6.6 mm. (a) Show that the field of view is $88\times66\ \mu m$. (b) Show that the sampling by the horizontal scan lines is at a spatial frequency high enough to satisfy the sampling theorem.

7.18 We wish to replace a telescope eyepiece with a CCD array that has 11 μm pixels. The CCD array lies in the focal plane of the objective lens. (a) What is the largest aperture (smallest F number) objective that we should use? (b) If we want a 15-cm (diameter) objective lens or mirror, what is the focal length? (c) Suppose you fixed a $10\times$ eyepiece to this objective. Would the magnifying power be useful or empty? Explain.

7.19 A sharp edge in a video microscope might rise from a low value of intensity to a high value within five or six pixels. Suppose, for example, that a very idealized edge is represented by the matrix

.. 50   50 100 150 200 200 . . . .

We want to filter noise from this image by using the rule that any pixel's intensity will be set equal to the average of its two nearest neighbors' if and only if its intensity differs in absolute value from that average by more than a certain value. What is the lowest value we can choose and not round the shoulder or the toe of the edge?

7.20   A bright line on a dark field may be represented digitally by

...0 0 0 1 1 1...1 1 1 0 0 0....

Convolve this edge with the kernel $(-1\ 0\ 1)$ and show that the result is a positive spike at the left edge and a negative spike at the right edge. What is the width of the spikes? Are they centered about the locations of the edges? Is there a more realistic representation of the edges, a representation that will give rise to spike that is symmetrical about the edges? Perform a similar analysis with the kernel $(-1\ 1)$ and show that this kernel will not allow the spikes to be symmetric about the edge locations.

8. Lasers

A laser consists of a fluorescing material placed in a suitable optical *cavity* that is generally composed of two mirrors facing each other. Although fluorescent light is not directional, some of the emission from the material strikes the mirrors and returns through the source (Fig. 8.1). If the mirror configuration is correct, and if the fluorescing medium is optically homogeneous, multiple reflections are possible.

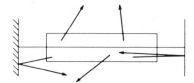

Fig. 8.1. Basic laser

Light passing through the fluorescing substance may be amplified by a process known as *stimulated emission*. If the material is properly prepared, stimulated emission can exceed *absorption* of the light. When sufficient amplification takes place. the character of the emission changes completely. In place of diffuse, nondirectional emission, a powerful, highly directional beam propagates along the axis defined by the two mirrors.

When such emission occurs, the assembly, called a *laser*, is said to *oscillate* or, colloquially, to *lase*. Laser emission is often very highly coherent both in space and time.

8.1 Amplification of Light

Suppose we locate an amplifying rod (or tube of gas or liquid) between two mirrors, as in Fig. 8.1. One mirror is partially transparent; both are aligned parallel to one another and perpendicular to the axis of the rod. The optical length of the *cavity* thus formed is d, the reflectance of the partially transmitting mirror is R, and the (intensity) *gain* of the rod is G.

Initially, the only light emitted is that arising from fluorescence or *spontaneous emission*. As we have noted, the fluorescent emission is not directional, but some of this light will travel along the cavity's axis. The following heuristic argument shows how amplified fluorescence brings about laser emission.

Consider a wave packet (see Sects. 4.2 and 5.6) that is emitted along the axis by a single atom. The packet undergoes many reflections from the mirrors. After each round trip in the cavity, it is amplified by G^2 and diminished by R. If the net round-trip gain exceeds 1,

$$G^2 R > 1 , \tag{8.1}$$

then the wave grows almost without limit. Only waves that travel parallel to the axis experience such continuous growth; the result is therefore a powerful, directional beam. The useful output of the laser is the fraction that escapes through the partial reflector or *output mirror*.

There is a second condition necessary for lasing. Consider again the wave packet emitted along the axis. Its coherence length is great compared with the optical length of the cavity. In a sense, the atom therefore emits the packet over a finite time. Because of multiple reflections, the packet returns many times to the atom before the emission is completed. If the packet returns out of phase with the wave that is still being emitted by the atom, it will interfere destructively with that wave and effectively terminate the emission. We can, if we wish, say that such a wave has been reabsorbed, but the effect is as if the wave never existed.

(The idea of reabsorption in this way may seem mysterious, but is easily seen with the help of an analogy. We liken the atom to a spring that oscillates at a certain frequency. If there is little damping, the spring will oscillate for many cycles. We can stop the oscillation in a few cycles by driving the spring with a force that oscillates at the frequency of the spring, provided that the force be applied just out of phase with the motion of the spring. In the same way, we stop the atom from emitting by driving it with an electric field just out of phase with the field being generated by the atom.)

The only waves that exist are, therefore, those for which constructive interference occurs,

$$m\lambda = 2d . \tag{8.2}$$

As a result of the large number of reflections, only wavelengths quite close to $2d/m$ exist. This is analogous to the sharpness of multiple-beam interference fringes. In general, there are many such wavelengths within the fluorescent linewidth of the source, so the value of d is not at all critical.

It is convenient now to consider the cavity separately from the amplifying medium. We may later combine their properties to account for the properties of the emitted light. In the following sections, we shall treat the properties of the amplifier and the cavity in somewhat greater detail.

8.1.1 Optical Amplifier

Consider a slab of material with thickness d and area A (Fig. 8.2). The material has N_1 absorbers per unit volume, and each absorber presents a cross-sectional area σ to an incident beam of light.

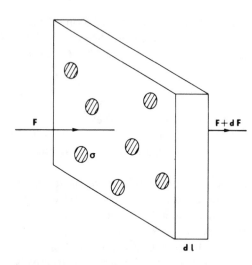

Fig. 8.2. Absorption by a slab

It is occasionally convenient to think of a light beam not as a wave or a series of wave packets, but as a stream of particles known as *photons*. Unadorned, the view is so naive it has been called the buckshot theory of light. It is nevertheless a useful heuristic device.

We allow a *flux* of F photons per unit area and unit time to fall on the slab. A photon is absorbed by the slab if it hits one of the absorbers and is transmitted if it does not. Therefore the fraction of F that is absorbed by the slab is equal to the fraction of the area A obscured by the absorbers. That is,

$$\frac{dF}{F} = -\frac{dA}{A}, \tag{8.3}$$

where dA is the total area of the absorbers in the slab. The volume of the slab is $A\,dl$, so $N_1\,A\,dl$ is the number of absorbers. Therefore,

$$dA = N_1 \sigma A dl, \tag{8.4}$$

if there are few enough absorbers that one never overlaps another. In this case,

$$\frac{dF}{F} = -N_1 \sigma dl. \tag{8.5}$$

If we integrate this equation over a rod whose length is l, we find

$$F_1(l) = F_0 e^{-N_1 \sigma l}, \tag{8.6}$$

where F_0 is the initial flux. The result is equivalent to *Lambert's law* in classical optics.

Now suppose the absorbers are quantum-mechanical systems (atoms, ions or molecules) with at least two sharp energy levels, as shown in Fig. 8.3. The

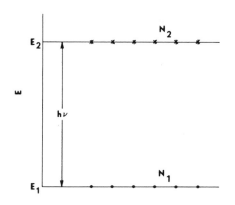

Fig. 8.3. Energy levels of a two-level laser

photon energy or *quantum energy* $h\nu$ coincides with the energy difference between the two levels. The internal energy of the system is raised from level 1 to level 2 when a photon or *quantum* of light is absorbed. Similarly, an *excited* system that is initially in level 2 may *emit* a quantum and thereby drop to level 1.

It is possible to show that a quantum can interact with an excited system in such a way that it forces the emission of another quantum. This is the process known as *stimulated emission*. It is sufficient for our purposes to say that the newly emitted quantum (which is really a wave packet) travels in company with the original wave packet and remains in phase with it.

The stimulated-emission cross section is precisely equal to the absorption cross section. If there are N_2 excited systems per unit volume, then, in the notation of Fig. 8.2,

$$\frac{dF}{F} = + N_2 \sigma dl \ , \tag{8.7}$$

by the same reasoning as before, and

$$F_2(l) = F_0 e^{+N_2 \sigma l} \ . \tag{8.8}$$

If we assume that some of the systems in the material exist in one state, and some in the other, we conclude that

$$F(l) = F_0 e^{(N_2 - N_1)\sigma l} \ . \tag{8.9}$$

The possibility of net amplification exists, provided that we can prepare a material with

$$N_2 > N_1 \ . \tag{8.10}$$

Finally, let us define the total number N_0 of systems,

$$N_0 = N_1 + N_2 \ . \tag{8.11}$$

The quantity $N_0\sigma$ is usually called the (passive) *absorption coefficient* α_0, in terms of which we can rewrite $F(l)$ as

$$F(l) = F_0 e^{-\alpha_0(n_2-n_1)l} ,\tag{8.12}$$

where $n_2 = N_2/N_0$ and $n_1 = N_1/N_0$. n_2 and n_1 are the *normalized populations* of the respective levels, and $n_1 + n_2 = 1$. The *single-pass gain* of the rod is therefore expressed as

$$G = e^{\alpha_0(n_2-n_1)l} .\tag{8.13}$$

The quantity $(n_2 - n_1)$ is called the *normalized population inversion n*; it varies between -1 and $+1$. Gain exceeds 1 when $n_2 - n_1 > 0$. In this case, net amplification is possible.

8.2 Optically Pumped Laser

In this section, we describe, for the sake of example, what is known as an optically pumped, three-level laser (Fig. 8.4). Energy levels 1 and 2 are the levels involved in the actual laser transition, as in Fig. 8.3. The third level is required to achieve a population inversion.

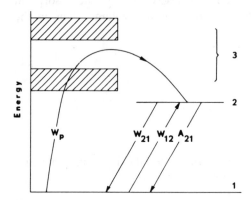

Fig. 8.4. Three-level laser

In an *optically pumped laser*, the active material is irradiated with a strong light, usually a *flashlamp*, known as a *pump*. The laser strongly absorbs the pump light in a band (level 3) around $h\nu_p$. However, this band is never populated because there is a nearly instantaneous, nonradiative transition from level 3 to level 2. If the actual atoms, ions or molecules can be pumped into level 2 faster than they decay (by spontaneous emission) to level 1, the required population inversion can be established.

8.2.1 Rate Equations

We may write the *rate equation* for the time development of N_2 as

$$\frac{dN_2}{dt} = (W_{21} + A_{21})N_2 - (W_p + W_{12})N_1 , \tag{8.14}$$

where W_p is the pumping rate per atom, W_{21} and A_{21} the stimulated and spontaneous emission rates from level 2 to 1, and W_{12} the rate of absorption from level 1. Because N_3 is 0, all atoms exist in either level 1 or 2, and dN_1/dt is just the negative of dN_2/dt. Because the stimulated-emission and absorption cross sections are equal,

$$W_{12} = W_{21} . \tag{8.15}$$

A steady-state solution to the rate equation is found by setting

$$\frac{dN_2}{dt} = 0 , \tag{8.16}$$

from which we find that

$$(W_p + W_{12})n_1 = (W_{12} + A_{21})n_2 , \tag{8.17}$$

in terms of normalized populations n_1 and n_2. Using the normalized population inversion n, we find that

$$n = \frac{W_p - A_{21}}{W_p + A_{21} + 2W_{12}} . \tag{8.18}$$

For amplification, $n > 0$, or

$$W_p > A_{21} . \tag{8.19}$$

We must pump to level 2 faster than n_2 is depleted by spontaneous emission. W_p depends both on the energy density of pump light and on the absorption cross section of the material. Thus, (a) an intense pumping source, (b) strong absorption of pump light, and (c) a long-lived upper level (small A_{21}) are desirable, if not necessary.

We estimate the pumping power required to achieve gain by writing W_p in terms of the photon flux F_p and cross section σ_p for absorption of pump light,

$$W_p = F_p \sigma_p . \tag{8.20}$$

The pump power per unit area at v_p is $F_p \cdot h v_p$. Further, A_{21} is just the reciprocal of the average spontaneous-emission lifetime τ of the upper laser level. Thus,

$$h v_p / \sigma_p \tau \tag{8.21}$$

is the required power density incident on the laser. For ruby, $\sigma_p \sim 10^{-19}$ cm^2, $\tau \sim 3$ ms and $h\nu_p \sim 2$ eV. We therefore require about 10^3 W·cm^{-2} or perhaps 50 kW incident at ν_p on a rod with total area 50 cm^2. The corresponding values for a four-level laser (in which the lower level is not the ground state) may be two orders of magnitude less. Because not all pump frequencies are effective in exciting the laser, the total input power required is greatly in excess of 50 kW; this immediately suggests short-pulse operation with a flashlamp for pumping. The first laser was such a flashlamp-pumped ruby laser, but we temporarily postpone further discussion of such systems.

8.2.2 Output Power

We can estimate the total output power emitted by a continuous-wave laser (or the average power emitted by certain lasers that produce irregular or repetitively pulsed outputs). We take ruby as our example. Suppose that the pump power absorbed at ν_p is equal to P. The total output power is approximately

$$P_0 = P \cdot \nu_{21}/\nu_p \,, \tag{8.22}$$

because, in the steady state, each excitation to the pumping level results in a single emission from level 2.

Let us assume that the laser begins to oscillate when the population inversion $n_2 - n_1$ only slightly exceeds 0. Then, approximately half of the systems will be in the upper level; this condition will persist as long as pumping is continued. Thus, the population N_1 per unit volume of the lower level is about

$$N_1 \cong N_0/2 \,, \tag{8.23}$$

and the total pumping power absorbed is

$$P = (N_0 V/2) W_p h\nu_p \,, \tag{8.24}$$

where V is the volume of the active medium. Further, we have seen that W_p is equal to A_{21} in steady state. Thus, the output power is

$$P_0 = (N_0 V/2) A_{21} h\nu_{21} \,. \tag{8.25}$$

A typical ruby rod may be 10 cm long with end faces 1 cm^2 in area. V is thus 10 cm^3, and the other parameters as in the preceding section. Thus, we find $P_0 \sim 10$ kW. This estimate is only slightly high for a system that may emit 1 or 2 J in a period of 1 ms.

8.2.3 Q-switched Laser

Applications often require high-peak-power (as opposed to high-energy) operation. In a *Q-switched* or *giant-pulse laser*, the laser is prevented from oscillating until n has been allowed greatly to exceed the usual threshold value n_t. When

n reaches a very large value, oscillation is allowed to occur, for example, by opening a shutter placed between the laser medium and the total reflector.

The process is detailed in Fig. 8.5. The top curve is drawn as if the reflectance R_{eff} of one mirror were allowed to vary from a low value to a high value. When R_{eff} is small, n grows large. When the shutter before the mirror is switched open, the power grows rapidly and continues to grow as long as the round-trip gain exceeds 1. The population of the upper level decreases rapidly because of the high energy density. The round-trip gain is just equal to 1 when $n = n_t$ and falls below 1 thereafter. *Peak power* is emitted, therefore, when n is just equal to n_t.

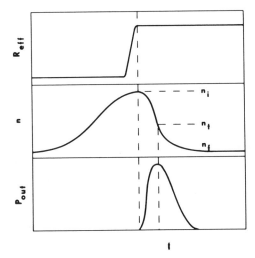

Fig. 8.5. Q-switched laser

Assuming virtually instantaneous switching, we can estimate some of the properties of the output by similar arguments. Suppose the normalized population inversion is n_i before the shutter is switched and decreases to n_f at the end of the pulse. Then the total energy emitted during the pulse is

$$E = (1/2)(n_i - n_f)N_0 V h v_{12}. \tag{8.26}$$

The factor 1/2 appears because the population difference changes by 2 units every time a quantum is emitted (because n_1 increases, whereas n_2 decreases after each downward transition). In ruby, n_t may be 0.2 and n_i, at most, 0.6 to 0.8. If the pulse is roughly symmetrical, n_f will be about 0.5. Using these figures for a small ruby rod, we estimate the output energy to be about 5 J. In fact, E is more nearly 1 J for such a laser.

We may estimate the duration of the Q-switched pulse by examining the decay of a pulse in an isolated cavity. Such a pulse oscillates between the mirrors and makes a round trip in time $t_1 = 2d/c$. Each time it strikes the output mirror, it loses $(1 - R)$ of its energy. In unit time, it therefore loses the fraction

$(1 - R)/t_1$ of its energy. The time

$$t_c = t_1/(1 - R) \tag{8.27}$$

is accordingly known as the *cavity lifetime*.

If the population inversion switches rapidly from n_1 to n_f, we can conjecture that the decay time of the Q-switched pulse is about equal to the cavity lifetime t_c. Thus, a symmetrical pulse would have a full duration of about $2t_c$. If we choose a cavity length of about 50 cm and take $R \sim 50\%$, we find that the pulse duration $2t_c$ is 10 or 15 ns. If anything, the estimate is slightly low.

The peak power can now be estimated. Assuming a nearly triangular shape for the pulse, we find

$$P_m = E/2t_c \tag{8.28}$$

for the case where $E \sim 1$ J and $2t_c \sim 10$ ns, the peak power is of the order of 100 MW.

8.2.4 Mode-Locked Laser

We begin by considering a laser oscillating in a large number N of different wavelengths known as *spectral modes*, all for simplicity taken to have equal amplitudes A. If we could place a detector inside the cavity at $x = 0$ and record the electric field $E(t)$ as a function of time, we would find

$$E(t) = A \sum_{n=0}^{N-1} e^{i(\omega_n t + \delta_n)} , \tag{8.29}$$

where ω_n is the angular frequency of the nth mode and δ_n, its relative phase. We shall find in the following section that the modes differ in frequency by $\Delta\omega$, where

$$\Delta\omega = \omega_n - \omega_{n-1} = 2\pi(c/2d) . \tag{8.30}$$

Usually, the modes are not related and the relative phases δ_n have different, random values. The modes are incoherent with one .another, and the total intensity is found by adding the intensities of the modes; that is,

$$I = NA^2 . \tag{8.31}$$

The intensity will have only small fluctuations, which occur whenever two or three modes happen to be precisely in phase.

Suppose we are able to make the modes interact so that they all have the same relative phase δ; that is

$$\delta_n = \delta . \tag{8.32}$$

Such a laser is known as a *mode-locked laser*. The intensity must now be found by adding the electric fields, rather than the intensities,

$$E(t) = A \, e^{i\delta} \sum_{n=0}^{N-1} e^{i\omega_n t} \, . \tag{8.33}$$

We may for convenience write ω_n as

$$\omega_n = \omega - n\Delta\omega \, , \tag{8.34}$$

where ω is the angular frequency of the highest-frequency mode.
Then the expression for $E(t)$ becomes

$$E(t) = A \, e^{i(\omega t + \delta)} \left[1 + e^{-i\phi} + e^{-2i\phi} + \cdots + e^{-(n-1)i\phi} \right] \, , \tag{8.35}$$

where $\phi = \Delta\omega t = \pi \, ct/d$. The term in brackets is the geometric series that we evaluated in connection with the diffraction grating. Thus, we may write that

$$I(t) = A^2 \, \frac{\sin^2 \dfrac{N\phi}{2}}{\sin^2 \dfrac{\phi}{2}} \, , \tag{8.36}$$

which is readily shown to have maxima of

$$I_m = N^2 A^2 \, , \tag{8.37}$$

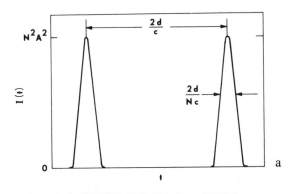

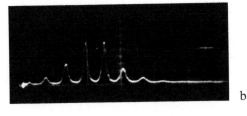

Fig. 8.6. (a) Mode-locked laser. (b) Output of a mode-locked ruby laser, 20 ns/cm

as in the case of the diffraction grating. Similarly, we may write that the output of the laser consists of short pulses separated by a time lapse of $2d/c$, or exactly one round-trip transit time (Fig. 8.6). The full duration of the pulses is $2d/cN$, which is about equal to the reciprocal $1/\Delta v$ of the fluorescence linewidth of the laser.

Thus, the output of a mode-locked laser consists of a sequence of short pulses, each with a peak power N times the average, or approximately N times the power of the same laser with the modes uncoupled. If N is about 100, the peak power of a mode-locked, Q-switched laser can be 1000 MW or more.

This result is easily interpreted as a short wave packet that bounces back and forth between the mirrors. The short pulses emitted by the laser appear each time the wave packet is partially transmitted by the output mirror. We will find this physical picture useful in describing several mechanisms for mode locking lasers, particularly in connection with argon-ion and neodymium:glass lasers.

8.3 Optical Resonators

An optical resonator consists of two reflectors facing each other, as in the Fabry-Perot interferometer. The reflectors need not be plane mirrors, but as long as they are aligned so that multiple reflections may take place, the analysis of interference by multiple reflections in Chaps. 5 and 6 will suffice.

8.3.1 Axial Modes

To begin, consider a plane wave that originates inside a Fabry-Perot resonator (Fig. 8.7), with amplitude A_0. After a large number of reflections has occurred, the total field inside the cavity is

$$E = A_0(1 + r^2 e^{-i\phi} + r^4 e^{-2i\phi} + \cdots) , \qquad (8.38)$$

for the wave traveling to the right. The notation is that of Chap. 5; r is the amplitude reflectance of both mirrors, and $\phi(= 2\,kd)$ is the phase change associated with one round trip. The wave is assumed to travel normal to the mirrors.

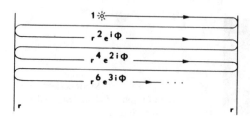

Fig. 8.7. Fabry-Perot resonator

Following Chap. 5, we calculate the sum of the geometric series and find that the intensity I inside the cavity is

$$I = I_0 \frac{1}{1 + F \sin^2 \frac{\phi}{2}} \, , \tag{8.39}$$

where $F = 4R/(1 - R)^2$ and R is the reflectance of one mirror. This is precisely the expression for the transmittance of a Fabry-Perot interferometer. I has value I_0 when

$$m\lambda = 2d \, , \tag{8.40}$$

the familiar condition for constructive interference. For all other values of λ, the intensity inside the cavity is small, so the wave is either absorbed or transmitted by the mirror and oscillation cannot take place. It is convenient to think of the output mirror as having an effective reflectance R_{eff} equal to R only when $m\lambda = 2d$. Specifically, we write

$$R_{\text{eff}} = \frac{R}{1 + F \sin^2 \frac{\phi}{2}} \, . \tag{8.41}$$

In most lasers, the second mirror has a reflectance of 100%.

Because R_{eff} has precisely the same functional form as the transmittance of a Fabry-Perot interferometer, we can adopt all the results of that discussion. Specifically, we found the interferometer to transmit fully at a series of wavelengths separated by the free spectral range

$$\Delta\lambda = \lambda^2/2d \tag{8.42}$$

or, in terms of frequency,

$$\Delta v = c/2d \, . \tag{8.43}$$

In the laser, therefore, the effective reflectance is equal to R only at discrete frequencies separated by Δv.

Figure 8.8 illustrates the effect of the cavity on the laser output. The top curve shows gain G as a function of frequency. The central curve plots R_{eff} vs frequency. R_{eff} is large (in this example) at only five points in the frequency interval within which G is large. As the bottom curve shows, laser emission occurs at these five discrete frequencies only. Each frequency is known as a *spectral mode* or an *axial mode*.

The modes are much more nearly monochromatic than the curve of R_{eff} would suggest. In many cases, their spectral width is determined only by minute vibrations that cause the optical length of the cavity to change slightly over short times. Their sharpness is a consequence of the very high net

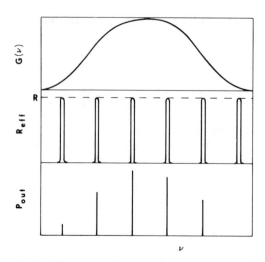

Fig. 8.8. Oscillation in several spectral modes

amplification that occurs after a great number of reflections. After each round trip, the central frequency is amplified more than the extreme frequencies. This results in a sharpening of the spectrum. When the cavity has gain, the number of round trips is very large, so the spectrum becomes extremely sharp.

8.3.2 Transverse Modes

These are most easily understood in terms of a cavity such as a *confocal cavity*. The confocal cavity shown in Fig. 8.9 has two identical mirrors with a common focus at w_0. Roughly speaking, we can define a *transverse mode* as the electric field distribution that is associated with any geometrical ray that follows a closed path. (We will not consider cavities that do not allow a ray to follow a closed path; these are known as *unstable resonators*.) Naturally, the ray will not describe the precise field distribution because of the effects of diffraction.

The simplest mode in a confocal cavity is described by a ray that travels back and forth along the axis. This is the 00 mode. Because of diffraction, the actual intensity distribution is that outlined. The output of a laser oscillating in this mode is a spherical wave with a Gaussian intensity distribution. The *beam width* is usually expressed as the radius w at which the beam intensity falls to $1/e^2$ of its maximum value.

The next-simplest mode is shown in Fig. 8.9b. This mode will oscillate only if the aperture (which is often defined by the laser tube) is large enough. The output is a spherical wave with the intensity distribution shown. Higher-order modes correspond to closed paths with yet-higher numbers of reflections required to complete a round trip. Transverse-mode patterns are labeled according to the number of minima that are encountered when the beam is scanned horizontally (first number) and then vertically (second number). The modes

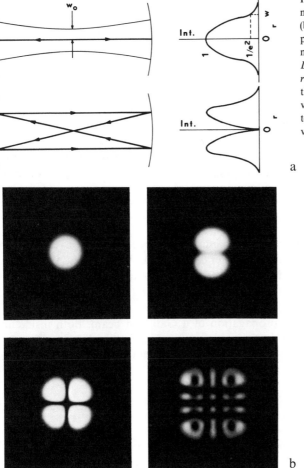

Fig. 8.9. (a) Transverse modes in a confocal cavity. (b) Laser transverse-mode patterns. *Upper left*, 00 mode. *Upper right*, 01 mode. *Lower left*, 11 mode. *Lower right*, coherent superposition of two or more transverse modes. [*Photos* courtesy of D. C. Sinclair, University of Rochester]

b

shown all have rectangular symmetry; such rectangular modes characterize nearly all lasers, even those with cylindrical rods or tubes.

In practice, the higher-order modes have greater loss (due to diffraction) than the 00 mode. If a laser oscillates in a certain high-order mode it also emits all modes with lower order. Such a *multimode laser* provides considerable power compared with that of a 00 or *single-mode laser*. Nevertheless, a single-mode laser is often desirable. In many gas lasers, the diameter of the laser tube is chosen small enough that diffraction loss prohibits oscillation of any mode other than the 00 mode. See Sect. 8.3.5.

8.3.3 Gaussian Beams

We have already noted that a laser oscillating in the 00 mode emits a beam with Gaussian intensity distribution. Higher-order modes also exhibit Gaussian intensity distributions, multiplied by certain polynomials (Hermite poly-nomials). Thus, beams emitted by most lasers are *Gaussian beams*.

In our earlier treatment of diffraction, we assumed that a uniform wavefront passed through the diffracting screen. We found, for example, that the far-field intensity distribution had an angular divergence of $1.22\lambda/D$ for a circular aperture. That result is not appropriate when Gaussian beams are used because then the intensity distribution is not uniform.

A detailed treatment of propagation of Gaussian beams is left to the references. Here we discuss the general results. The field distribution in any curved-mirror cavity is characterized by a *beam waist* such as that shown in Fig. 8.9; in a symmetrical cavity, the beam waist is located in the center of the cavity. The intensity distribution in the plane of the waist is Gaussian for the 00 mode; that is,

$$I(r) = \exp(-2r^2/w_0^2)\,, \tag{8.44}$$

where r is the distance from the center of the beam. For convenience, the intensity is normalized to 1 at the center of the beam. When $r = w_0$, the intensity is $1/e^2$ times the intensity at the center. Higher-order modes are characterized by the same Gaussian intensity distribution, but for the Hermite polynomials mentioned above.

Figure 8.10 shows the propagation of a Gaussian beam. Both inside and outside the cavity, it retains its Gaussian profile. That is, at a distance z from the beam waist, the intensity distribution is given by (8.44) with w_0 replaced by $w(z)$, where

$$w(z) = w_0\left[1 + \left(\frac{\lambda z}{\pi w_0^2}\right)^2\right]^{1/2}. \tag{8.45}$$

At large distances z from the beam waist, the term in parentheses becomes much larger than 1. In this case, the Gaussian beam diverges with angle θ, where

$$\theta = \lambda/\pi w_0\;; \tag{8.46}$$

this is the far-field divergence of a Gaussian beam.

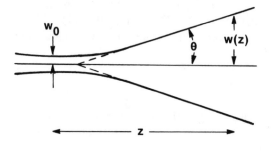

Fig. 8.10. Propagation of a Gaussian beam

If a Gaussian beam with beam width w is brought to a focus with a lens whose diameter is at least $2w$, the radius of the focal spot is $\lambda l'/\pi w$, which is somewhat smaller than the corresponding Airy-disk radius, $0.61\lambda l'/w$. In addition, the diffraction pattern is not an Airy disk, but has a Gaussian intensity distribution with no secondary maxima (unless the lens aperture vignettes a significant portion of the incident beam).

The radiation converges toward a beam waist and diverges away from it. Therefore, the wavefront must be plane at the beam waist. At a distance z from the waist, the radius of curvature $R(z)$ of the wavefront is

$$R(z) = z\left[1 + \left(\frac{\pi w_0^2}{\lambda z} \right)^2 \right]. \tag{8.47}$$

Only at great distances z from the waist does the beam acquire a radius of curvature equal to z.

Thus far, our comments have been general and apply to any Gaussian beam with a beam waist w_0 located at $z = 0$. To apply the discussion to a laser cavity, we must know the size and position of the beam waist. In a confocal cavity whose mirrors are separated by d, w_0 is given by

$$w_0 = (\lambda d/2\pi)^{1/2} , \tag{8.48}$$

and the waist $(z = 0)$ is located in the center of the cavity, as it is in any symmetrical, concave-mirror cavity.

When the cavity is not confocal, it is customary to define *stability parameters* g_1 and g_2 by the equations

$$g_1 = 1 - (d/R_1) , \tag{8.49a}$$

and

$$g_2 = 1 - (d/R_2) , \tag{8.49b}$$

where R_1 and R_2 are the radii of curvature of the mirrors.

To find the size and location of the beam waist, we argue that the radius of curvature of the beam wavefront at the positions of the mirrors must be exactly equal to the radii of the mirrors themselves. If this were not so, then the cavity would not have a stable electric-field distribution; in terms of our earlier ray picture, the rays would not follow a closed path.

We therefore know the radius of curvature of the wavefront at two locations, which we may call z_1 (the distance from the waist to mirror 1) and z_2 (the distance from the waist to mirror 2). Setting $R(z_1) = R_1$ and $R(z_2) = R_2$, we may solve (8.47) for z_1 and thereby locate the waist. The result is

$$z_1 = \frac{g_2(1 - g_1)}{g_1 + g_2 - 2g_1 g_2} d , \tag{8.50}$$

in terms of the stability parameters.

Similarly, we may solve for the beam waist w_0, which is, in general,

$$w_0 = \left(\frac{\lambda d}{\pi}\right)^{1/2} \left(\frac{g_1 g_2 (1 - g_1 g_2)}{g_1 + g_2 - 2g_1 g_2}\right)^{1/4} . \tag{8.51}$$

We may now use the results of Gaussian-beam theory to find the radius $R(z)$ or the spot size $w(z)$ at any location z. In particular, at mirror 1,

$$w(z_1) = \left(\frac{\lambda d}{\pi}\right)^{1/2} \left(\frac{g_2}{g_1(1 - g_1 g_2)}\right)^{1/4} , \tag{8.52}$$

and at mirror 2,

$$w(z_2) = (g_1/g_2)^{1/2} \, w(z_1) . \tag{8.53}$$

Sometimes it is necessary to *match* the mode of one cavity to that of another; that is, the mode emitted by the first cavity must be focused on the second so that it becomes a mode of that cavity as well. For example, mode matching may be necessary when a spherical, confocal Fabry-Perot interferometer is used with a laser.

The simplest way to match two cavities a and b is to locate the point where their spot sizes $w(z)$ are equal. Then, calculate the beams' radii of curvature $R_a(z)$ and $R_b(z)$ at that point. A lens whose focal length f' is given by

$$\frac{1}{f'} = \frac{1}{R_a} - \frac{1}{R_b} \tag{8.54}$$

will match the radii of curvature of the two modes at the point.

Both radius and spot size must be matched to ensure effective mode matching. If both parameters are not matched, for example, power may be lost to higher-order modes in the second cavity. When it is not possible to match two cavities in the simple way described here, it may be necessary to expand or reduce the beam size with a lens before attempting mode matching with a second lens.

8.3.4 Stability Diagram

The expressions for $w(z_1)$ and $w(z_2)$ contain a fractional root of $(1 - g_1 g_2)$. Unless the product $g_1 g_2$ is less than 1, the spot sizes on the mirrors become infinite or imaginary. Laser cavities for which $g_1 g_2$ exceeds 1 are unstable; those for which the product is just less than 1 are on the border of stability, because the spot size may exceed the mirror size and bring about great loss. Thus the stability criterion for lasers is

$$g_1 g_2 = \left(1 - \frac{d}{R_1}\right)\left(1 - \frac{d}{R_2}\right) < 1 . \tag{8.55}$$

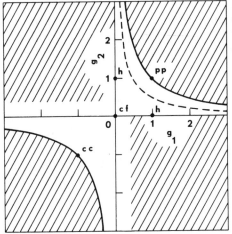

Fig. 8.11. Stability diagram. *h*, hemispherical cavity. *cc*, concentric. *cf*, confocal. *pp*, plane parallel. *Dashed curve* shows the equation $g_1g_2 = 1/2$

The limiting case, $g_1g_2 = 1$, is a hyperbola, as shown in Fig. 8.11. Stable resonators lie between the two branches of the hyperbola and the axes; unstable resonators, outside the two branches. Resonators that are least sensitive to changes within the cavity (such as thermally induced focusing effects in solid or liquid lasers) lie on the hyperbola $g_1g_2 = 1/2$, which is shown as a dashed line in Fig. 8.11.

8.3.5 Coherence of Laser Sources

Although a laser is usually thought of as a coherent source, only a laser that oscillates in a single axial and transverse mode emits highly coherent light. A multimode laser, although much more intense, may be no more coherent (in space or time) than a suitably filtered thermal source.

Let us look first at temporal coherence. A single-mode laser has virtually infinite temporal coherence, owing to its narrow spectrum. On the other hand, a multimode laser may have a spectral width that is nearly as great as that of the fluorescence emission from which it is derived. It will therefore have about the same coherence length as the corresponding thermal source.

Inexpensive He-Ne lasers are made to oscillate for stability in two modes. An interference experiment performed with the OPD equal to the length of such a laser will result in nearly zero coherence. For many purposes, the coherence length of this laser is therefore much less than the length of the laser itself.

The spatial-coherence area of a single-mode laser covers the entire beam. A double-slit experiment will result in high-contrast fringes even when the slits are separated by much more than the beam diameter $2w$.

When the laser oscillates in more than one transverse mode, the spatial coherence is radically altered. In general, the higher-order transverse modes oscillate at slightly different frequencies from the 00 mode; for example, in confocal geometry the frequency of *odd modes* (10, 21, 30, etc.) differs by approximately $c/4d$ from that of the 00 and other *even modes*. Other effects cause the frequency difference between spectral modes to differ slightly from the nominal value of $c/2d$. Because of these we may associate each spectral or transverse mode with a different set of radiating atoms. The modes are therefore incoherent with one another.

As a result, a laser that oscillates in many transverse modes closely approximates the spatial coherence of a thermal source. The statement is surprisingly accurate even for oscillation in the lowest two modes (00 and 10) only and shows the importance of single-mode oscillation. Figure 8.12 shows the degree of spatial coherence of a He-Ne laser oscillating in single-mode, double-mode and multi-mode fashion; this is defined as the visibility of the fringes of a double-slit experiment [cf. (6.11)]. The beam diameter $2w$ is a few millimeters. Only in the case of single-mode operation does the coherence in the brightest part of the beam differ significantly from the coherence calculated for the appropriate thermal source.

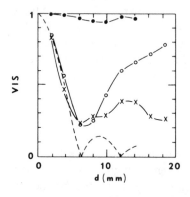

Fig. 8.12. Degree of spatial coherence or fringe visibility of a gas laser. ●, 00 mode. ○, two modes. ×, multimode. *Dashed line* shows the coherence function calculated for the appropriate thermal source. [After Young et al.; Opt. Commun. **2**, 253 (1970)]

8.4 Specific Laser Systems

8.4.1 Ruby Laser

The first ruby lasers employed a *xenon flashtube* in a helical configuration, with a ruby rod located along the axis of the helix. The end faces of the rod were polished flat and parallel to one another, and coated with gold. One of the coatings was thin enough to allow partial transmission and served as the output mirror.

Most pulsed lasers today have external mirrors. The rod itself has antireflection coatings on the ends (unless the ends are cut at *Brewster's angle*, for which one polarization has nearly zero reflectance). A linear flashlamp is located next to the rod. The rod and lamp are oriented at the foci of a long, highly polished *elliptical cylinder* to ensure that nearly all the emission from the flashlamp is focused on the rod. Finally, dielectric reflectors are located on adjustable mounts outside the housing of the rod and lamp.

Ruby is made of *aluminium oxide* (Al_2O_3) with a small concentration of *chromium oxide* (Cr_2O_3) impurity. (Pure Al_2O_3 is called *sapphire*.) Laser rods are nominally single crystals of ruby, which contain about 0.03% chromium oxide by mass. Many of the Cr^{+3} ions occupy locations in the crystal normally reserved for Al^{+3}; these Cr^{+3} ions are responsible for the emission of light by the crystal. The relevant energy levels of the ruby (strictly, of these ions) are indicated in Fig. 8.4. Laser action occurs because of the transition from level 2 to 1. The energy of this transition is about 1.8 eV, the corresponding wavelength, 694 nm.

Figure 8.13 describes the output in a highly schematic way. A large, high-voltage capacitor is discharged through the flashlamp. The lamp therefore flashes for a duration that may be between 0.1 and 10 ms, depending upon the circuit design. This is sketched in the upper curve.

The middle curve of Fig. 8.13 indicates the development of the normalized population inversion n. n grows steadily until it reaches the *threshold inversion* n_t, which refers to the population inversion required to produce net amplification. When n reaches n_t, the power density in the cavity becomes so great that an enormous number of stimulated emissions drives n below n_t. The laser thus

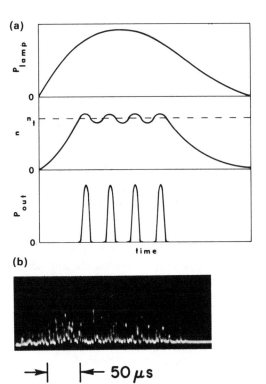

(a)

P_{lamp}

n

n_t

P_{out}

time

(b)

$\rightarrow$| |$\leftarrow$ 50 μs

Fig. 8.13. (a) Output characteristics of pulsed ruby laser. (b) Spiking in the output of a pulsed ruby laser

produces a short pulse and stops, as in the bottom curve. Because the flashlamp is still active, n grows to n_t and the process begins again.

The output of a pulsed ruby laser is an irregular series of pulses. (The irregularity is related to multimode oscillation.) Typically, the capacitor bank may store 2000 J with a drop of 2000 V across the flashlamp. This laser may emit 2 J or more each time the lamp is pulsed. If the pulse width is 1 ms, this corresponds to an average power of a few kilowatts. The spikes have a duration of a few microseconds, so the *peak power* is tens or hundreds of kilowatts. The output of the laser is in general multimode, both with regard to spectral and transverse modes.

The pulsed ruby laser has been used successfully for precise welding and drilling of metal, for drilling industrial diamonds, for repairing detached retinas in ophthalmology, and for holography and photography of moving objects. It has now been replaced by other lasers in most of those areas.

The ruby laser is often *Q switched* for high peak power. If the parameters are properly adjusted, then only a single pulse will be emitted. The pulse may have a peak power up to 100 MW and a full width of 10 or 20 ns. The pulse may be amplified by directing the light through a second apparatus, devoid of external reflectors, and pumped synchronously with the laser itself. The laser or the amplifier will be damaged when the power density exceeds 200 MW/cm^2.

Q switching is effected in one of three ways. The total reflector can be rotated rapidly, so that it is only instantaneously aligned parallel with the output reflector. If we drive the mirror with a synchronous motor or otherwise detect its position, we can time the flashlamp to fire so that the mirror is aligned only when n approaches its maximum value. The laser will emit a giant pulse, after which the mirror will no longer be aligned.

Other means of *Q* switching place *electro-optic shutters*, such as Kerr or Pockels cells, or *acousto-optic shutters* between the laser crystal and one mirror. In each case, the shutter is effectively closed until n reaches its maximum value. The shutter is synchronized with the flashlamp to open at the proper time.

Finally, a *passive shutter* consists of an optical-quality cell containing a *saturable dye*. Saturable dyes, like ruby, have energy-level diagrams similar to that shown in Fig. 8.4. We choose a dye for which the energy difference labeled $h\nu_p$ corresponds to that of the laser transition $h\nu$. We adjust the concentration of the dye so that oscillation can just take place with the dye cell in the cavity. The laser then pumps the dye molecules into level 2, where they may remain for a short time. With few molecules in level 1, the dye is relatively transparent to laser light and *Q* switching has been effected. This very simple and elegant means of *Q* switching can also help force the laser to oscillate in a single axial mode, a condition that may be desirable for many scientific purposes.

8.4.2 Neodymium Laser

This is another, optically pumped laser material and is an example of a *four-level laser*, shown in Fig. 8.14. In such a laser, the lower level lies far above the ground

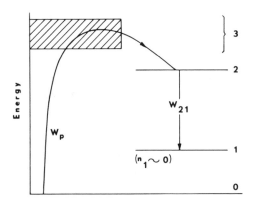

Fig. 8.14. Four level laser

state of the system and is generally unpopulated. Thus, $n_1 = 0$, so the normalized population inversion n is just equal to n_2. Any population in level 2 gives rise to an inversion with $n > 0$. Because it is not necessary to pump a four-level laser from $n = -1$ to $n = 0$ before achieving gain, such lasers are much more efficient than three-level lasers. If level 1 decays to level 0 rapidly enough, $n_1 = 0$ under all conditions, and the four-level laser never exhibits absorption of the laser light itself.

The active Nd^{+3} ion may be incorporated into several hosts, in particular, certain glasses and a crystal known as YAG (for yttrium aluminum garnet). The Nd^{+3} ion surrounds itself with several oxygen atoms that largely shield it from its surroundings. Hence, Nd : glass and Nd: YAG lasers oscillate at about the same frequency. Their output is in the near infrared, at 1.06 μm.

The Nd:YAG laser is most often used in a *quasicontinuous* fashion – that is, repetitively pulsed at a high rate. The peak output power is of the order of kilowatts. Nd:glass lasers, on the other hand, are normally operated in single pulses, just as the ruby laser is. The glass laser may be pulsed or Q switched and is highly resistant to damage from high power densities. This is in large part due to the fact that glasses, unlike crystals, can be made almost entirely free of internal stress and strain. Nd : glass is also free of the microscopic particles of metal and Cr_2O_3 that characterize rubies, and, whether pumped or not, the glass laser does not absorb the laser emission because the lower laser level is not the ground state. In addition, high-quality glass laser rods can be made in almost any diameter, to allow high power with relatively low power density. For these reasons, the Nd: glass laser may be combined with one or more amplifiers to produce very high peak powers.

The fluorescence linewidth of the Nd : glass laser is quite great, and the laser is rarely if ever operated in a single spectral mode. Rather, the Q-switched glass laser is often induced to operate in a *mode-locked* fashion, emitting a train of pulses, each with a duration of the order of 10 ps. Nd:YAG is also commonly mode locked, but the pulses are 5 or 10 times longer. A single one of these pulses may be isolated with an electro-optic switching technique and passed through several amplifying stages. Pulses with peak power in excess of 10^6 MW have been generated in this way and are used in laser-fusion programs and in fluorescence spectroscopy, for example.

A mode-locked laser may incorporate a weak saturable dye in the cavity. If a short pulse travels through the dye, the dye may bleach and allow the pulse to pass relatively unobstructed. Now, suppose that the shutter, once bleached, were to close again in a time short compared to the round-trip transit time t_1 of the laser. Laser oscillation that involves an intense, short pulse traversing the cavity would experience little loss. Competing, long-pulse oscillation at lower power would, on the other hand, experience considerable loss because the shutter would, on the whole, be closed. It is therefore possible to establish conditions under which oscillation occurs as a short pulse that bounces between the mirrors. This is identical with the *mode-locked laser* described earlier.

Because the total energy in a mode-locked train of pulses is roughly the same as the total energy in a single, longer pulse, the peak power of the mode-locked train is much in excess of the peak power of a conventionally Q-switched laser.

Finally, for low-power, continuous-wave applications, the Nd:YAG laser may be pumped with an array of semiconductor-diode lasers (Sect. 8.4.8). Frequency-doubling (Sect. 9.3.1) this laser yields a useful source of green light at the wavelength of 532 nm. Indeed, the diode-pumped YAG laser displays greater stability than the He-Ne laser, which suffers from fluctuations in intensity because of the instability of gas discharges.

8.4.3 Organic-Dye Lasers

The other important class of optically pumped laser is the *organic-dye laser*. The active medium is an organic dye dissolved in a suitable liquid. The dye molecules are basically three-level systems, with energy-level schemes like that in Fig. 8.15. The energy levels are shown as *bands*; these bands result from the energy of vibration of the dye molecule. The width of the bands gives rise to a broad fluorescence spectrum and hence the existence of tunable lasers. Unfortunately,

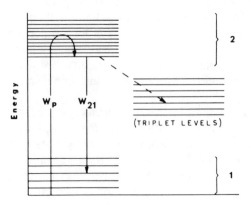

Fig. 8.15. Dye laser

level 2 has a lifetime that is on the order of microseconds, rather than milliseconds. As a result, a very great pumping power is required to maintain a population inversion.

In addition, the organic dyes have a fourth level known as a *triplet level*. The triplet level does not contribute to laser action. Molecules in level 2 decay to the triplet level in about 1 μs. The triplet level is very long lived, so the molecules will not soon return to the ground state, level 1. Laser action therefore stops when a significant fraction of the molecules is in the triplet level.

To minimize the effect of the triplet state, flashlamp-pumped dye lasers use special, low-inductance capacitors and, sometimes, specially constructed flashlamps to permit discharging the capacitor across the flashlamp in a few microseconds or less. The dye is thereby made to lase efficiently before molecules are lost to the triplet level. In addition, certain triplet-state-quenching additives help somewhat to bring the dye from the triplet level back to level 1.

Other dye lasers are continuously pumped with an argon-ion laser. The triplet level is still a problem, but is usually surmounted by flowing the dye solution through the laser cavity. If the active volume is small enough and the rate of flow great enough, molecules are physically removed from the cavity before an appreciable fraction of the molecules is lost to the triplet level.

Despite these difficulties, dye lasers are of considerable importance, mainly because of their tunability. Each dye has a broad fluorescence spectrum, and the laser is easily tuned across the spectrum with a grating or prism located inside the cavity. In addition, the great number of available dyes allows coherent radiation to be produced at any wavelength in the visible spectrum. Dye lasers, usually pumped by argon-ion lasers, have been mode locked to achieve pulses with durations well under 1 ps. These lasers are useful for studies in optical communications and detectors, as well as photochemistry and fast physical phenomena.

The *Ti:sapphire laser* is a solid-state laser pumped by an argon-ion laser. Its broad fluorescence spectrum, 680 nm–1.1 μm, and the convenience of solid-state lasers make it a useful replacement for many dye lasers. Ti:sapphire lasers may also be mode-locked to achieve pulses well under 1 ps in duration.

8.4.4 Helium-Neon Laser

The helium-neon or He-Ne laser is not optically pumped, but electrically pumped. The active medium is a gas mixture of about 5 parts helium to each part of neon, at a pressure of about 400 Pa (3 Torr). Pumping takes place because of a glow discharge established in the tube. Helium is excited to a certain level by electron impact. The energy is transferred rapidly to a neutral neon atom which has an energy level that is very slightly below that of the helium atom. This is the upper laser level; the most important laser transition takes place at 633 nm.

Although the He-Ne laser is not optically pumped, its behavior below threshold can be described adequately by *rate equations* (Sect. 8.2.1). In the case of the ruby laser, we observed very large fluctuations, or spiking, in the output. This was attributed to a sort of oscillation of the normalized population inversion n about the threshold value n_t. In the He-Ne laser, n does not undergo such oscillations, but rather takes on precisely the value n_t. As a result, the output of the He-Ne laser is continuous and stable.

He-Ne lasers are continuously pumped, usually with a dc power supply. Typically, they emit powers ranging from 0.3 to 15 or more milliwatts in the 00 transverse mode. The spectral width Δv of the 633-nm neon line is about 1500 MHz; the gain of the laser tube may exceed 1 over a bandwidth of more than 1000 MHz (see Fig. 8.8). The length of most common He-Ne laser cavities is such that the output will exhibit either one mode or two, depending on the precise length of the cavity. As the cavity length changes minutely owing to thermal expansion, the laser output remains constant to perhaps 1%.

Many He-Ne lasers employ hemispherical cavities with mirrors external to the plasma tube. The gain of the laser is extremely low; *Brewster-angle windows* are essential to eliminate reflection loss. Even so, the output mirror is likely to have a reflectance of over 99% with a 15- or 20-cm plasma tube. Because of the Brewster window, the output is polarized with the electric-field vector in the plane that includes the axis of the laser and the normal to the Brewster window.

Other He-Ne lasers, especially inexpensive models, are manufactured with mirrors cemented directly to the laser tube. Lacking a Brewster window, these lasers contain no element designed to control the polarization and are often called unpolarized or randomly polarized. In fact, however, alternate spectral modes are alternately polarized. That is, if a mode that has an even mode number m in (8.40) is polarized horizontally, then modes that have odd mode numbers $m \pm 1$ are polarized vertically. Since the laser generally oscillates in two spectral modes whose powers vary with cavity length, the polarization of the output beam is not constant but varies from somewhat horizontal to somewhat vertical as the even or the odd mode gains ascendancy. The only way to obtain a steady, polarized output from such a laser is to place a polarizer at 45° and sacrifice half the output power.

A single-mode laser, on the other hand, requires delicate temperature or length control to ensure that the cavity mode coincide precisely with the center of the spectral line. Otherwise, the output power will vary greatly as changes in cavity length reduce the gain by altering the resonant frequency.

8.4.5 Ion Lasers

The *argon-ion laser* can be made to oscillate at several wavelengths at the blue and green end of the visible spectrum. The important transitions take place between energy levels of the Ar^{+1} spectrum. A high-current arc discharge will produce enough singly ionized argon atoms to bring about the required gain. A close relative of the argon laser is the *krypton-ion laser*, which produces a strong red line, among others.

Because of the energy required to ionize an atom and then raise the ion to an excited state, the efficiencies of all arc-discharge lasers are extremely low.

Nevertheless, once a population inversion is maintained, these lasers have very high gain and can provide continuous-output powers up to several watts.

The most important argon-laser line has a wavelength of 514.5 nm. Other lines may be chosen by rotating a grating or prism inside the cavity. The laser will oscillate in all its lines simultaneously if broadband reflectors are used and the grating is removed. The krypton laser can be made to look nearly white when several lines across the visible spectrum are made to oscillate simultaneously.

Like the Nd:glass laser, the argon-ion laser is often mode locked for short, high-power pulses. Mode locking of the argon-ion laser is usually accomplished by means of a *loss modulator*, rather than a saturable dye, placed in the cavity. The loss modulator is a device, either acousto-optic or electro-optic, that periodically reduces the overall gain of the cavity by introducing a time-varying loss into the system. The loss is generally sinusoidal in time, and the frequency is half the frequency difference between adjacent spectral modes, or $c/4d$.

As with passive, dye mode locking of the Nd:glass laser, mode locking of the argon-ion laser is most easily understood in terms of a pulse developing in the cavity, rather than in terms of the spectral modes of the cavity. To begin, consider a short pulse originating in the cavity and passing through the loss modulator at a time when its loss is 0. (The pulse may be a small fluctuation in the spontaneous emission that takes place before the onset of laser action.) The round-trip transit time of the cavity is $2d/c$. After this time interval, the loss modulator has gone through one-half period, and its loss is again 0. The pulse slips through with relatively little loss; all other small fluctuations occur at different times and pass through the loss modulator when its loss is comparatively high. Thus the first small pulse that passes through the loss modulator with the right timing is amplified and ultimately gives rise to a mode-locked output such as we have discussed previously.

8.4.6 CO_2 Laser

The molecular-CO_2 laser oscillates at 10.6 μm in the infrared. The important transition occurs between vibrational energy levels of the CO_2 molecule. CO_2 lasers are operated continuously, pulsed, or Q-switched. Even a small, continuous CO_2 laser is capable of emitting a fraction of a watt and can heat most materials to incandescence in a short time. (Because the beam must be blocked, it is important to use materials that do not release dangerous contaminants, particularly beryllia, into the air.) CO_2 lasers are in use today for cutting metal and fabric and for welding metals.

The electric discharge that excites most gas lasers is a glow discharge or an arc that is maintained by an anode and a cathode at the ends of a long, thin plasma or discharge tube. A few lasers are excited by an rf (radio frequency) discharge. All such lasers operate with gas pressures well below atmospheric pressure.

There is another class of gas laser known as *transversely excited atmospheric-pressure lasers*, or *TEA lasers*, for short. The TEA laser is always pulsed, and, as the name implies, it is excited by an arc discharge at roughly atmospheric pressure. The current in the arc flows at right angles to the axis of the laser.

Many CO_2 lasers are also TEA lasers. They require relatively simple gas-handling systems and are therefore inexpensive and easy to construct. They may be repetitively pulsed and, like other CO_2 lasers, display high peak power or high average power.

The danger to the tyro who attempts to build such a laser cannot be overestimated.

8.4.7 Other Gas Lasers

Another laser system with increasing importance is the helium-cadmium laser, which oscillates continuously at 442 nm, in the blue. Many of the difficulties of vaporizing sufficient cadmium metal and preventing it from plating on electrodes and cooler portions of the tube have been largely overcome. The He-Cd laser is relatively inexpensive and can compete favorably with the He-Ne and argon-ion laser for low-power applications, especially where short wavelength is desirable.

Other gas lasers include the water-vapor and HCN lasers, both low-power, far-infrared lasers. Hydrogen- and deuterium-fluoride lasers oscillate at various wavelengths between 3 and 5 μm, in the infrared. They are transversely excited and are attractive as high-power sources at wavelengths shorter than 10.6 μm. The pulsed nitrogen laser is a source of high power in the uv portion of the spectrum, at 337 nm. Finally, *excimer lasers* are lasers that use halides of the noble gases as their active media. These are pulsed and produce high peak powers in the uv spectrum.

Dozens of other materials have been made to exhibit laser action at hundreds of wavelengths; Table 8.1, Principal laser lines, mentions only those that are common or commercially available. See also Volume 2 of the Springer Series in Optical Sciences.

8.4.8 Semiconductor Lasers

Also known as a *diode laser, junction laser,* or *injection laser,* the semiconductor laser is important in optical communications and optical-computer design. It is close relative of the light-emitting diode, which has already seen considerable application in alphanumeric and other displays, optical range finding, and short-range communication.

A semiconductor laser is a light-emitting diode with two of the faces cleaved or polished so they are flat and parallel. The other two faces are roughened. Because of the high refractive index of the material, the polished faces have sufficient reflectance to allow oscillation. 840-nm light is emitted from a slab along the junction and because of diffraction has a beam divergence of 5–10°.

Table 8.1. Principal laser lines

Species	Host	Wavelength	Usual modes of oscillation
Cr^{+3}	Al_2O_3 (ruby)	694 nm	pulsed, Q switched
Nd^{+3}	glass	1.06 μm	pulsed, Q switched, mode locked
Nd^{+3}	YAG	1.06 μm	cw, repetitively pulsed, mode locked
Er^+	silica fiber	1.55 μm	diode-laser pumped
Ti^{+3}	sapphire	680 nm–1.1 μm	cw, mode-locked, pulsed
Ne	He	633 nm; 1.15, 3.39 μm	cw
Cd	He	325 nm, 442 nm	cw
CO_2	—	10.6 μm	cw, Q switched, repetitively pulsed
Ar^+	—	488 nm, 515 nm	cw, pulsed, mode locked
Kr^+	—	647 nm	cw, pulsed
GaAs	GaAs substrate	840 nm	pulsed, cw
GaAlAs	GaAs	850 nm	pulsed, cw
GaP	GaAs	550–560 nm	pulsed, cw
GaInAsP	InP	0.9–1.7 μm	pulsed, cw
Rhodamine 6G	ethanol, methanol, water	570–610 nm	short pulse, cw, mode locked
Sodium fluorescein	ethanol, water	530–560 nm	short pulse, cw
Water vapor	—	119 μm	pulsed, cw
HCN	—	337 μm	pulsed, cw
HF, DF	—	2.6–4 μm	pulsed
N_2	—	337 nm, 1.05 μm	pulsed, pulsed
Excimer (KrCl KrF, XeCl, XeF)	—	222, 248, 308, 351 nm	pulsed
Cu vapor	—	511, 518 nm	pulsed

Like light-emitting diodes, semiconductor lasers are basically pn junctions of gallium arsenide, although more complicated structures involving gallium aluminum arsenide and gallium aluminum arsenide phosphide have evolved. Continuous, room-temperature semiconductor lasers are now available and are used for optical communications at 1.3 and 1.55 μm as well as 840 nm (Sect. 12.1).

8.5 Laser Safety

Probably the most dangerous aspect of many lasers is the power supply. Still, the beam emitted by some lasers can be harmful to the eyes or even the skin.

Therefore, it is worthwhile to dwell on the dangers posed by lasers and other intense sources.

Radiation can harm the eye in several ways, depending on the wavelength. Ultraviolet light below 300 nm or so can "sunburn" the cornea or, at higher intensity, the skin. Slightly longer-wavelength uv radiation penetrates the cornea at least in part and is absorbed in the lens. It may cause a *cataract*, or opacity of the lens. Visible and near-ir radiation through about 1.4-μm wavelength penetrates fairly efficiently to the retina, where it may be focused to a small spot and cause either photochemical or thermal damage. Radiation between the wavelengths of 1.4 and 3 μm penetrates to the lens and may cause cataracts; longer-wavelength infrared radiation is absorbed near the surface of the cornea and can cause damage there. Even microwave radiation, because of its ability to heat tissue, has been implicated in the formation of cataracts.

Laser sources have been placed into four classes defined by the lasers' ability to cause damage to the eye. These are usually designated by Roman numerals from I to IV. Class I lasers are believed to be unable to cause damage even when shone directly into the eye for an extended period of time. Class II lasers emit low-power visible radiaton that can probably not cause damage within 0.25 s if shone directly into the eye; the duration, 0.25 s, is assumed to be the time required for an aversion response, in this case, a blink. In the visible and near-ir regions, Class II lasers are those that emit between approximately 1 μW (depending on wavelength) and 1 mW. There are fairly large safety factors built into the classification, but a glance down a Class II laser should not be encouraged.

Class III lasers are those that can create a hazard in less than 0.25 s; Class IV lasers are those that can create dangerous levels of radiation by diffuse reflection. Many Class IV lasers are also fire hazards or emit so much power that they can vaporize whatever is used to block the beam and thereby put dangerous chemicals (such as beryllium compounds) into the air. All lasers except Class I lasers must have labels that state the laser's classification. In addition, operators of Class III and IV lasers are required to employ a variety of safety measures to protect themselves as well as more casual passers-by.

To put the laser-classification scheme into perspective, let us compare the retinal irradiance levels that result from a direct look at the sun and a direct look down a 1-mW He-Ne laser. The sun subtends about 10 mrad and delivers an irradiance about equal to 100 mW $\cdot$ cm^{-2} at the earth's surface. The focal length of the eye is about 25 mm, so the image of the sun has a diameter of about 25 mm times 10 mrad, or 0.25 mm. If the eye is bright-adapted, the pupil diameter is approximately 2 mm, and the total power incident on the pupil is about 3 mW. The irradiance is equal to this power divided by the area of the image, or 6 W $\cdot$ cm^{-2}. Had the eye been dark-adapted immediately before the exposure, the pupil diameter might have been as much as 8 mm. Then the retinal irradiance could have been as much as 100 mW $\cdot$ cm^{-2}.

Now let us assume that a 1-mW laser with a 2-mm beam diameter (1-mm beam width w) is aimed directly into the eye. With a 2-mm pupil, the eye is diffraction-limited, so we assume diffraction-limited imagery in this case. If the wavelength of the laser is 633 nm, the radius of the beam waist on the retina is $\lambda f'/\pi w$, or about 5 μm. The average irradiance inside a circle with this radius is about 1 kW $\cdot$ cm^{-2}, or about 200 times more than the irradiance brought about by the sun; the total focused powers are roughly equal. Whether power or irradiance is the important quantity depends on the mechanism of damage. In any case, looking down the bore of a common, 1-mW He-Ne laser is at least roughly comparable to looking directly at the sun. Those who suffer from eclipse blindness can attest that the full power of the sun is not necessary to bring about permanent damage in a relatively short time.

8.5.1 Sunglasses

Long-term exposure to relatively high irradiance can cause damage to the retina, whether or not the light is coherent. The blue and ultraviolet spectral regions can cause photochemical damage at irradiances far below the threshold for visible burns. Light of these wavelengths may be implicated in *senile macular degeneration*, a disorder that causes elderly people to lose their visual acuity.

The danger from the short-wavelength radiation is potentially increased with some sunglasses if they are on average dark enough to increase the pupil diameter and yet are manufactured so that they transmit harmful wavelengths selectively. Therefore, sunglasses should probably be designed for greater attenuation of wavelengths below about 550 nm; they should at any rate not exhibit high transmittance at wavelengths just shorter than visible light. Unfortunately, some sunglasses, among them some polarizing sunglasses, exhibit high transmittance just below 400 nm. A rule of thumb, which is not necessarily accurate, states that yellow or brown lenses are less likely to transmit short-wavelength radiation than are neutral or gray lenses. [Glasses that have colored lenses (especially blue lenses) and are not specifically described as "sunglasses" should probably be avoided entirely.]

There is also speculation that ambient levels of infrared radiation may be harmful to the eye, but this is less well established. Many sunglasses transmit near-infrared radiation efficiently, and this also might cause a hazard if the pupil dilates significantly. Another rule of thumb, again not necessarily applicable to all cases, suggests that metal-coated sunglasses are the least likely to have high-transmittance windows at wavelengths other than those of visible light.

Problems

8.1 Suppose that a three-level laser has a fourth level that can partially deplete level 2. Write a rate equation that includes the effects of level 4.

Suppose that the laser is to be pulsed and that level 4 is long lived compared to the pulse duration. Assuming that steady-state conditions are approximately valid, solve the rate equations and show that the threshold is increased and that the population inversion (and therefore the gain) is decreased by the presence of level 4.

8.2 (a) Show that the duration of the pulses emitted by a mode-locked laser is about equal to the reciprocal of the linewidth Δv of the laser transition.

(b) Explain why the pulse duration may be longer than $1/\Delta v$ if not all the modes are locked to the same phase.

8.3 *Depth of Focus of a Gaussian Beam.* This may be defined as the axial distance over which the beam width remains less than, say, $2w_0$.

(a) Calculate the depth of focus on the basis of this definition. [Answer: $\sqrt{3}\pi w_0^2/\lambda$.]

(b) A lens can be shown to transform a plane wave at one focal point into a plane wave at the other. Explain what happens when a Gaussian-beam waist is located at the primary focal point F of a lens. Consider two cases: when the beam waist at F is small compared with f, and when it is large. [Hint: Calculate the radius of curvature of the beam when it hits the lens.]

(c) Suppose that we try to collimate a Gaussian beam by locating a small waist at F. Over what range may the beam be considered collimated?

8.4 A certain 1-mW He-Ne laser is made with a nearly hemispherical cavity about 30 cm long. The beam diameter on the surface of the curved mirror is 1 mm. A spatial filter is to be made using a $10 \times$ microscope objective and a pinhole. Find the diameter of the pinhole.

8.5 Calculate the approximate length of a He-Ne laser that will oscillate in one spectral mode when the cavity resonance happens to coincide with the wavelength of maximum gain, but will oscillate in two modes when the cavity resonance is far from that wavelength. Calculate the maximum length of a temperature-stabilized laser that is designed to oscillate in only one spectral mode at precisely the center of the gain curve of the laser.

8.6 A He-Ne laser cavity is 30 cm long. Show that the degree of coherence is nearly zero for path lengths equal to 30 cm.

8.7 The higher-order transverse modes have greater diffraction loss than the 00 mode. A laser will therefore oscillate in a single transverse mode only when diffraction is large enough to be a factor. Consider a plane-parallel cavity that contains a circular aperture near one end. If the diameter of the aperture is D and the length of the cavity is d, when will diffraction become apparent? Develop a criterion for estimating the aperture size necessary to force 00-mode oscillation. What is this diameter for $L = 1$ m and $\lambda = 1.06$ μm (Nd:YAG)?

8.8 We need to know the spot size w_0 of a Gaussian beam at the focal point of a $40 \times$ microscope objective. The lens is well corrected, and its entrance-pupil diameter D is much larger than the beam waist $w = 1$ mm. The spot at the focal point is too small to measure directly. Explain in detail how to measure w_0. ($\lambda f'/\pi w$ is not sufficiently accurate for our purpose because neither f' nor w is known accurately enough.)

8.9 A He-Ne laser cavity is 1.5 m long, $\lambda = 633$ nm, and the fluorescence linewidth is 1500 MHz. Calculate the frequency difference Δv between adjacent spectral (axial) modes. Explain why the coherence length of the laser is about equal to that of the incoherent fluorescence line.

8.10 A laser could be mode locked by vibrating one of the end mirrors at frequency $c/4d$. Give a qualitative explanation in terms of cavity modes. In terms of Doppler shift. [This is an example of a mode-locking mechanism that is more easily visualized physically in terms of modes than in terms of a pulse propagating back and forth in the cavity. In practice, the mirror cannot be vibrated fast enough, but the laser may be mode locked by changing the optical length of the cavity (as opposed to its physical length) with an electro-optic device.]

9. Electromagnetic and Polarization Effects

Under this heading we discuss electromagnetic theory of light, polarization, birefringence, harmonic generation, electro- and acousto-optics, and related topics. Light consists of time-varying electric and magnetic fields. These fields are vectors, and their directions are almost always perpendicular to the direction of propagation of the light. When the electric-field vector E of a light wave lies in one plane only, the light is said to be *plane-polarized*. The magnetic-field vector H is then perpendicular to both the direction of propagation and the electric-field vector, as shown in Fig. 9.1. Because the fields propagate together and maintain a constant 90° phase difference with one another, it is usually sufficient to describe the wave with either the electric vector or the magnetic vector. It is conventional to choose the electric vector, largely because the interaction of matter with the electric field is stronger than that with the magnetic field. Therefore, the wave shown in Fig. 9.1 is said to be vertically polarized because the electric-field vector lies in a vertical plane. Unfortunately, in classical optics, the *plane of polarization* is defined perpendicular to the electric-field vector. We shall use electric-field vector throughout.

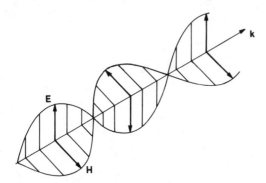

Fig. 9.1. Electromagnetic wave propagating in free space

9.1 Reflection and Refraction

9.1.1 Propagation

To understand reflection, refraction, harmonic generation, and related topics physically, we will find it helpful to discuss propagation in optically *dense* media, that is, transparent media whose refractive index is greater than 1.

Consider, for simplicity, a plane-polarized electromagnetic wave that propagates through vacuum and across an interface with a transparent, dielectric medium. The electric field associated with the wave induces dipoles in the material; the induced dipole moment per unit volume P is

$$P = XE,\qquad(9.1)$$

where X, the *polarizability*, is a property of the material and is, for most practical purposes, a constant.

The dipoles oscillate with the same frequency as the field, but with not necessarily the same phase. We know from classical electrodynamics that such oscillating dipoles radiate at their oscillation frequency. Therefore, the material radiates electromagnetic waves with the same frequency as the incoming field.

A portion of this radiation propagates back into the vacuum and is said to be the reflected wave. Light is reflected whenever it strikes a boundary between media that have different refractive indices.

The remainder of the radiation emitted by the induced dipoles propagates into the material, where it interferes with the original wave. Because of the phase difference between P and E, the velocity of the total field in the medium is reduced by a factor of n, the refractive index.

9.1.2 Brewster's Angle

The plane that contains the incident, reflected, and refracted waves is known as the *plane of incidence*. Consider a wave that is incident on a surface and polarized so that its electric-field vector lies in the plane of incidence. There will be some angle of incidence i_B, called Brewster's angle, for which the refracted and reflected waves propagate at right angles to one another, as in Fig. 9.2. The reflected wave, however, is driven by the oscillating dipoles induced in the material. Oscillating dipoles radiate primarily in the direction perpendicular to their axes; they radiate no energy along their axes. Therefore, in this polarization, there is no wave reflected at Brewster's angle.

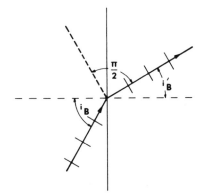

Fig. 9.2. Brewster's angle

Using Snell's law, combined with the fact that the reflected and refracted waves are perpendicular, we find that *Brewster's angle* i_B is

$$i_B = \tan^{-1} n .\tag{9.2}$$

For a glass-air interface, i_B is about $57°$.

Brewster's angle is important for several reasons. First, many lasers use *Brewster windows* inside the cavity to reduce reflection loss. Brewster windows are important in low-gain lasers such as He-Ne lasers, where a few percent loss can completely inhibit laser action, and in high-power lasers, where antireflection coatings would be destroyed by the intense beam.

If unpolarized light is shone on a surface at Brewster's angle, the refracted wave will be partially polarized because the reflectance at Brewster's angle is 0 only for waves whose electric-field vectors lie in the plane of incidence. Waves whose electric-field vectors are perpendicular to the plane of incidence exhibit about 15% reflectance from a glass surface. A beam that passes through a number of plates of glass at Brewster's angle will be nearly 100% plane-polarized with its electric field in the plane of incidence. Polarizers made of a pile of plates are useful in laser applications where other polarizers would be damaged by the laser.

9.1.3 Reflection

Consider, for simplicity, a plane-polarized wave incident on an air-to-glass interface. The index of refraction of the air is very nearly 1; the index of the glass is n (about 1.5 for common optical glass). The fraction of light reflected depends on the angle of incidence, the direction of the electric-field vector of the incident light, and the index n. A calculation based on electromagnetic theory may be found in any text on electricity and magnetism; here we discuss the results.

If the incident light is polarized with its electric vector parallel to the plane of incidence, the *amplitude reflectance* is

$$r_{||} = \frac{\tan(i - i')}{\tan(i + i')}\tag{9.3a}$$

and the *amplitude transmittance is*

$$t_{||} = \frac{2 \sin i' \cos i}{\sin(i + i') \cos(i - i')} .\tag{9.3b}$$

When $i + i' = \pi/2$, $r_{||} = 0$. This is *Brewster's law.*

If the incident light is polarized with its electric vector perpendicular to the plane of incidence, then the amplitude reflectance is

$$r_{\perp} = \frac{-\sin(i - i')}{\sin(i + i')} ,\tag{9.3c}$$

and the amplitude transmittance is

$$t_\perp = \frac{2 \sin i \cos i'}{\sin(i + i')}.$$ (9.3d)

The preceding four equations are known as *Fresnel's laws*.

The minus sign in the equation for the amplitude reflectance of light polarized with its electric vector perpendicular to the plane of incidence signifies a *phase change on reflection* of π.

The (intensity) *reflectance* $R_{||}$ or $R_\perp$ is calculated by squaring $r_{||}$ or $r_\perp$. Figure 9.3 shows a plot of reflectance as a function of angle of incidence for reflection from glass whose index of refraction is 1.5. The most prominent features of the graph are the existence of the Brewster angle at 57° and the rapid approach to 100% reflectance at grazing incidence ($i = \pi/2$). The reflectance of both polarizations is about 4% for angles of incidence between 0 and 30°. The dashed line shows the reflectance of unpolarized light; it is the average of the two solid curves.

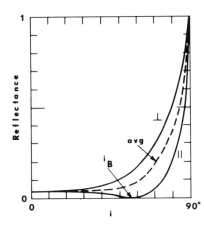

Fig. 9.3. Reflectance from the surface of a dielectric medium whose index of refraction is 1.5

Problem. Show that the reflectance at normal incidence is

$$R = \frac{(n - 1)^2}{(n + 1)^2}$$ (9.4)

for either polarization.

The (intensity) *transmittance* $T_{||}$ or $T_\perp$ is calculated by squaring the amplitude transmittance and multiplying by n, because the intensity in a medium is n times the square of the electric field strength (Sect. 5.1).

Because of the effects of the second medium on the electric and magnetic fields, the sum of amplitude reflectance and transmittance is not 1. The relationships among the incident, reflected, and transmitted amplitudes must be calculated by applying the boundary conditions of electromagnetic theory.

Similarly, the sum of the intensity reflectance and transmittance is not 1. This is because real beams of light have finite cross sections; the cross section is diminished after refraction at the interface. Therefore, the intensity (strictly speaking, irradiance or power per unit area) of the beam in the medium is increased in proportion to the amount by which the cross section is diminished (Problem 9.1).

9.1.4 Interface between Two Dense Media

Fresnel's laws may also be applied to the case of reflection at the interface between media whose indices are n and n'. If we define the *relative index of refraction* μ by the equation

$$\mu = n'/n \,, \tag{9.5}$$

we may replace n by μ in any of the Fresnel equations. In particular, the reflectance at normal incidence is

$$R = \left(\frac{\mu - 1}{\mu + 1}\right)^2 \tag{9.6}$$

in the general case.

9.1.5 Internal Reflection

This is the case where the light propagates across the boundary from glass to air, not from air to glass. Fresnel's laws apply to internal reflection, provided only that i and i' be interchanged. Figure 9.4 shows reflectance as a function of angle of incidence for glass for which $n = 1.5$. As we saw in Chap. 1, when i exceeds the *critical angle* i_c, the refracted ray does not exist. The reflectance for both polarizations is 1 for all angles of incidence between i_c and $\pi/2$. (There is no

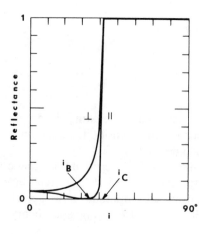

Fig. 9.4. Internal reflection from the surface of a dielectric medium whose index of refraction is 1.5

discontinuity in the reflectance curves, however: both $R_\parallel$ and $R_\perp$ approach 1 rapidly near the critical angle.) When i exceeds i_c, we speak of *total internal reflection.*

Even though total internal reflection is indeed total, electric and magnetic fields penetrate into the low-index material. The field strength decreases rapidly with distance from the boundary; the electric field falls practically to 0 within a few wavelengths of the interface.

The wave in the low-index medium is known as an *evanescent wave* and is one of the few examples of a longitudinal wave in electromagnetics. Nonetheless, because the wave penetrates slightly into the low-index medium, the propagating wave can be continued past the interface by bringing a second glass surface into near contact with the first. The phenomenon is known as *frustrated total internal reflection.* The transmittance depends on the separation between the two surfaces. Variable-reflectance mirrors and shutters and components for integrated optics have been made using the principle of frustrated total internal reflection.

9.1.6 Phase Change

Electromagnetic theory shows that the totally reflected ray undergoes a phase shift after reflection from the interface. The magnitude of this phase shift, or *Goos-Hänchen shift*, depends on the angle of incidence, on the relative index of refraction μ, and on the polarization of the incident wave. If the wave is polarized with its electric-field vector perpendicular to the plane of incidence, the phase shift $\Phi_\perp$ is given implicitly by the equation

$$\tan^2(\Phi_\perp/2) = \frac{\sin^2 i - (1/\mu^2)}{\cos^2 i} \tag{9.7a}$$

where $\mu = n/n' > 1$ and $\sin i_c = 1/\mu$. When the electric-field vector is polarized parallel to the plane of incidence, the phase shift $\Phi_\parallel$ is

$$\tan(\Phi_\parallel/2) = (1/\mu^2)\tan(\Phi_\perp/2) . \tag{9.7b}$$

These equations may be rewritten in terms of the complement θ of the angle of incidence,

$$\tan^2(\Phi_\perp/2) = \frac{\cos^2\theta - \cos^2\theta_c}{\sin^2\theta} , \tag{9.8a}$$

and

$$\tan(\Phi_\parallel/2) = (1/\mu^2)\tan(\Phi_\perp/2) , \tag{9.8b}$$

where θ_c is the complement of the critical angle. The angle θ varies between 0 and θ_c; when θ exceeds θ_c, i is less than i_c, and the incident ray is mostly refracted.

The case where μ is close to 1 has interest for studies of optical waveguides. The angle θ_c is small in that case, so we may use the approximation that, for small x, $\cos x \cong 1 - (x^2/2)$ to show that

$$\tan^2(\Phi_\perp/2) \cong (1/\alpha^2) - 1 , \tag{9.9}$$

where $\alpha = \theta/\theta_c$. Because μ is close to 1, $\Phi_\parallel \cong \Phi_\perp$. Defining the normalized variable α gives the equation more general applicability because Φ no longer depends explicitly on θ_c.

Figure 9.5 shows the phase shift as a function of angle for three values of μ. The value 1.01 corresponds to a typical optical waveguide; 1.5 to an air-glass or air-vitreous silica interface; and 3.6 to an air-gallium arsenide interface (Chap. 12). The horizontal axis is the normalized variable θ/θ_c, which varies between 0 and 1; when θ/θ_c is greater than 1, the reflection is no longer total. Normalizing the independent variable in this way makes it possible to draw a family of curves on the same scale and to develop an approximation that is independent of μ over a certain range.

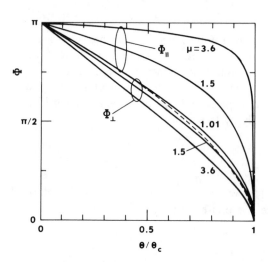

Fig. 9.5. Phase shift on reflection

Previous equations give Φ implicitly; for the case where $\mu \sim 1$, Φ may be approximated by the empirical relation,

$$\Phi = \pi(1 - \alpha)^p . \tag{9.10}$$

When $p = 0.6$, the empirical curve approaches the exact curves asymptotically and is almost exact as long as $\theta/\theta_c < 0.6$. This is shown by the dotted curve in the figure. The maximum error is about 2.5°. (The value $p = 0.56$ does not approach the curve asymptotically but gives a slightly smaller maximum error.) These curves are useful in the design of dielectric waveguides for integrated optics.

When $\mu = 1.01$, the curves that correspond to the two polarizations coalesce to the approximation (9.10); the difference between the two exact curves is about equal to the thickness of the line in the figure.

The phase shift on reflection is related to the existence of the evanescent wave. Because of the phase shift, a ray appears to penetrate slightly into the less dense medium before emerging and being reflected. This is so because the phase shift is equivalent to an advance of phase of a fraction of a wavelength. This argument suggests (correctly) that power propagates in the less dense medium. Although the electric-field strength diminishes approximately exponentially in the direction perpendicular to the boundary and the intensity is 0 more than a few wavelengths past the boundary, power does propagate in the second medium in the direction parallel to the boundary. This power may be important in *optical waveguides* (Chaps. 10–12); the evanescent wave may be used for coupling power into a waveguide, or, on the other hand, may contribute to loss of power from the waveguide.

9.1.7 Reflection from Metals

Most polished-metal surfaces have relatively high reflectance in the visible and near ir. The reflectance decreases significantly below 300 or 400 nm.

At normal incidence, silver and aluminum have reflectances greater than 0.9 throughout the visible spectrum. These metals owe their high reflectance to the presence of free electrons, which are readily set into oscillation by the incident electric field. The wave penetrates only a few wavelengths into the metal; nearly all of the light is reflected. The electrons cannot respond to the high frequency of uv radiation, and the metals have lower reflectance in that portion of the spectrum.

The reflectance of metals varies with angle of incidence, much as that of dielectrics. Light polarized with its electric-field vector parallel to the plane of incidence passes through a *principal angle* that is analogous to Brewster's angle; that is, the reflectance is a minimum, but not 0, at the principal angle. The reflectance of both polarizations approaches 1 as the angle of incidence approaches $\pi/2$.

9.2 Polarization

We have already discussed the fact that light may be plane-polarized. There are other states of polarization as well. Consider the light wave of Fig. 9.6. It consists of two plane-polarized waves out of phase with one another by an amount ϕ. That is, at $z = 0$,

$$E_x = A_x e^{i\omega t}, \quad \text{and} \tag{9.11}$$

$$E_y = A_y e^{i(\omega t + \phi)}, \tag{9.12}$$

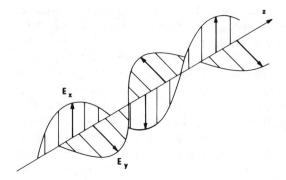

Fig. 9.6. Elliptically polarized light wave resolved into two plane-polarized components that differ in phase

where the A's are amplitudes. Light polarized in this way is termed *elliptically polarized*, because in any plane normal to the z axis, the tip of the electric-field vector describes an ellipse once during each period of the wave. It is not instructive to dwell on this point, but rather to describe elliptically polarized light in terms of its components parallel and perpendicular to any convenient axis.

When A_x and A_y are equal and ϕ is $\pi/2$, the ellipse becomes a circle; the light is then termed *circularly polarized*.

When A_x and A_y are equal, but the phase difference ϕ is a random variable that changes rapidly with time, the light is termed *unpolarized*. Much natural light, such as sunlight or blackbody radiation, is unpolarized or nearly unpolarized. Natural light that is reflected at glancing incidence from a smooth surface may be *partially polarized* as a result of the difference of reflectance between the two orthogonal planes of polarization.

9.2.1 Birefringence

Certain crystals are anisotropic; that is, their physical properties, such as index of refraction, vary with direction. The anisotropy is determined by the crystal structure of the material. The simplest kind of anisotropic crystal has a single axis of symmetry; in optics, the axis is known as the *optic axis*.

Figure 9.7 shows an anisotropic crystal whose optic axis is inclined at an angle to the surface of the crystal; the optic axis lies in the plane of the page. The light beam falling on the surface has one component polarized in the plane containing the optic axis. The direction of its electric vector is indicated by the slashes perpendicular to the direction of propagation. Because the crystal is anisotropic, its refractive index varies with direction of propagation.

We appeal to Huygens's construction to determine the behavior of the wave inside the crystal. At each point on the surface of the material, a Huygens wavelet begins to propagate, as we have seen in Fig. 5.10. However, because of the anisotropy of the crystal, the wavelet may be not a sphere, but an ellipsoid of revolution known as the *wave ellipsoid*. Arguments based on symmetry show that the optic axis must be one of the axes of the ellipsoid. In Fig. 9.7, it is shown

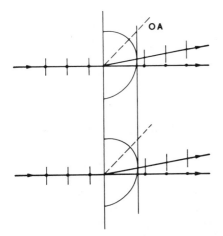

Fig. 9.7. Propagation in a birefringent crystal

as the major axis; in this case, we term the optic axis a *fast axis*, because a wave travels faster along the optic axis than perpendicular to it. Calcite is a common crystal that has a fast axis; quartz and mica have *slow axes*.

The ellipses in Fig. 9.7 represent the intersections of the Huygens wavelets with the plane of the page. Two such wavelets are shown. The wavefront that propagates is the common tangent to the wavelets. Thus, the beam propagates through the crystal at an angle to the surface, even though it was incident normal to the surface. (The apparent violation of Snell's law is resolved if we realize that the electric-field vector and the wavefront both remain parallel to the surface, but that the direction of propagation is not normal to the wavefront. Snell's law is derived by applying the boundary conditions of electromagnetic theory and can be said to describe the behavior of the wavefront, not the direction of propagation.)

Any beam of light polarized with its electric-field vector in the same plane as the optic axis behaves in this peculiar fashion and is termed an *extraordinary ray*.

If a wave is polarized with its electric-field vector perpendicular to the optic axis, it displays no extraordinary behavior and is called an *ordinary ray*. A ray polarized with its electric-field vector perpendicular to the plane of Fig. 9.7 will continue to propagate through the crystal in its original direction. Its wave surface is a sphere whose radius is equal to either the major or the minor axis of the wave ellipsoid of the extraordinary ray.

If a narrow beam of unpolarized light were shone on the crystal of Fig. 9.6, two beams polarized orthogonally to one another would emerge. For this reason, the phenomenon is called *birefringence*.

9.2.2 Wave Plates

Consider a thin slab of birefringent material cut so that its optic axis OA lies in the plane of the surface, as shown in Fig. 9.8. For convenience, we take the optic

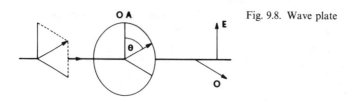

Fig. 9.8. Wave plate

axis to be vertical. Suppose that plane-polarized light falls at normal incidence on the slab. Its electric vector makes an angle θ with the optic axis.

We may resolve the electric vector into components perpendicular and parallel to the optic axis. The component E parallel to the optic axis propagates through the crystal as an extraordinary ray; the component O perpendicular to the optic axis, as an ordinary ray. Suppose, for example, that the optic axis is a fast axis, as in calcite. Because it is incident normal to the optic axis, the extraordinary ray propagates through the crystal unchanged in direction, but somewhat faster than the ordinary ray.

Suppose that the index of refraction is n_o for ordinary rays and n_e for extraordinary rays. If the thickness of the crystal is d, then the extraordinary ray will propagate through an optical thickness $n_e d$, and the ordinary ray, $n_o d$. The extraordinary ray will lead the ordinary ray by $(n_o - n_e)d$ after leaving the crystal.

As a special case, let θ be $\pi/4$. Then the ordinary and extraordinary waves will have the same magnitude. Further, suppose that $(n_o - n_e)d$ is equal to $\lambda/4$ at some specific wavelength λ. Then the phase difference between the two waves will be $\pi/2$, and the light transmitted by the crystal will be circularly polarized. The crystal is called a *quarter-wave plate*.

As another special case, let the phase difference between the rays be π. Then the crystal is known as a *half-wave plate*. Suppose a wave with unit electric-field strength falls on the half-wave plate. The wave is plane polarized, its electric-field vector making angle θ with the optic axis of the crystal. The component of the incident electric field parallel to the optic axis is $\cos\theta$; the component perpendicular to the optic axis is $\sin\theta$.

The half-wave plate causes a phase shift of π between the two waves. Therefore, the components become $\cos\theta$ and $-\sin\theta$, where we arbitrarily associate the phase shift with the ordinary ray. The emerging wave is plane-polarized; its electric-field vector makes angle θ with the optic axis, but it now lies on the opposite side of the optic axis from the incident wave. The half-wave plate can therefore be used to rotate the plane of polarization through an angle 2θ. Quarter- and half-wave plates behave precisely as they are intended at one wavelength only, because the dispersion of the ordinary ray is slightly different from that of the extraordinary ray. Unless otherwise stated, values of the index difference for different materials are usually measured for sodium light, 590 nm.

Problem. A *polarizer* is a device that transmits only light whose component of electric-field vector is parallel to an imaginary line we call the *axis* of the polarizer.

Suppose that a plane-polarized light beam is shone on a polarizer so that its electric field vector makes an angle θ with the axis of the polarizer. Show that the transmitted intensity is $I_0 \cos^2\theta$, where I_0 is the incident intensity. This result is known as *Malus's Law*.

9.2.3 Glan-Thompson and Nicol Prisms

Birefringent prisms may be used to plane-polarize elliptically polarized or unpolarized light. Such prisms take advantage of the difference between the indices of refraction for the extraordinary and ordinary rays. Because of this index difference, the ordinary ray has a slightly different critical angle from the extraordinary ray. For example the indices of refraction of crystal quartz at 700 nm (approximately the wavelength of the ruby laser) are $n_e = 1.55$ and $n_o = 1.54$; the corresponding critical angles are $40.2°$ and $40.5°$.

Suppose we wish to make a polarizing prism of quartz and specify that the light must enter the prism at right angles to the face of the prism. The prism on the left side of Fig. 9.9 is a right-angle prism with one of the acute angles between $40.2°$ and $40.5°$; the optic axis is perpendicular to the plane of the page. Unpolarized light enters the prism. The component with polarization parallel to the optic axis behaves as an extraordinary ray when it enters the prism and is totally reflected. (The optic axis OA is shown as a dot to indicate its direction normal to the page; the vertical component of polarization is similarly shown as a series of dots along the ray.) The component of polarization in the plane of the page (indicated by the slashes) passes through the prism but emerges nearly parallel to the interface between the prisms because its angle of incidence is very near to the critical angle.

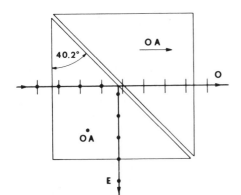

Fig. 9.9. Glan-Thompson prism

The second prism is provided to redirect the refracted ray to its original direction of propagation. The optic axis of the second prism is parallel to the directon of propagation, to ensure that the ray remains an ordinary ray.

A pair of prism elements such as those in Fig. 9.9 is called a *Glan-Thompson prism*. For laser work, these prisms are generally made of quartz and have a small air space between the two elements. The principle must not be confused with frustrated total internal reflection, however; the air space must be large enough that that total reflection of the extraordinary ray will not be frustrated.

Because of the small index difference between the extraordinary ray and the ordinary ray, the Glan-Thompson prism has a very small angular tolerance. It must be used in highly collimated light so that the angle of incidence at the interface is between 40.2° and 40.5°. When the angle of incidence is less than 40.2°, both polarizations are transmitted; when it is larger than 40.5°, both polarizations are reflected. Therefore, the prism is effective for a range of angles not exceeding one or two tenths of a degree.

Calcite is another common, birefringent material. The index difference $(n_o - n_e)$ of calcite is considerably greater than that of quartz. In sodium light (590 nm), $n_o = 1.66$ and $n_e = 1.49$. Common optical cement has an index of refraction of about 1.55; this makes it possible to fabricate a polarizing prism of calcite by cementing the two elements together. Cementing alleviates mechanical problems inherent in constructing an air-spaced Glan-Thompson prism; unfortunately neither calcite nor cemented interfaces are durable enough for high-power laser applications.

The first birefringent polarizers were Nicol prisms, which are made of two cemented calcite elements. The principal difference between the Glan-Thompson prism and the Nicol prism is the angle of incidence: The incoming beam is normal to the face of the former whereas it enters at a small angle to the face of the latter. With the exception of laser polarizers, birefringent prisms are comparatively uncommon today.

9.2.4 Dichroic Polarizers

Certain crystals, such as tourmaline, are not only birefringent; they absorb one plane of polarization and transmit the other. Such crystals are called *dichroic crystals*.

Polarizing films are made of sheets of an organic polymer containing submicroscopic, dichroic crystals whose axes are aligned parallel to one another. The process whereby the films are made is complicated, but involves stretching the organic polymer to orient the molecules and then treating the polymer chemically to create dichroic crystals with their axes parallel to the oriented, polymer molecules.

Polarizing films are almost uniformly gray throughout the visible and transmit highly polarized light. The angle of incidence is not critical. The films are inexpensive and have all but replaced Nicol prisms in the laboratory. Unfortunately, because of their strong absorption, they cannot be used with high-power lasers.

9.2.5 Optical Activity

Certain substances, among them crystal quartz and sugar solutions, will rotate the plane of polarization of a linearly polarized beam. The direction of rotation may be either clockwise of counterclockwise (as seen by the observer looking backward toward the source). Clockwise rotation is called right-handed; counterclockwise, left-handed. Materials that rotate the plane of polarization are termed *optically active*.

A plate of crystal quartz with its optic axis parallel to the incident beam of light will rotate sodium light approximately 20° per millimeter of thickness; the effect is roughly inversely proportional to wavelength. Crystal quartz may be either right-handed or left-handed, depending upon the sample. (This is possible because there are two possible crystal structures, one the mirror image of the other.)

9.2.6 Liquid Crystals

These are organic compounds that flow like liquids at certain temperatures, but whose molecules nevertheless show more long-range order than the molecules of ordinary liquids. Liquid crystals are commonly used in electronics in the displays of digital watches, lap-top computer screens, and pocket television receivers.

The liquid crystals used in these displays are usually *nematic* liquid crystals. These are compounds whose molecules are aligned in the direction parallel to their long axes, but are not aligned in layers perpendicular to the axes of the molecules. They are often likened to matches in a box: If you shake the box, the matches align parallel to one another, but there is no particular order in the other two dimensions. The matches can be poured, but if you confine them, as in a narrow box, they will realign parallel to the long axis of the box. Nematic liquid crystals behave analogously; whereas they flow like liquids, they can be made to align by confining them, for example, between two plates of glass separated by 5 or 10 μm.

In bulk, nematic liquid crystals display only short-range order. For displays, they must be sandwiched between two specially prepared glass plates. These plates are prepared so that their surfaces have a preferred direction, that is, so that their surfaces are anisotropic. This preparation may be effected in the laboratory by rubbing a surface in one direction with a cloth. For mass production, a coating is sputtered or evaporated onto the surface, but the surface is oriented so that the droplets or molecules strike the surface at glancing incidence as they are deposited. Either treatment prepares an anisotropic surface that causes the liquid-crystal molecules to align parallel to the preferred direction on the surface.

In a display, the two glass surfaces are oriented parallel to one another and 5–10 μm apart, but one of the plates is rotated so that the preferred directions of the surfaces are aligned perpendicular to one another. When the space between the surfaces is filled with liquid crystals, the molecules that are in contact with one surface orient along the preferred direction. Similarly, the molecules that are in contact with the other surface orient parallel to that preferred direction. Since these directions are orthogonal, the molecules in the volume between the two surfaces follow a helix that begins at one glass surface and ends at the other.

Such an arrangement is called a *twisted nematic liquid crystal* and is optically active. Specifically, the plane of polarization of a light beam incident on the sandwich at normal incidence is rotated by 90°, since that is the angle of twist of the helix. We may, however, change the angle of rotation by applying an electric field perpendicular to the plane of the sandwich. As the field is increased, the molecules gradually change their orientation so that they become more and more nearly parallel to the electric field, or perpendicular to the glass plates. When the molecules are fully perpendicular to the plates, the sandwich loses its anisotropy, at least for beams that are incident perpendicular to the plane of the sandwich; this is so because the molecules display no preferred orientation in the directions perpendicular to their axes, and the beam is incident parallel to the optic axis of the liquid-crystal layer.

To make a display, we simply enclose the sandwich between crossed polarizers. The axes of the polarizers are parallel to the preferred orientations of the glass plates. We may orient the polarizers for either transmission or extinction. In either case, as we change the electric field across the sandwich, the transmittance of the sandwich changes gradually from clear opaque, or the reverse. To make a visual display, we use a network of fine wires or of transparent electrodes to create the needed electric field as a function of position across the plane of the sandwich. A *liquid-crystal television* may be used in the object plane of an optical processor (Sect. 7.2).

Liquid crystals that align themselves parallel to their short axes, like coins in a bottle, are called *smectic liquid crystals*; these are not as important in the optics of displays. *Ferroelectric liquid crystals* are those whose molecules display a permanent dipole moment. If the dipole moments are aligned in an electric field, the liquid crystals become anisotropic and therefore birefringent. A sandwich of ferroelectric liquid crystals may be used in displays, but generally these displays do not exhibit a continuous range of transmittance like displays that use twisted nematic crystals. That is, the alignment of the molecules may be reversed by reversing the electric field, but it is difficult to align the molecules at intermediate angles other than parallel or perpendicular to the field. For this reason, the elements in the display may be made either completely transmitting or completely opaque, but it is difficult to obtain a range of transmittance. Ferroelectric liquid crystals therefore are less useful for visual displays, but they show promise for optical processing (Sect. 7.2) and for digital computers based on optics rather than electronics.

9.3 Nonlinear Optics

We began this chapter with a discussion of propagation of light through an optically dense medium and noted that the induced dipole moment per unit volume P is, for most purposes, proportional to the incident electric field E. In fact, when the electric field becomes sufficiently large (as with a laser), P is no longer quite proportional to E. We may regard the relation $P = XE$ as the first term in a series expansion of some more general function. In that case, the function may be expressed

$$P = X_1E + X_2E^2 + X_3E^3 + \cdots, \tag{9.13}$$

where $X_2, X_3, \ldots$ are known as *nonlinear polarizabilities*.

Suppose that the incident field has the form $E = A \sin \omega t$. We use the trigonometric identities $\cos 2x = 1 - 2\sin^2 x$ and $\sin 3x = 3 \sin x - 4 \sin^3 x$ to show that

$$P = X_1 A \sin \omega t + \tfrac{1}{2} X_2 A^2 - \tfrac{1}{2} X_2 A^2 \cos 2\omega t - \tfrac{1}{4} X_3 A^3 \sin 3\omega t + \cdots \tag{9.14}$$

(where we have used the fact that $X_2 \ll X_1$).

The interesting terms in the expansion are the terms involving 2ω and 3ω. These terms represent dipole moments per unit volume induced in the medium but oscillating at frequencies two and three times the incident frequency. These oscillating dipoles in turn drive electric fields at frequencies 2ω and 3ω. The effects are known as *second-harmonic generation* and *third-harmonic generation*. (Higher-order harmonics can also be generated.) Second-harmonic generation is the more important; it is a means whereby infrared radiation may be converted to visible and visible to ultraviolet.

The term $\tfrac{1}{2} X_2 A^2$ gives rise to a dc field across the medium; this field has been observed, but the effect (known as optical rectification) is of comparatively little practical importance.

9.3.1 Second-Harmonic Generation

In most crystalline materials, the nonlinear polarizability X_2 depends on the direction of propagation, polarization of the electric field, and orientation of the optic axis of the crystal. In short, X_2 is not a constant, but a *tensor*, and the correct expression for the second-harmonic polarization is

$$P_i^{(2)} = \sum_{j,k} X_{ijk}^{(2)} E_j E_k, \tag{9.15}$$

where i, j, k represent x, y, or z. For example, if i is the x direction, $P_x^{(2)}$ has six terms that involve the products E_x^2, E_y^2, E_z^2, $E_x E_y$, $E_x E_z$, and $E_y E_z$, where the electric fields E are the components of the incident eleric field and have frequency ω. As a rule, most of the coefficients X_{ijk} are 0, and the incident electric field is plane-polarized, so there will be only one or two components to deal with.

We have introduced the tensor equation for $P^{(2)}$ to show that only certain crystals will exhibit second-harmonic generation. To show this, consider a crystal that is isotropic. In that case, X_{ijk} is a constant X, which is independent of direction. We now reverse the directions of the coordinate axes; that is, let x become $-x$, y become $-y$ and z become $-z$. If we let the electric fields and the dipole moment per unit volume remain unchanged in direction, then their signs must change when we reverse the directions of the axes. That is,

$$-P_i^{(2)} = \sum_{j,k} X_{ijk}^2(-E_j)(-E_k) = +P_i^{(2)}. \tag{9.16}$$

Therefore, $P_i^{(2)} = 0$, and $X = 0$. Second-harmonic generation cannot take place in an isotropic medium. Further, by the same reasoning, it cannot take place in a crystal whose structure is symmetrical about a point. In the language of solid-state physics, only crystals that lack inversion symmetry exhibit second-harmonic generation. (Third-harmonic generation is possible in crystals that exhibit inversion symmetry.)

9.3.2 Phase Matching

Figure 9.10 shows a fundamental wave at frequency ω driving a second-harmonic wave at frequency 2ω. The direction of propagation is the z direction, and the length of the material is l. The amount of second-harmonic radiation $dE^{(2)}$ produced within a slab with width dz located at z is

$$dE^{(2)}(z) \propto P^{(2)}(z)dz. \tag{9.17}$$

$P^{(2)}$ is the second-harmonic dipole moment per unit volume induced at frequency 2ω; it is proportional to the square of the incident electric field E. Therefore,

$$dE^{(2)}(z) \propto e^{2i(k_1 z - \omega t)}dz; \tag{9.18}$$

that is, the spatial variation of the second-harmonic polarization is characterized by a wave number $2k_1$ as well as frequency 2ω.

The second-harmonic radiation, on the other hand, propagates with wave number k_2, where, in general, k_2 is not equal to $2k_1$ because of dispersion. (Recall that $k = 2\pi n/\lambda$.) Consequently, at the end of the crystal, where $z = l$, the second-harmonic radiation produced by the slab located at z is

$$dE^{(2)}(l) \propto dE^{(2)}(z)e^{ik_2(l-z)}dz, \tag{9.19}$$

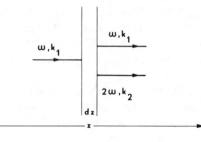

Fig. 9.10. Propagation of fundamental and second-harmonic waves in a crystal

where $(l - z)$ is the distance from the slab to the end of the crystal. Combining the last two equations, we find that

$$dE^{(2)}(l) \propto e^{i(2k_1 - k_2)z} e^{i(k_2 l - 2\omega t)} dz . \tag{9.20}$$

These equations have been derived by assuming that the second-harmonic power is small compared to the incident power. In that case, the incident power is nearly unchanged as the beam propagates through the crystal.

The last equation is easily integrated to yield

$$E^{(2)}(l) \propto \frac{\sin(2\pi \Delta n l/\lambda)}{(2\pi \Delta n/\lambda)} \tag{9.21}$$

where λ is the vacuum wavelength of the incident radiation and $\Delta n = (n_2 - n_1)$.

$E^{(2)}(l)$ is a maximum when the argument of the sine is $\pi/2$, or when

$$l = \lambda/4\Delta n . \tag{9.22}$$

This value of l is often called the coherence length for second-harmonic generation; for ordinary materials it may be no more than a few micrometers. Increasing l beyond the coherence length will not result in any increase of $E^{(2)}$.

In a crystal such as potassium dihydrogen phosphate (KDP), the incident wave may be introduced into the crystal as an ordinary ray. The second-harmonic wave will be an extraordinary ray. The wave ellipsoid (sphere) for an ordinary ray with frequency ω crosses the wave ellipsoid for an extraordinary ray with frequency 2ω, as shown in Fig. 9.11. Therefore, if the incident wave is made to propagate through the crystal at precisely the angle θ_m shown in the figure, Δn will be very nearly 0. The incident and second-harmonic waves are said to be *phase-matched*; the coherence length for second-harmonic generation may be made quite large by phase matching.

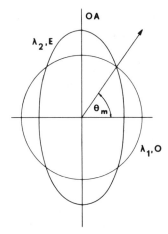

Fig. 9.11. Phase matching in second-harmonic generation

The efficiency of conversion to second-harmonic power is proportional to incident power density, because the second-harmonic polarization is proportional to E^2. Conversion efficiencies of 15–20% are typical at input-power densities of the order of $100 \text{ MW} \cdot \text{cm}^{-2}$.

9.3.3 Optical Mixing

We return to the equation for the nonlinear polarization P and concentrate on the term $X_2 E^2$. Until now, we have assumed that the term E^2 means the product of the electric field strength with itself. This need not be the case; we may, if we wish, consider the nonlinear polarization that results from the interaction of two fields with different frequencies ω_1 and ω_2. The term we are interested in is $P^{(2)}$, where

$$P^{(2)} = X_2(A_1 \sin \omega_1 t + A_2 \sin \omega_2 t)^2 \ . \tag{9.23}$$

A_1 and A_2 are the amplitudes of the two waves. Expansion of the square reveals terms with frequencies $2\omega_1$ and $2\omega_2$, as expected, as well as a term

$$2A_1 A_2 X_2 \sin \omega_1 t \sin \omega_2 t \ . \tag{9.24}$$

Using trigonometric identities for the sum and difference between two angles, we find that this term may be written as

$$A_1 A_2 X_2 [\cos(\omega_1 - \omega_2)t - \cos(\omega_1 + \omega_2)t] \ . \tag{9.25}$$

The nonlinear polarization and, therefore, the emitted light contain frequencies $\omega_1 + \omega_2$ and $\omega_1 - \omega_2$.

The sum and difference frequencies can be observed experimentally; the process is known as *parametric amplification*. As with second-harmonic generation, phase matching is extremely important, although the requirement is more stringent because of the number of frequencies involved. In second-harmonic generation, we found that it was necessary to find a direction in the crystal such that waves with frequencies ω_1 and ω_2 encountered the same index of refraction. In terms of wavenumber k, we had to find a direction such that k_1 was equal to k_2.

In the case of parametric amplification, three waves must be phase matched, rather than two. Therefore, if we wish to obtain a frequency ω_3, where

$$\omega_3 = \omega_1 \pm \omega_2 \ , \tag{9.26}$$

we must also satisfy the equation

$$k_1 \pm k_2 = k_3 \ , \tag{9.27}$$

where the three waves are assumed to be collinear.

To obtain high power from a parametric amplifier, it is sometimes placed inside an optical resonator. The laser that pumps the parametric amplifier is mode matched to the resonator, the mirrors of which have high reflectances for ω_1 and ω_2, but high transmittance for the laser frequency ω_3. If the pumping by the laser is sufficient, the parametric amplifier in the resonator goes into oscillation, much as a laser does, at both frequencies ω_1 and ω_2. The process is therefore known as *parametric oscillation*.

The parametric oscillator can be tuned either by rotating the crystal so that different frequencies are phase matched or by changing the temperature of the crystal so that its properties change slightly. Thus, the parametric oscillator provides a tunable source of coherent radiation.

Parametric amplification can also be used to convert from a low frequency to a higher one. In this application, two collinear beams with frequencies ω_1 and ω_2 are shone on a nonlinear crystal. If the crystal is phase matched for these frequencies and for the sum frequency ω_3, a third wave whose frequency is ω_3 will be generated. The first two waves may be filtered out if necessary. This process is known as *frequency upconversion*.

One of the main applications of frequency upconversion is to detect radiation from ir or far-ir lasers, at whose frequencies fast or sensitive detectors are often not available. The radiation from the laser is converted to near-ir visible radiation that is easily detected.

9.4 Electro-optics, Magneto-optics and Acousto-optics

9.4.1 Kerr Effect

When certain liquids and glasses are placed in electric fields, their molecules tend to align themselves parallel to the direction of the electric field. The greater the field strength, the more complete the alignment of the molecules. Because the molecules are not symmetrical, the alignment causes the liquid to become anisotropic and birefringent. Such electric-field-induced birefringence in isotropic liquids is called the *Kerr electro-optic effect* or *Kerr effect*.

The optic axis induced by the field is parallel to the direction of the field; for constant electric-field strength, the liquid behaves exactly as a birefringent crystal with indices n_o (the index of the material in the absence of the field) and n_e.

The amount of field-induced birefringence $(n_o - n_e)$ is proportional to the square of the electric-field strength and to the wavelength. Therefore, the index difference between the ordinary and extraordinary rays is

$$\Delta n = K\lambda E^2 , \tag{9.28}$$

where K is the *Kerr constant* and E is the applied electric field. Nitrobenzene has an unusually large Kerr constant, 2.4×10^{-10} cm·V^{-2}. Glasses have Kerr constants between about 3×10^{-14} and 2×10^{-23}. The Kerr constant of water is 4.4×10^{-12}.

A *Kerr cell* is a cell containing nitrobenzene or other liquid between two flat, parallel plates spaced by several millimeters or more. The potential difference between the plates is typically 10 or 20 kV. If the cell is located between crossed polarizers, it may be used as a fast shutter known as an *electro-optic shutter*.

In an electro-optic shutter, the direction of the electric field is 45° to the directions of the polarizer axes. When the electric-field strength is 0, the polarizers transmit no light. Ideally, when the electric field is applied to the cell, its magnitude is such that $(n_o - n_e)d = \lambda/2$. In this case, the Kerr cell acts as a half-wave plate and rotates the plane of polarization by 90°. The voltage necessary to make the cell a half-wave plate is known as the *half-wave voltage* of the cell.

If the half-wave voltage is applied to the cell as a fast pulse, the cell acts as a fast shutter; shutter speeds in the 10-ns range can be obtained routinely in this fashion. The Kerr effect itself is extremely fast; shutter speeds are limited by the difficulty of generating fast electronic pulses in the kilovolt range.

9.4.2 Pockels Effect

This is an electro-optic effect that is observed in certain crystals such as potassium dihydrogen phosphate (KDP). It differs from the Kerr effect in that the Pockels effect is linear in applied electric field, whereas the Kerr effect, as we have seen, is quadratic in applied electric field. More important, the half-wave voltage of typical *Pockels cells* is at least an order of magnitude less than that of Kerr cells.

Suppose that a Pockels cell is made by applying an electric field to a crystal parallel to the crystal's optic axis. The direction of propagation of a light beam is also parallel to the optic axis. (This may be accomplished either by using partially transparent electrodes on the faces of the crystal or by fixing electrodes with holes in their centers to the faces of the crystal.) When the electric field is absent, any ray that propagates parallel to the optic axis is an ordinary ray; the refractive index is independent of direction of polarization.

The electric field deforms the crystal and induces a second optic axis OA′ in the plane perpendicular to the field. The direction of OA′ depends on the structure of the crystal and need not concern us. Because of this additional anisotropy induced by the electric field, rays polarized with their electric-field vectors parallel to OA′ experience a refractive index that differs from that of rays polarized perpendicular to OA′. That is, when the electric field is applied, the crystal acts on the light ray as a birefringent crystal whose optic axis lies in the plane perpendicular to the direction of propagation.

The difference between the indices of refraction for rays polarized perpendicular and parallel to OA′ is

$$(n_o - n_e)' = pE , \tag{9.29}$$

where E is the applied field and p is a proportionality constant. p is approximately 3.6×10^{-11} mV^{-1} for KDP, 8×10^{-11} for deuterated KDP (KD*P), and 3.7×10^{-10} for lithium niobate.

Because the electric field is applied parallel to the optic axis, the preceding case is known as the *longitudinal Pockels effect*. The Pockels effect can also be observed when the electric field is applied perpendicular to the optic axis; this case is called the *transverse Pockels effect*. Commercially available Pockels cells may be either longitudinal or transverse cells.

The transverse Pockels effect has certain advantages over the longitudinal effect. First, the electrodes lie parallel to the beam and do not obscure or vignette it. Second, the index difference $(n_o - n_e)'$ depends on electric-field strength in the crystal, not on the voltage between the electrodes. If the length of the crystal in a longitudinal Pockels cell is increased and the voltage maintained constant, the electric field in the crystal will decrease proportionately. $(n_o - n_e)'$ will therefore decrease, and the phase difference or *retardation* between the two polarizations will remain independent of the length of the crystal.

On the other hand, in a transverse Pockels cell, the electrodes need be separated by the diameter of the beam and no more. The electric-field strength in the crystal depends on the separation between the electrodes and not on the length of the crystal. Consequently, lengthening the crystal and maintaining the spacing between the electrodes will result in increased retardation.

Low-voltage Pockels cells will always be the transverse-field type. High-speed Pockels cells, which require low capacitance and therefore small electrodes, will often be the longitudinal-field type.

Pockels cells, like Kerr cells, may be used as high-speed electro-optic shutters; because of their lower voltage requirements, they have nearly replaced Kerr cells for this application. A Pockels cell is usually the active element in an electro-optically Q-switched laser.

9.4.3 Electro-optic Light Modulation

A Pockels cell may be used to modulate a beam of light. For amplitude modulation, the apparatus is similar to that of an electro-optic shutter: The Pockels cell is placed between crossed polarizers, and a time-varying voltage is applied to the electrodes. The optic axis OA' is oriented at 45° to the axis of the polarizers.

The retardation induced by the electric field is proportional to the voltage between the electrodes. If we call the half-wave voltage V_π, then the retardation is $\pi V/V_\pi$. From our study of wave plates, we conclude that the transmittance of the modulator as a function of voltage is

$$\sin^2 \frac{\pi}{2} \frac{V}{V_\pi}. \tag{9.30}$$

The $\sin^2$ is not a linear function in the neighborhood of 0; hence the output of an electro-optic modulator will not in general be a linear function of the modulating voltage, except for small voltages. Fortunately, the $\sin^2$ is a nearly linear function in the neighborhood of 45°. If the retardation is approximately 90°, the modulator will be a linear device. That is, the Pockels cell is not modulated about an average value of 0 V, but about an average dc bias of $(1/2)V_\pi$. The peak-to-peak modulating voltage is substantially less than V_π.

Biasing the Pockels cell at its *quarter-wave voltage* of $(1/2)V_\pi$ causes it to behave as a quarter-wave plate. Including a quarter-wave plate in the electro-optic modulator will therefore have the same effect as biasing the modulator. Consequently, the electro-optic modulator may be operated about 0 V and will yield a linear output, provided that a quarter-wave plate is incorporated into the device.

Light may also be phase modulated with a Pockels cell. In this application, which is equivalent to frequency modulation, the plane of polarization of the incident light is either parallel or perpendicular to the second optic axis OA′. In this case, the light remains linearly polarized, but its phase changes with electric field by an amount

$$\phi = (2\pi/\lambda)pEt, \tag{9.31}$$

where t is the thickness of the crystal.

The electro-optic phase modulator has potential value in optical communications. It may also be used in a mode-locked laser cavity, where the index modulation that takes place is equivalent to vibrating one of the mirrors (see Chap. 8).

9.4.4 Acousto-optic Beam Deflection

An acousto-optic beam deflector consists of a block of quartz or other material through which an ultrasonic wave propagates. The wave has wavelength λ_s. Because it is a longitudinal or compression wave, it causes the index of refraction of the material to vary sinusoidally with wavelength λ_s.

To analyze the interaction of light with sound, we regard the sound wave in the medium as a series of planes from which the incident light beam is reflected. The treatment is identical with that of Bragg diffraction from crystal planes. The incident light beam strikes the planes at angle θ, and a fraction of the beam is reflected because of the spatially varying index of refraction. For a certain angle θ only, the beams reflected from adjacent planes interfere constructively. From Fig. 9.12, we see that constructive interference occurs when the optical-path difference between the two waves is equal to one optical wavelength λ. Therefore,

$$\sin \theta = \lambda/2\lambda_s . \tag{9.32}$$

For typical sound frequencies, θ may be a few degrees. Under proper conditions, nearly all of the incident light may be diffracted, or *deflected*, into angle θ. Equation (9.32) is called the *Bragg condition*.

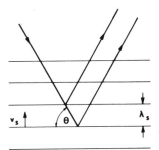

Fig. 9.12. Bragg reflection in an acousto-optic light modulator

The deflected light is shifted slightly in frequency from the incident light. This is a result of the propagation of the sound wave either toward or away from the source with a velocity component equal to $v_s \sin \theta$, where v_s is the velocity of sound in the medium.

When a source is viewed by reflection from a moving mirror, the source appears to have a velocity twice that of the mirror. Consequently, the Doppler shift observed corresponds to a velocity twice that of the mirror. (The Doppler shift Δv is given by $\Delta v/v = v/c$, where v is the component of velocity parallel to the line between source and observer.) The Doppler shift of the deflected beam is therefore

$$\Delta v/v = (2v_s/c) \sin \theta . \tag{9.33}$$

Using the relationship between θ and λ_s, we find that

$$\Delta v = v_s . \tag{9.34}$$

The light is Doppler shifted by a frequency equal to the sound frequency.

Acousto-optic deflectors are used for light modulation as well as beam deflection. An *acousto-optic light modulator* is an acousto-optic beam deflector followed by a spatial filter that consists of a lens and pinhole. The spatial filter rejects the deflected beam and passes the undeflected beam. To modulate the beam, we propagate the sound wave across the crystal, thereby removing power from the undeflected beam. Because of the presence of the spatial filter, this is equivalent to modulating the incident beam.

Like the electro-optic light modulator, the acousto-optic modulator is not linear; rather, the deflected power is proportional to the square of the sine of the acoustic amplitude. For linearity, the device would have to be used in the neighborhood of 0.5 transmittance.

Acousto-optic deflectors are useful in data processing and in computers. In particular, they may be made to scan a plane in a computer memory by varying the sound frequency.

9.4.5 Faraday Effect

When lead glass and other glasses are placed in strong magnetic fields, they become optically active. The amount of rotation induced by the magnetic field is equal to

$$VBl ,\qquad\qquad\qquad (9.35)$$

where B is the magnetic field strength, l is the length of the sample, and V is a constant known as the *Verdet constant*. If l is measured in millimeters and B in teslas, then V is equal to 0.004 for fused quartz, 0.11 for dense flint glass and 0.0087 for benzene.

A Faraday rotator in combination with a polarizer may be used as an optical isolator, a device that permits light to pass in one direction, but blocks it in the other direction.

Problems

9.1 Use conservation of energy to prove that (intensity) transmittance and reflectance at an interface are related by

$$R + nT(\cos i')/(\cos i) = 1 ,$$

not $R + T = 1$. (Examine the transmission of a beam with a finite width through an interface.)

9.2 Late-afternoon sunlight is reflected from a nearly horizontal surface such as the rear window of an automobile. How will polarizing sunglasses help to reduce glare more than simply attenuating sunglasses? Which plane of polarization should be transmitted by the polarizing lenses?

9.3 A right isosceles prism is made from a birefringent material. The optic axis of the prism lies in a plane parallel to one of the two equal faces of the prism. Light strikes the other face at normal incidence (that is, parallel to the optic axis). Describe the propagation of each polarization after it is reflected from the hypotenuse (cf. Glan-Thompson prism). What if the prism is not isosceles and the rays do not hit the hypotenuse at 45°?

9.4 *Optical Isolator.* Show that, when plane-polarized light is transmitted through a quarter-wave plate and is reflected back through the quarter-wave plate, the result is plane-polarized light, polarized at right angles to the incident light. This is the principle of an *optical isolator*, which is a device that consists of a polarizer, a quarter-wave plate and a plane mirror, and may be used to prevent unwanted reflections.

9.5 (a) Show how the electric-field vector of a plane-polarized beam can be rotated with relatively little loss by several polarizers in sequence. Determine a configuration that will rotate the electric-field vector through 45°, while transmitting at least 90% of the light. (It will be convenient to use approximations when solving the problem; eight-place calculations are not necessary.)

(b) Show that a pile of polarizers with infinitesimal angle between them can rotate a beam through a finite angle with a transmittance of 1.

9.6 A wave plate is sandwiched between crossed polarizers. The optic axis of the wave plate is inclined at $45°$ to the axes of the polarizers. The wave plate causes a phase shift of ϕ between the components parallel and perpendicular to its own axis. Show that the transmittance of the system is $\cos^2(\phi/2)$ when the incident beam is polarized parallel to the first polarizer. What is the transmittance when the axes of the polarizers are parallel?

9.7 (a) A Kerr cell is filled with nitrobenzene. The electrodes are spaced 5 mm apart, and their length is 2 cm. A potential difference of 10 kV is applied to the electrodes. If the cell is used as an electro-optic shutter, what is the transmittance of the shutter?

(b) What voltage is required to achieve the same transmittance with a 1-cm Pockels cell that uses the longitudinal Pockels effect in deuterated KDP?

10. Optical Waveguides

Optical fibers have long been used to carry beams over short, flexible paths, as from a screen to a detector. If the fibers in a bundle are oriented carefully with respect to one another, the bundle may be used to transmit an image. For example, a short *fiber-optic face plate* may be used in place of a lens to carry an image a very short distance. The spacing between the individual fibers is the resolution limit of such an imaging device.

Perhaps more important, newly developed, low-loss *optical waveguides* are valuable in optical communications and other closely related fields. Optical waveguides are compact, flexible, relatively lossless, and insensitive to electromagnetic interference.

10.1 Rays in Optical Fibers

An optical-fiber waveguide is made of a glass *core* surrounded by a lower-index *cladding*. The core of the fiber shown in Fig. 10.1 has uniform index of refraction n_1. The index of refraction changes abruptly to n_2 at the *core-cladding boundary*. If $n_2 < n_1$, the fiber is able to trap a light beam in the core by total internal reflection (Sect. 9.1.5). A ray will be guided only if angle i exceeds the critical angle i_c. If θ is the greatest angle of incidence for which a ray will be totally internally reflected, we see from the figure that

$$n_0 \sin \theta = n_1 \sin \theta' = n_1 \cos i_c \qquad (10.1)$$

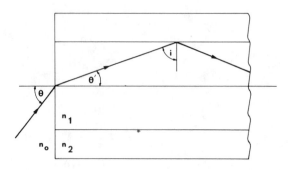

Fig. 10.1. Optical waveguide

where n_0 is the index of refraction of the medium in which the ray originates, and n_1 is the index of the core.

The critical angle is given by

$$\sin i_c = n_2/n_1, \tag{10.2}$$

where n_2 is the index of refraction of the cladding. Using the last two equations, along with the identity $\cos^2 i_c = 1 - \sin^2 i_c$, we find that

$$n_0 \sin \theta_m = (n_1^2 - n_2^2)^{1/2} \sim [2n_1(n_1 - n_2)]^{1/2}. \tag{10.3}$$

$n_0 \sin \theta_m$ is known as the *numerical aperture* (NA) of the waveguide and is defined in precisely the same way as the numerical aperture of a microscope objective. For good collection efficiency, the waveguide should be illuminated by a source whose numerical aperture does not exceed that of the waveguide. The value of n_0 may be 1, but quite possibly the waveguide will be cemented to a light-emitting diode or other component with a cement whose index is close to that of the core. As an example, an optical waveguide for which $n_1 = 1.5$ and $n_1 - n_2 = 0.01$ has a numerical aperture of about 0.17. When $n_0 = 1$, this corresponds to an acceptance cone with a 10° half-angle.

Because of the limited range of angles that will propagate along the fiber, coupling a source into a waveguide is a problem. For example, in optical communications, light-emitting diodes are among the most useful sources. These sources radiate into a full hemisphere, so the waveguide can accept only a small fraction of the power radiated by the diode. If we assume that the diode is a Lambertian source (see Chap. 4), we may show that the power radiated into a cone whose half angle is θ_m is proportional to $\sin^2 \theta_m$. Therefore, if the core is at least as large as the diode, the coupling efficiency between the waveguide and the diode is equal to $\sin^2 \theta_m$. (It is less if the diode is larger than the core.) For the waveguide in the example above, the coupling efficiency is about 3%.

For optical communications, the electrical frequency response of the waveguide is important. Inspection of Fig. 10.1 will show that rays inclined at the critical angle i_c travel a factor of $1/\sin i_c$ further than axial rays. Because of the extra path length that these rays travel, they are delayed relative to the axial rays by an interval $\Delta\tau$ given by the relation

$$\frac{\Delta\tau}{\tau} = \frac{1}{\sin i_c} - 1, \tag{10.4}$$

where τ is the time for a short pulse to propagate down the waveguide. $\Delta\tau$ is the amount by which the pulse is broadened by propagating down the waveguide. The reciprocal of $\Delta\tau$ is the greatest modulation frequency that can be transmitted along the waveguide and therefore defines the widest electrical bandwidth that can be employed. In our example, this is ~ 30 MHz for a 1-km waveguide, enough for a few video channels. This bandwidth decreases as the length of the waveguide is increased.

In an actual waveguide, more power is transmitted by rays near the axis than by rays inclined at nearly the critical angle because the latter suffer greater loss. In addition, there is significant scattering in a very long waveguide; as a result, power is interchanged from oblique rays to axial rays and vice versa. Both these factors reduce the variation in transit time and tend to make the bandwidth somewhat larger than our prediction; in many waveguides, after a few hundred meters of propagation, $\Delta\tau$ becomes proportional to the square root of length, rather than length.

In practice, optical waveguides differ substantially from older optical fibers. Because of cracks and scattering losses, the transmittance of the very best glass optical fibers would have been of the order of 10^{-4} (-40 dB) for a fiber 1 km long. Optical waveguides made of high-purity silica glass may exhibit transmittances well under 50% (-3 dB) for a 1-km section. In addition, optical waveguides have very thick cladding layers, in part to reduce interference, or "crosstalk", between neighboring waveguides.

Graded-index waveguides display relatively little transit-time variation; these waveguides are made with a refractive index that decreases gradually with distance away from the axis of the fiber. Because of this index variation, rays that are inclined at an angle to the axis travel, on the average, through a medium that has a lower index of refraction than the index of refraction along the axis. If the index profile of the waveguide is controlled properly, the waveguide will have minimal transit-time variation and will be able to transmit very large electrical bandwidths.

When the diameter of the core of a waveguide is less than 10 or 20 μm, ray optics is not sufficiently accurate to describe the propagation in the waveguide, and wave optics must be employed. In particular, waveguides with cores only a few micrometers in diameter, called *single-mode waveguides,* can transmit only a very simple beam known as a single radiation mode. Single-mode waveguides are immune to the transit-time variations described above and can in principle transmit electrical bandwidths as great as 10^6 MHz over a 1-km path. Such a bandwidth corresponds to a very large information-carrying capacity. Unfortunately, single-mode waveguides are so small that it is difficult to transmit much power or to couple power efficiently into the waveguide. Nevertheless, room-temperature semiconductor lasers and efficient coupling techniques have made single-mode waveguides attractive for long-distance or undersea applications, and the great majority of the optical waveguides sold today are single-mode waveguides.

10.2 Modes in Optical Waveguides

A plane wave may be described as a family of rays perpendicular to the wavefront. Any such family of rays may be directed into the end face of an optical waveguide. If the angle of incidence on the entrance face is less than the acceptance angle of the fiber, the family of rays (or, what is equivalent, the plane wave) will be bound and will propagate down the waveguide. The rays will zigzag as in Figs. 10.1 or 10.2, reflecting alternately from each surface of the waveguide. In general, there will be no fixed phase relationship among reflected rays. At some locations along the axis, there might be constructive interference; at others, destructive. Thus, the intensity will generally vary with position along the axis.

Certain of the families of rays, however, propagate in a very special way: Rays interfere constructively with their parallel neighbors, and the intensity within the waveguide does not vary with position.

We may see the significance of these special families by referring to Fig. 10.2. The figure is an axial section of a planar waveguide that has infinite extent in the direction perpendicular to the page. The two rays shown strike the interface between the two media at angle i. The dashed lines AA' and BB' perpendicular to the rays represent wavefronts. The optical-path difference between the two rays is

$$\text{OPD} = 2n_1 d \cos i \; ; \qquad\qquad (10.5)$$

not coincidentally, the result has the same form as the expression we derived earlier for the Fabry-Perot interferometer.

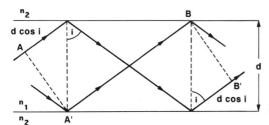

Fig. 10.2. Rays in an optical waveguide

A phase difference ϕ results from this difference of optical path; it is equal to $2n_1 kd \cos i$, where k is the vacuum wavenumber. In addition, electromagnetic theory shows that there is a phase shift Φ on reflection (Sect. 9.1). We may position the reference wavefronts so that one of the rays undergoes no reflections between the wavefronts, while the other ray is reflected twice. The total phase difference ϕ_t between the two rays is the sum of those due to path difference and to reflection; that is,

$$\phi_t = 2n_1 kd \cos i - 2\Phi \; . \qquad\qquad (10.6)$$

According to Fig. 9.5, $\Phi = 0$ when i is near the critical angle, and $\Phi = 90°$ at glancing incidence. When i is less than the critical angle, the ray is refracted, and the mode is not guided. When $\phi_t = 2m\pi$, the rays interfere constructively; this happens only for certain angles of incidence.

Pairs of rays inclined equally to the axis like those shown in Fig. 10.2 correspond to *waveguide modes* if the condition for constructive interference is satisfied. If only one mode is *excited* by proper choice of angle of incidence, the electric-field distribution inside the waveguide remains undistorted as the radiation propagates down the waveguide.

In terms of wave optics, each waveguide mode corresponds to constructive interference perpendicular to the waveguide axis. That is, the component $n_1 k \cos i$ perpendicular to the waveguide axis exhibits constructive interference between the two interfaces. The waveguide mode is equivalent to a standing wave perpendicular to the axis and a traveling wave along the axis. Thus, the electric field distribution perpendicular to the axis is the result of interference between two plane waves; we therefore expect a cosine-squared function for the intensity, and this is indeed the case.

Figure 10.3 sketches the electric-field distribution inside the waveguide for several values of m. The electric field does not fall precisely to 0 at the interfaces; this is related to the phase change on reflection and shows that power propagates in the cladding as well as in the core.

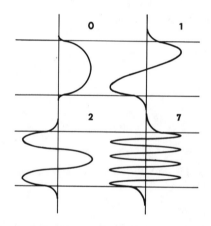

Fig. 10.3. Modes in a slab waveguide

The lower-order modes, those that have small values of m, have large angles of incidence, according to (10.6). That is, they correspond to rays that travel along the waveguide at nearly grazing incidence. The lowest-order mode, $m = 0$, has an amplitude distribution that is a single cycle of a cosine, as sketched in Fig. 10.3. Higher-order modes are characterized by oscillating field distributions. The highest-order bound mode is the one that strikes the interface just beyond the critical angle. Rays that are refracted at the interface correspond to *unbound modes*. Reflection at the boundary is not especially important, and there is a continuum of such modes, rather than the discrete set that characterizes the bound modes.

A ray may enter a waveguide at an angle that does not satisfy the relation $\phi_t = 2m\pi$. In such cases we cannot describe the electric field distribution as one waveguide mode; rather, it will be a complicated superposition of many modes. In mathematical terms, the modes form part of a *complete set*, that is, a set of elementary functions that may be used to describe an arbitrary function. Any field distribution that is bound to the waveguide can be described by any complete set. However, the waveguide modes are a convenient set to use because they correspond to a simple, physical model.

In a planar waveguide, the bound modes correspond to pairs of rays that have the same inclination to the axis; in a circular waveguide, these pairs of rays must be replaced by complicated cones of rays. The physical interpretation of the modes as standing waves is similar but more difficult to visualize with circular symmetry. In addition, there are standing waves in the azimuthal direction; consequently, two numbers are needed to describe a mode in a circular waveguide, just as two numbers are needed in the planar guide in which both transverse dimensions are finite.

10.2.1 Propagation Constant and Phase Velocity

The wavenumber k may be regarded as a vector whose direction is the direction of propagation. This vector is called the *wave vector*. In a medium with index n_1, the magnitude of the wave vector is kn_1. For propagation inside a waveguide, we are interested in the component β parallel to the axis of the waveguide,

$$\beta = n_1 k \sin i = n_1 k \cos \theta , \qquad (10.7)$$

where θ is the complement of i (or the angle between the ray and the waveguide axis, as shown in Fig. 10.4). β is known as the *propagation constant* and plays the same role inside the waveguide that the wavenumber k plays in free space. According to (10.6) and (10.7), θ and i will vary with wavelength.

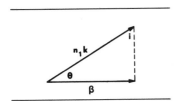

Fig. 10.4. Wavevector and propagation constant

The angle of incidence i varies between i_c and $\pi/2$. When $i = i_c$, $\beta = n_1 k \sin i_c$, or kn_2. When $i = \pi/2$, $\beta = kn_1$. Therefore,

$$kn_2 < \beta < kn_1 . \qquad (10.8)$$

That is, the value of the propagation constant inside the waveguide always lies between the values of the wavenumbers in the two bulk materials. If we use the relationship that $c = \omega/k$ (where c and k are the values in vacuum), we may rewrite this relationship in terms of phase velocities,

$$c/n_1 < v < c/n_2 \ . \tag{10.9}$$

The phase velocity of bound modes lies between the phase velocities of waves in the two bulk materials. This leads directly to the earlier result (10.4).

Finally, the analysis of Sect. 5.2 shows that the phase velocity in the direction parallel to the axis of the waveguide is $c/n_1 \cos \theta$. This suggests that we define an *effective index of refraction,*

$$n_e = n_1 \cos \theta. \tag{10.10}$$

Because θ is a function of mode number m, n_e is also a function of m. Consistently with (10.9), the value of n_e for bound modes is always between n_2 and n_1. When n_e is less than n_2, the beam is not trapped by the waveguide.

10.2.2 Prism Coupler

Planar waveguides or waveguides with rectangular cross sections may be deposited on a dielectric substrate such as glass, semiconductor, or electro-optic crystal. The high-index waveguide is located between the substrate and the air. Many times, we may want to introduce the light into the waveguide through the interface with the air, rather than through the edge of the waveguide; this approach presents fewer alignment difficulties and does not require special preparation of the edge of the guide.

One way of *coupling* the light into the guide through the top is the *prism coupler* shown in Fig. 10.5. The prism has a higher index of refraction than the waveguide and is not quite in contact with the surface of the guide. If it were in contact, the light would simply refract into the waveguide and reflect back out after a single reflection from the lower interface. By *weakly coupling* the prism to the waveguide, however, we can divert a significant fraction of the incident

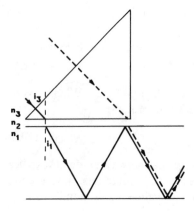

Fig. 10.5. Prism coupler

beam into the waveguide. Weak coupling is effected by keeping a space of a wavelength or so between the prism and the waveguide; power is transmitted into the waveguide as a result of the penetration of the evanescent wave into the second medium.

The beam in the prism must have finite width. Power will be coupled efficiently into the waveguide only if the direction of the ray that is refracted into the waveguide corresponds to a mode of the waveguide. According to Snell's law, $n_3 \sin i_3 = n_1 \sin i_1$; multiplying both sides of this equation by the vacuum wavenumber k, we find that the horizontal components of the wave vectors in the two media (1 and 3) must be equal. The horizontal component of the wave vector inside the waveguide is the propagation constant β; therefore,

$$\beta = k_3 n_3 \sin i_3 \ . \tag{10.11}$$

If i_3 is adjusted so that the value of β is that of a waveguide mode, light coupled into the waveguide at the left side of the figure will propagate down the waveguide and be precisely in phase with light coupled into the prism further along. In other words, there will be constructive interference. The power inside the waveguide can therefore be made to increase with *coupling length*. If the condition on i_3 is not satisfied, there will be no such constructive interference, and the light will be coupled back and forth between the prism and the waveguide, with comparatively little power ultimately ending up inside the waveguide. In short, the wave will be largely reflected out of the waveguide, precisely as the transmittance of a Fabry-Perot interferometer is nearly 0 when the condition, $m\lambda = 2d$, is not satisfied.

When the electric-field strength inside the waveguide becomes sufficiently large, light will pass from the waveguide back into the prism. To prevent this from happening and to ensure optimum coupling, the beam in the prism is cut off sharply at the edge of the prism. The actual coupling coefficient is adjusted experimentally by adjusting the pressure between the prism and the waveguide until the maximum coupling has been achieved.

This discussion may appear to suggest that the coupling efficiency can never exceed 50%. In fact, it may reach 80% or so. This is so because the electric field inside the waveguide is not constant but is less at the surface of the guide than at the center. Coupling from the prism to the waveguide can continue until the evanescent wave associated with the wave in the guide becomes large. This will happen only after a sizeable fraction of the incident power has been trapped inside the waveguide.

10.2.3 Grating Coupler

Light may also be coupled into a planar waveguide with a *grating coupler*. Such a device is shown in Fig. 10.6; it functions in a similar way to the prism coupler.

A grating, usually a phase grating, is etched onto the surface of the waveguide. The index of refraction of the waveguide is n_1; the material above the

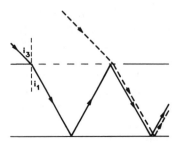

Fig. 10.6. Grating coupler

waveguide is usually air. (The index of refraction n_2 of the medium below the grating is irrelevant here, except that n_2 must be less than n_1 for bound modes to exist.) The grating equation (6.2) becomes for this case

$$m\lambda = d' \sin i_3 - n_1 d' \sin i_1, \tag{10.12}$$

where d' is the groove spacing and the angles are positive as shown. Here m is the order of interference, and we take it to be 1. The factor n_1 appears on the right side of the equation because the wavelength inside the medium is λ/n_1 or, what is equivalent, because the optical-path length inside the medium is n_1 times the geometrical-path length. If we multiply both sides of this equation by $2\pi/\lambda$ and rearrange slightly, we find that

$$k_3 \sin i_3 = k_1 n_1 \sin i_1 + 2\pi/d'. \tag{10.13}$$

To this point, i_1 is arbitrary. If we wish to excite a mode of the waveguide, then $k_1 n_1 \sin i_1$ must be equal to the wavenumber β of an allowed mode. That is, we may excite a mode provided that we choose i_1, i_3, and d' so that

$$k_3 \sin i_3 = \beta + 2\pi/d', \tag{10.14}$$

where β is the propagation constant of a mode of the waveguide. Only when this equation is satisfied will the dashed ray in Fig. 10.6 enter the waveguide precisely in phase with the solid ray. As with the prism coupler, light can also be coupled out of the waveguide, so the length of the grating must be chosen carefully; coupling efficiency is not as easily adjusted as with prism couplers.

The grating may also be an ultrasonic wave excited on the surface of the substrate. It may be used for coupling light out of the waveguide as well as into it. It may additionally be used for modulation of the light that remains inside the waveguide. For example, the power of the ultrasonic wave may be amplitude modulated or pulsed. Whenever a pulse is applied to the ultrasonic wave, the power of the light inside the waveguide decreases accordingly. In this way, as we shall see, a planar waveguide may be used to modulate a beam for communication or other purposes.

In general, a grating or prism coupler may be used to excite only one mode of the waveguide at a time. If the alignment is not nearly perfect, constructive interference will not take place, and power will be coupled only very inefficiently into the waveguide. Still, we should not conclude that a waveguide will propagate light only if the light excites one mode, nor that light entering the end of a waveguide at the wrong angle will be rejected by the waveguide. In general, this is not so but is the result of using grating or prism couplers. In fact, if a beam is shone into the end of a multimode waveguide at an arbitrary angle less than the acceptance angle, the light will not be rejected by the waveguide but will propagate as a superposition of several modes. This is so because diffraction by the edges of the waveguide will spread an incident plane wave into a range of angles and allow excitation of more than one mode.

10.2.4 Modes in Circular Waveguides

A mode in a slab waveguide corresponds to a pair of rays zigzagging in concert down the guide at an angle that satisfies (10.6) with $\phi_t = 2m\pi$. The modes of a rectangular waveguide are very similar to slab-waveguide modes. We simply think of components of the propagation constant, one in the plane parallel to the horizontal faces of the waveguide and one in the plane parallel to the vertical faces. Each component must satisfy, separately, an equation exactly similar to (10.6). Therefore, the modes of a rectangular waveguide are very nearly products of slab-waveguide modes that correspond to the height and width of the rectangular waveguide. Each mode must be specified by two numbers, m and m', instead of just m.

The modes of a circular waveguide or fiber are substantially more complicated. Each mode corresponds to a complicated cone of rays. When the cone is coaxial with the waveguide, the modes correspond closely to the modes of a slab waveguide; an equation like (10.6) pertains, with d replaced by the diameter of the waveguide.

Unfortunately, most of the cones are not coaxial with the waveguide. Similarly, most of the rays never pass through the axis of the waveguide. Such rays are known as *skew rays*, to distinguish them from the *meridional rays*, those rays that do pass through the fiber axis. Skew rays spiral around the axis of the waveguide, never intersecting it. Skew rays may be bound to the fiber, but some skew rays are loosely bound or *leaky rays*. Leaky rays turn out to be very important for certain measurements, and we deal with them in more detail in Sect. 10.3.4.

10.2.5 Number of Modes in a Waveguide

The mode number m of a particular mode in a slab waveguide may be found from (10.6), with $\phi_t = 2m\pi$. The phase shift on reflection Φ is always less than π and may be ignored for large values of m. Mode number increases from 0, when the wave vector is nearly parallel to the axis, to a maximum value M when i is nearly equal to the critical angle. Because every mode corresponds to a positive integer or 0, $M + 1$ is the number of modes allowed in the slab (not including a factor of 2 for polarization).

When $m = M$, i is very nearly equal to i_c; therefore, $\Phi = 0$ and (10.6) becomes

$$2n_1 k_1 d \cos i_c \cong 2\pi M. \tag{10.15}$$

If we use the fact that $\sin i_c = n_2/n_1$, we may rewrite this equation as

$$M \cong (2d/\lambda)\sqrt{n_1^2 - n_2^2}\,, \tag{10.16}$$

or, from (10.2) for numerical aperture NA,

$$M \cong (2d/\lambda)NA\,. \tag{10.17}$$

We require two sets of mode numbers m_1 and m_2 to count the modes in a waveguide with a square cross section. For each value of m_1, there are $M + 1$ values of m_2. Therefore, the total number of modes in such a waveguide is approximately M^2, when $M \gg 1$.

For circular waveguides, we define a parameter,

$$V = (2\pi a/\lambda)\sqrt{n_1^2 - n_2^2} = (2\pi a/\lambda)NA\,, \tag{10.18}$$

known as *normalized frequency* or just V *number*. a is the radius of the waveguide. When V is large, the total number of modes in a circular waveguide is of the order of V^2. For example, the number of modes in a fiber that has a uniform core is about $V^2/2$.

10.2.6 Single-Mode Waveguide

If M is 0, then only the lowest-order mode can propagate down a slab waveguide or one that has a square cross section. The higher-order modes are said to be *cut off*. When the mode for which $m = 2$ is precisely at cutoff, $M = 1$, and we find from (10.17) that the condition

$$(2d/\lambda)NA \lesssim 1\,, \tag{10.19}$$

guarantees single-mode operation. A fiber or waveguide that can operate only in the lowest-order mode is known as a *single-mode waveguide* or a *monomode waveguide* or fiber. A single-mode waveguide can be constructed by making either d or NA sufficiently small.

A circular waveguide may likewise be forced to operate in the lowest-order mode by adjusting the core radius a or the numerical aperture; the analogous relationship is

$$(2\pi a/\lambda)NA \lesssim 2.4\,, \tag{10.20}$$

where 2.4 is the first zero of the Bessel function that describes the second-lowest-order mode of a circular waveguide. Here also the equation is only approximate because the phase shift on reflection has been ignored.

If we use the preceding equation to solve for the fiber radius a, we find that

$$a \lesssim 1.2\lambda/\pi NA . \tag{10.21}$$

Equation (8.46) gives the divergence of a Gaussian beam whose spot size is w_0. In paraxial approximation, the divergence θ is equal to the numerical aperture, so the spot size of a Gaussian beam, written in terms of the numerical aperture, is $\lambda/\pi NA$. Thus, we find that a fiber will be a single-mode fiber when its radius is about the same as the spot size of the Gaussian beam that has the same numerical aperture as the fiber. In fact, the beam that emerges from the end of the fiber has a very nearly Gaussian profile, and its beam divergence is approximately that of a Gaussian beam with the same spot size. We will find these facts important when we discuss certain measurements and connector problems.

A single-mode fiber propagates what amounts to a single cone of meridional rays that all have the same inclination to the fiber axis. All the rays travel the same distance. Therefore, there is no *multimode distortion* such as we encountered in connection with (10.4). Single-mode fibers can transmit very high electrical bandwidths that are in principle limited only by the dispersion of the silica from which they are made.

10.3 Graded-Index Fibers

A *graded-index fiber* is an optical waveguide whose index of refraction decreases with distance from the fiber axis. The rays in such a fiber travel in curved paths, as shown in Fig. 10.7. A ray that intersects the axis with a relatively large slope travels a greater distance than the axial ray. However, it experiences a lower index of refraction and hence a higher velocity over a part of its path. Thus, the two rays may spend approximately the same time traversing the fiber. Fibers are manufactured in this way to compensate for multimode distortion. To distinguish between graded-index fibers and those that have a uniform core, we call the latter *step-index fibers*.

10.3.1 Parabolic Profile

Let us derive an expression for the *index profile* of a graded-index fiber in paraxial approximation. We begin with a meridional ray that crosses the axis of the fiber with angle of incidence i. Because the index of refraction of the fiber decreases monotonically with distance from the axis, the ray will follow the curved path shown in Fig. 10.7. At the *turning point*, the slope of the ray path changes sign, and the ray bends toward the axis as a result of total internal reflection. At a distance L from the starting point, the ray again crosses the axis.

To trace the ray, we imagine that the fiber is made up of layers, like the proverbial onion. The index of refraction of each successive layer is less than that

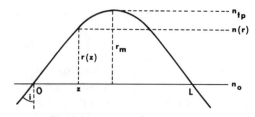

Fig. 10.7. Ray path in graded-index fiber

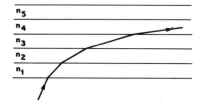

Fig. 10.8. Ray path in layered material

of the last. If we trace the ray through several layers, as in Fig. 10.8, we find that

$$n_1 \sin i_1 = n_2 \sin i_2 ,$$
$$n_2 \sin i_2 = n_3 \sin i_3, \text{ etc. },$$
$$\text{(10.22)}$$

or

$$n \sin i = \text{constant} , \qquad (10.23)$$

where i is the angle of incidence (measured with respect to the normals to the layers). This is a generalization of Snell's law. It applies whether or not the medium is layered or has continuous index of refraction, as we see by letting the thickness of the layers approach 0.

We return to Fig. 10.7 and apply the generalized Snell's law to the ray that intersects the axis with angle of incidence i. The ray follows a path that we describe with a function $r(z)$, where z is the distance along the fiber and r is the distance from the axis. The ray turns back toward the axis when it reaches the radius r_m. Equation (10.23) shows immediately that

$$n(r_m) = n_0 \sin i \qquad (10.24)$$

for any value of i.

Now, let us demand that all guided rays follow curved paths that intersect periodically. The period L should be independent of i, so all the rays are focused periodically to the same points; this is the condition that guarantees that they travel the same optical path length (Fermat's principle, Sect. 5.5).

Since the trajectories are to be periodic, we try a sinusoidal function,

$$r(z) = r_m \sin(\pi z/L) , \qquad (10.25)$$

for the ray height $r(z)$. The slope of this curve is

$$dr(z)/dz = (r_m \pi/L)\cos(\pi z/L) , \qquad (10.26)$$

which is equal to $r_m \pi/L$ when $z = 0$. The slope of the curve at 0 is also (in paraxial approximation) equal to $\cos i$. Therefore, using $\sin i = (1 - \cos^2 i)^{1/2}$ in (10.22), we find that

$$n(r_m) = n_0[1 - (\pi r_m/L)^2]^{1/2} . \qquad (10.27)$$

r_m may take any value, depending how we choose i. If the index profile of the fiber does not have axial variations, $n(r_m)$ is the function we seek, and we may drop the subscript m. If we use the approximation that, for small x, $(1 - x)^a \cong 1 - ax$, we may write

$$n(r) \cong n_0 [1 - (1/2)(\pi r/L)^2] . \tag{10.28}$$

Finally, we require that the fiber have a uniform cladding; that is, that the index of refraction be a constant $n(a) = n_2$ for all values of $r > a$. If we define Δn by the equation

$$\Delta n = n_0 - n_2 , \tag{10.29}$$

we find immediately that

$$n(r) = n_0 - \Delta n(r/a)^2 . \tag{10.30}$$

The index profile described by this equation is known as a *parabolic profile*. It has been derived for meridional rays in paraxial approximation only. The result is in fact a special case of the *power-law profile*,

$$n(r) = n_0 [1 - 2\Delta (r/a)^g]^{1/2} \tag{10.31}$$

where

$$\Delta = (n_0^2 - n_2^2)/2n_0^2 . \tag{10.32}$$

The *delta parameter* that appears in (10.31) is not the same as Δn; when Δn is small, $\Delta n \cong n_0 \Delta$. The exponent g is known as the *profile parameter*. When all factors are taken into account, the best value for g may differ by 5–10% from the value 2 that we derived using the paraxial approximation.

10.3.2 Local Numerical Aperture

The numerical aperture of a graded-index fiber may be defined as $(n_0^2 - n_2^2)^{1/2}$. Rays that enter the fiber on the axis and within this numerical aperture will be guided; those outside this numerical aperture will not. Off the axis, where the index of refraction is $n(r) < n_0$, the situation is somewhat different. We may use the generalization of Snell's law (10.23) to show that an off-axis, meridional ray will be guided only if

$$n(r) \sin \theta < n_2 . \tag{10.33}$$

This leads us to define a *local numerical aperture*,

$$N = [n^2(r) - n_2^2]^{1/2} , \tag{10.34}$$

for meridional rays incident on the fiber end face at radius r. Because $n(r)$ is always less than n_0, we conclude that the graded-index fiber traps light less efficiently than a step fiber that has the same numerical aperture. The light-gathering ability falls to 0 at the core-cladding interface. This is reflected by the fact that the number of modes of a fiber with a parabolic profile is half the number of modes of a step fiber that has the same numerical aperture.

10.3.3 Leaky Rays

When light is incident on a curved surface the laws of geometrical optics break down; even if the angle of incidence exceeds the critical angle, light may be transmitted beyond the boundary, although not as efficiently as by ordinary refraction. The phenomenon is reminiscent of frustrated total internal reflection. Rays that strike the curved interface but are partly transmitted are called *leaky rays* and, in a fiber, are distinguished from ordinary *bound rays* and *refracted rays*. When mode theory is used, these rays are associated with *leaky modes*.

Figure 10.9 shows a ray that is incident on a curved surface whose radius of curvature is R. The angle of incidence i is greater than the critical angle i_c. Still, there is power in the second medium because of the evanescent wave. The amplitude of this wave decreases approximately exponentially with radial distance from the interface. Nevertheless, the evanescent wave carries power in the direction tangent to the interface, and there is a phase associated with the evanescent wave.

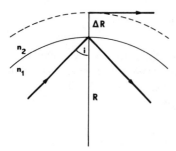

Fig. 10.9. Leaky ray

The oscillatory wave in the first medium, so to speak, drags the evanescent wave along with it, and the evanescent wave propagates power along the boundary. The evanescent wave must therefore keep pace with the oscillatory wave. Farther and farther from the boundary, the evanescent wave must propagate faster and faster to keep pace. Eventually, its speed will reach the velocity of light c/n_2 in the second medium. Relativity theory precludes velocities greater than light, so we conclude that the evanescent wave most likely turns into a propagating wave at the radius at which its velocity reaches c/n_2. This interpretation is borne out by electromagnetic theory.

We may get some idea of the physics behind leaky rays by writing Snell's law in the form

$$\sin i_2 = \sin i_1 / \sin i_c , \tag{10.35}$$

where $\sin i_c = n_2/n_1$. The cosine of i_2 is

$$\cos i_2 = [1 - (\sin i/\sin i_c)^2]^{1/2} . \tag{10.36}$$

When $i_1 < i_c$, the cosine is real. As i approaches i_c, $\cos i$ approaches 1. Finally, when i exceeds i_c, the cosine becomes imaginary; we conclude that a propagating wave no longer exists in the second medium. At a distance ΔR from the interface, however, the evanescent wave again becomes a propagating wave. When this happens, the imaginary angle i_2 will become real again, and $\cos i_2 = 1$. Therefore, when the evanescent wave becomes a propagating wave, $i_2 = \pi/2$; the radiation is emitted parallel to the surface, but at a distance ΔR beyond the surface. Because there is a gap between the incident wave and the transmitted wave, leaky rays are sometimes called *tunneling rays*. The greater that gap, the less power is lost to tunneling, since the transmittance must be somehow closely related to the amplitude of the evanescent wave at the distance ΔR from the interface.

The electromagnetic theory of leaky rays is quite complicated. However, we may make an educated guess as to the relation among R, ΔR, and i. First, we know that the evanescent wave extends to ∞ when $R = \infty$ and the surface is flat. Therefore, we guess that ΔR is proportional to R.

Next, we assume that the transition from refracted rays (with $i < i_c$) to leaky rays (with $i > i_c$) is continuous. When the rays are refracted, the radiation in the second medium originates at the interface. When i is precisely i_c, the distinction between a refracted ray and a leaky one is not sharp. Therefore, the transmitted propagating wave ought to originate from the interface when $i = i_c$. In short, ΔR ought to be 0 when $i = i_c$ and ought to increase with increasing i.

Consequently, we suspect three things: that $\Delta R \propto R$; $\Delta R = 0$ when $i = i_c$; and ΔR increases monotonically with increasing R. A very simple relationship that meets all three criteria is

$$\Delta R/R = (\sin i/\sin i_c) - 1 , \tag{10.37}$$

which proves to be rigorously correct.

10.3.4 Restricted Launch

In a fiber, Fig. 10.9 represents the projection of a ray onto a plane perpendicular to the axis of the fiber. Some rays enter the fiber with angle greater than the acceptance angle for meridional rays. Sometimes, however, the projections of these rays hit the core-cladding interface at angles greater than the critical angle. These rays are leaky rays and have higher loss than ordinary, bound rays. This can create a problem for certain kinds of measurements; for example, the leaky rays may be a factor in a short fiber but not in a long one. Attenuation per unit length, for instance, varies with the length of the fiber being measured, all other things being equal.

Many researchers try to overcome this problem by exciting bound modes only. For the case of a parabolic profile, the acceptance angle for meridional rays is given by an equation analogous to (10.3),

$$\sin^2 \theta_m(r) = n^2(r) - n_2^2. \tag{10.38}$$

Using (10.30), we may rewrite (10.38) as

$$\sin^2 \theta_m(r) = \sin^2 \theta_m - 2n_0 \, \Delta n(r/a)^2, \tag{10.39}$$

where θ_m is the acceptance angle for rays incident on the axis of the fiber. Rays incident at any angle that is less than the local acceptance angle are certain to be bound rays, whether or not they are meridional rays. Other rays are either leaky or refracted.

Equation (10.39) shows that $\sin^2 \theta$ is a linear function of $(r/a)^2$. Figure 10.10 is a plot that shows the relations among bound, leaky, and refracted rays for different values of θ and r. The axes are normalized so that the plot will have universal applicability. The diagonal line represents (10.39); all bound rays are represented on this graph by points that lie below this line. Points that lie above the line but inside the square represent either leaky or refracted rays. Points outside the square represent only refracted rays.

Figure 10.10 shows how to launch bound rays only: Restrict θ and r to values that lie below the diagonal line. We can accomplish this in many ways. For example, we could focus a point on the axis of the fiber and let $\theta \leqslant \theta_m$; in this case, we launch all the rays represented by the vertical axis between 0 and 1. Likewise, we could illuminate the fiber with a collimated beam whose radius is a; this case is represented by the horizontal axis between 0 and 1. However, if we wish to launch as many rays as possible (or excite as many modes as possible),

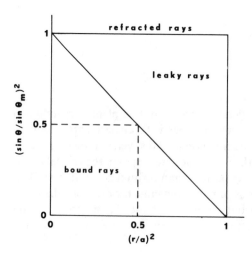

Fig. 10.10. Domains of bound, leaky and refracted rays

we will attempt to maximize the useable area below the line. This is accomplished by restricting both axes to the ranges between 0 and 0.5, as indicated by the dashed lines. The square so formed has the largest area of any rectangle that can be constructed wholly below the diagonal line. The conditions for launching only rays represented by points inside the square are

$$\sin \theta < 0.71 \sin \theta_m ,$$ (10.40a)

and

$$r < 0.71a ,$$ (10.40b)

where 0.71 is approximately the square root of 0.5.

These equations describe a *restricted launch* that is called the *70-70 excitation condition*. The condition is easily met by using a lens whose numerical aperture is equal to 0.71 times the fiber's numerical aperture to focus a diffuse source to an image whose radius is equal to 0.71a. In practice, to avoid exciting leaky rays at all, slightly smaller factors than 0.71 may be used.

A 70-70 excitation does not excite modes that are characterized by large values of r or $\sin \theta$. Such modes turn out in practice to be somewhat lossier than the lower-order modes that are excited by the 70-70 condition. A long fiber, say longer than 1 km, may therefore have lost much of the power in its higher-order modes, no matter how it was initially excited. Therefore, the 70-70 condition yields results that are meaningful for many measurements on long fibers.

10.3.5 Bending Loss and Mode Coupling

Figure 10.11 shows a bound ray near the critical angle of a waveguide that has an infinitesimal discontinuity. Beyond the discontinuity the ray is refracted because the angle of incidence in the second section of the waveguide is less than the critical angle. A low-order mode is represented by rays inclined well past the critical angle, that is, by rays that are nearly parallel to the axis of the fiber. The low-order mode will not become lossy unless the discontinuity is very sharp, but the high-order mode will invariably become lossy. A bend may be regarded as a series of infinitesimal discontinuities like the one shown. We conclude that high-order modes are likely to be lossier at bends than are low-order modes. Since all fibers have *microbends* that result from strain induced by spooling or

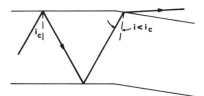

Fig. 10.11. Radiation loss at infinitesimal bend

cabling, the high-order modes will be selectively lost from long fibers, leaving only the relatively lower-order modes to propagate.

Similarly, skew rays that have primarily large values of r exhibit higher loss than other rays. This so because they travel a greater distance than more nearly axial rays and also because much of the power of the corresponding mode is propagated in the evanescent wave. There is a sharp discontinuity in the index of refraction at the core-cladding boundary of even a graded-index fiber. Any imperfections at this boundary cause substantial scattering, and modes that have more power near the boundary suffer greater loss than others.

Rays or modes that correspond to points near the values of 1 on either axis of Fig. 10.10 show higher loss than those nearer the center of the bound-mode area. Suppose that we excite a fiber uniformly, as by contacting the end with a light-emitting diode. The power distribution inside the fiber at the outset corresponds to uniform excitation of the modes in the triangular area below the slanted line in Fig. 10.10. After propagation down a substantial length of the fiber, both leaky rays and points near the corners of the triangle will have been attenuated or lost. The line will no longer delimit the mode distribution inside the fiber. Instead, the ends of the line will droop toward the square that represents the 70-70 condition.

On the other hand, if we use the 70-70 condition to excite the fiber, some rays will be scattered by imperfections but will still remain bound to the core; this is known as *mode coupling*. As a result of mode coupling, the square will often not adequately describe the mode distribution after a considerable propagation length. Rather, the dashed lines will rise in some fashion to fill more of the area under the line.

In a sufficiently long fiber, both launching conditions will result in the same distribution of power among the modes. Beyond the point where that happens, the modal-power distribution will not change (although the total power will decrease due to losses). Such a distribution of power among the modes is known as the *equilibrium mode distribution*; the propagation length required to reach the equilibrium mode distribution is called the *equilibrium length*. Even though many fibers have so little scattering that the equilibrium length is many kilometers, the concept of equilibrium mode distribution helps explain why the 70-70 condition gives representative results for many fibers under arbitrary launch conditions.

10.4 Connectors

Optical fibers cannot be joined haphazardly like wires or coaxial cables; rather, they must be butted against one another with their cores aligned precisely. Today, the majority of fibers are single-mode fibers, and their cores are typically about 5 μm in radius. An alignment error or offset of 1 μm can cause a very significant loss of guided power. Unless this *coupling loss* can be held within tight tolerances, either the designer of fiber communication systems will have to use higher-power sources or more-frequent repeaters than are optimal, or connectors will have to be adjusted by hand. Both options are costly, so considerable effort has been expended to reduce the loss of both fixed *splices* and demountable *connectors*.

Before they are joined, either in a permanent splice or a demountable connection, fibers are usually *cleaved*, or broken, so that the end of the fiber is flat and perpendicular to the axis. This is accomplished by first removing the coating from the fiber. The fiber is put under tension, and the outer surface is scratched with a diamond or other hard edge at the point where the cleaved end is needed. Sometimes, the fiber is scratched and then subjected to a shear force. If the tension is right, the fiber breaks smoothly. Two fiber ends may then be welded or fused in an electric arc, or pushed through a ceramic or glass capillary tube and held together with a refractive index-matching fluid or cement. Either method works only if the core is located very accurately at the center of the cladding, if the cleave is perpendicular to the axis, and, in systems that use a capillary tube or a vee groove, if the cladding diameter and ovality are held within precise limits. If, for example, the core of a single-mode fiber is offset from the center of the cladding by even 1 μm, the loss will be significant.

10.4.1 Multimode Fibers

We may derive some insight into the connector problem by considering a multimode slab waveguide. Assume that the waveguide has a step-index profile and its core is uniformly illuminated both across the face of the waveguide and within its numerical aperture.

Now consider two such waveguides cut off at right angles and butted against one another, as in Fig. 10.12. The left side of Fig. 10.12 illustrates an *axial misalignment* between two waveguides that are closely connected but not aligned quite correctly. Because we are assuming that the waveguides are uniformly illuminated, the transmittance when there is such an axial misalignment is

$$T_\delta = (1 - \delta/b) , \tag{10.41}$$

where b is the width of the slab (the diameter of the core in real fibers).

The right side of Fig. 10.12 shows an angular misalignment ϕ. Here, we assume that the angle is small enough that the waveguides are nearly in contact despite the angular misalignment. The acceptance half-angle θ of the waveguide is $\sin^{-1}(NA)$. Because the second waveguide is inclined at angle ϕ to the first, some of the rays that are emitted from the first waveguide will fall onto the second with angle of incidence greater than θ. Because we assume uniform illumination throughout a full angle of 2θ, we find by inspection that

$$T_\phi = 1 - (\phi/2\theta) \tag{10.42}$$

is the transmittance of the connection.

Fig. 10.12. Loss mechanisms in fiber-to-fiber couplers, *Left*, radial or transverse misalignment. *Right*, angular misalignment. [This and following two figures after M. Young, Geometrical theory of multimode optical fiber-to-fiber connectors, Opt. Commun. **7**, 253–255 (1973)]

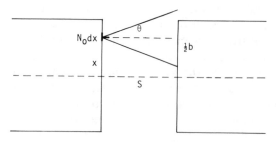

Fig. 10.13. Axial separation between two fibers

Axial separation between the waveguides is more complicated and cannot be calculated as easily. Figure 10.13 shows two waveguides separated by a distance s but otherwise correctly aligned. Dimension x in the plane of the face of the waveguide is measured from the axis. In agreement with our assumption of uniform filling of the numerical aperture of the waveguide, we assume that the radiance L of any differential element dx is

$$L = L_0, \text{ angles} < \theta \; ;$$

$$L = 0, \text{ angles} > \theta \; . \tag{10.43}$$

According to (4.2), the increment of power dP emitted by a differential element dx is

$$dP = L_0 dx \; , \tag{10.44}$$

where we ignore the second dimension in this one-dimensional approximation and use the fact that the cosine is 1 for small angles.

Whenever

$$x + s \tan \theta < b/2 \; , \tag{10.45}$$

all the power dP is transferred to the second fiber. When x exceeds

$(b/2) - s\tan\theta$, however, only the fraction

$$dP_{\text{eff}} = \frac{s\tan\theta + (b/2) - x}{2s\tan\theta}\, dx \qquad (10.46)$$

is transferred to the second fiber. This function is linear in x and has slope -1; it falls to 0.5 when x reaches its maximum value $b/2$.

The differential power transferred to the second fiber as a function of x is a constant as long as (10.45) holds; then it decreases linearly with increasing x until x reaches $b/2$. This is shown in Fig. 10.14, normalized so that the constant value is 1.

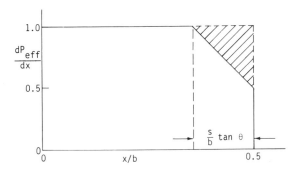

Fig. 10.14. Differential power coupled between two fibers as a function of normalized distance x/b between the two fibers

The transmittance of the connection is the integral of the function depicted in Fig. 10.14. It is equal to the area of the rectangle minus the upper-right corner, which is shown hatched. Therefore,

$$T_s = 1 - (s\tan\theta)/2b \qquad (10.47)$$

is the transmittance of a connection that displays axial separation s. In (10.47), $\tan\theta$ is very nearly equal to the numerical aperture of the waveguide.

Most multimode fibers today are graded-index fibers. In consequence, neither the core nor the numerical aperture accepts or emits light uniformly. Rather, the irradiance of the exit face is a maximum at the center of the core and decreases to 0 at the core-cladding interface. Similarly, the angular distribution of power is a maximum on the axis of the fiber and decreases to 0 at angle θ. Finally, fibers are circular in cross section, not slabs. Equations (10.41, 42) and (10.47) therefore slightly underestimate the losses if they are applied to optical fibers. For example, for graded-index fibers with a uniform mode distribution, the second (or loss) terms on the right side of (10.41) and (10.42) must be multiplied by 1.7, whereas the second term on the right of (10.47) must be multiplied by 2. For an equilibrium mode distribution, the loss is yet higher.

Still, we may compare the three loss mechanisms by calculating the loss for typical misalignments. A typical multimode fiber has a core diameter of 62.5 μm and a numerical aperture of 0.25. The cladding diameter may be uncertain by about 2 μm, so two fibers are apt to be joined with capillaries that are 2 μm oversized. The axial misalignment therefore may be of the order of 2 μm as well. Then, according to (10.41), the transmittance of a misaligned connection is about 0.97 (loss of 0.1 dB).

A typical cleaved fiber displays an end that is nearly perpendicular to the axis; a typical angular misalignment, caused, for example, by a slightly angled cleave, may be about 1°. If the numerical aperture of the fiber is 0.25, according to (10.42), this results in a transmittance of 0.97. Finally, it is easy to join two fibers with an axial separation of 1 μm; (10.47) shows that the transmittance of the connection is then 0.998 (0.01 dB). The most severe losses are therefore axial misalignment and angular misalignment.

We have so far assumed that the two fibers are identical; they may not be. If the diameter or the numerical aperture of the second fiber is less than that of the first, there will be loss. Consider first a diameter mismatch. Assume uniform illumination but, this time, circular cores. The transmittance of the joint is, by inspection, the ratio of the diameters of the two fibers, at least, if there is no misalignment. That is,

$$T_D = (D_2/D_1)^2 , \tag{10.48}$$

where the subscript 2 means the second (smaller) fiber and 1 means the first. With 62.5 μm cores, the transmittance is 0.97 if there is a 1 μm difference between the two core diameters. Very similar fibers must be joined if losses are to be held to a minimum.

Finally, the numerical apertures may be mismatched as well. Assume, as before, that power is emitted uniformly within the numerical aperture of the first fiber. Likewise, the second fiber has uniform acceptance within its numerical aperture, but its numerical aperture is less than that of the first fiber. The total power that the first fiber radiates into a cone is proportional to the square of the numerical aperture, according to (4.17); here the angles are so small that the difference between a Lambertian source and a uniform source is insignificant. The second fiber accepts only radiation that is incident within its numerical aperture. By an argument exactly similar to the argument that led to (10.48), we find that the transmittance is

$$T_{NA} = (NA_2/NA_1)^2 , \tag{10.49}$$

when the numerical aperture of the second fiber is less than that of the first. If the numerical apertures are about 0.25 and differ by 0.01, or 4%, the transmittance is about 0.92 (0.4 dB). This is the greatest loss we have encountered and shows again that very similar fibers must be joined to keep losses to a minimum.

10.4.2 Single-Mode Fibers

A single-mode fiber is characterized by its waveguide radiation mode, not by rays; coupling from one single-mode fiber to another requires mode matching

(Sect. 8.3.3). That is, a beam will couple efficiently into a fiber only if that beam displays an amplitude profile – electric-field amplitude as a function of radius – that is identical to that of the fiber. Otherwise, there will be loss, whether the beam comes from another fiber or is a laser beam focused by a lens into the core of the fiber. This is so because fiber modes are a resonance phenomenon (Sect. 10.2 and Fig. 10.2), where the electric field resonates transversely to the direction of propagation (between the walls, in the case of a slab waveguide). If two single-mode waveguides are joined imperfectly, the field of the first does not bring about a perfect resonance in the second, and coupling loss results.

The radiation mode of many single-mode fibers can be described by a Gaussian amplitude distribution (Sect. 8.3.3). To propagate a laser beam efficiently into such a fiber, that beam must be focused onto the entrance face of the fiber with approximately the same numerical aperture and, therefore, spot size as the fiber. If the numerical aperture of the beam is too large, the spot size is smaller than that of the fiber, and loss results. Similarly, if the numerical aperture of the beam is too small, the spot size is larger than that of the fiber, and loss also results. One method for adjusting the numerical aperture of the incident beam is to begin with a microscope objective whose numerical aperture is higher than that of the fiber and to expand the laser beam (say, with negative lenses) until the most efficient coupling is obtained. A semiconductor-laser beam is usually elliptical and has to be adjusted with cylindrical lenses until its profile is approximately circularly symmetric.

Modes in a fiber form a complete mathematical set. This means that the integral of the product of one mode with the complex conjugate of another is 0 if the modes differ and 1 if the modes are the same. The lowest-order mode in any fiber is therefore characterized by

$$E(r) = A e^{-r^2/w^2} \tag{10.50}$$

where

$$A = (2/\pi)^{1/2}/w \tag{10.51}$$

because of the condition that

$$2\pi \int_0^\infty E^2(r) r \, dr = 1 \ . \tag{10.52}$$

The loss introduced by a mismatch to the fiber's radiation mode may be estimated by calculating the *overlap integral* between the incident radiation mode and that of the fiber; this is the integral over all space of the product of the amplitude distributions,

$$t = 2\pi A_1 A_2 \int_0^\infty e^{-r^2/w_1^2} e^{-r^2/w_2^2} r \, dr \ . \tag{10.53}$$

Equation (10.53) may be integrated by making a change of variables $u = \sqrt{2}r/w$, where $1/w^2 = 1/w_1^2 + 1/w_2^2$. The integration results in the *amplitude* transmittance,

$$t = 2/[(w_2/w_1) + (w_1/w_2)], \qquad (10.54)$$

where 1 means the first fiber and 2 means the second fiber. We are interested, however, in power or intensity, so we square (10.54) to yield the transmittance,

$$T_w = 4/[(w_2/w_1) + (w_1/w_2)]^2. \qquad (10.55)$$

Unlike the result (10.48) for multimode fibers, (10.55) does not depend on which fiber is the smaller.

A typical single-mode fiber has a mode-field radius w of the order of 5 μm. If we assume a possible mismatch of 0.5 μm, then $(w_1/w_2) = 0.9$, and $T = 0.99$ (0.04 dB). The ray approximation (10.48) would yield 0.9, so it is completely inappropriate for single-mode fibers.

A similar calculation shows that if there is a radial alignment error δ between two identical fibers with mode-field radius w, the loss of the connection is

$$T_\delta = e^{-\delta^2/w^2} \qquad (10.56)$$

(Problem 10.12). If $w = 5$ μm and $\lambda = 1$ μm, the transmittance of the joint is 0.96, which represents a very significant loss when we consider that single-mode fibers can propagate only relatively little power compared with multimode fibers.

An angular misalignment ϕ gives rise to a transmittance

$$T_\phi = e^{-(\pi n \phi w/\lambda)^2}, \qquad (10.57)$$

where n is the index of refraction of the material between the ends of the fibers. If $\lambda = 1.3$ μm, $w = 5$ μm, $n = 1.46$, and $\phi = 1°$, then $T_\phi = 0.91$, a significant loss.

The formula for axial separation is complex but shows that separation between the ends of the fibers is not important. Radial and angular misalignment are therefore the most important. It is comparatively easy with a vee groove or a tight capillary to align a connection within a fraction of 1°, and fibers can easily be brought within 1 μm of each other. The diameter of the cladding of a fiber, however, is not known accurately within 1 μm, and the eccentricity between the core and the center of the cladding may also be in error by 1 μm. Achieving an axial offset of 1 μm is therefore difficult without manual adjustment of the transmittance of the connector.

10.4.3 Star Couplers

A device that connects the output of one fiber to many or, perhaps, of many fibers to many, is called a *star coupler*. The simplest star coupler is a hollow tube (or a dielectric rod) with a single fiber inserted at one end and along the axis of the tube. Light emitted from the fiber fills the tube and is reflected from the wall.

A bundle of fibers is inserted into the other end of the tube. If the tube is long enough, each fiber in the bundle will receive roughly the same power as its neighbors. A star coupler therefore allows a signal to be transferred from a single fiber to any number.

If the output bundle consists of N fibers, the overall core area in the bundle is N times the area of the core of a single fiber. Let us call the ratio of this overall core area to the cross section of the star coupler the *packing fraction*. Because of a low packing fraction, a star coupler is not very efficient at distributing power. A multimode fiber with a 125 μm cladding diameter, for example, might have a core diameter of 62.5 μm. Even if the fibers are closely packed, under 25% of the light in the star coupler will fall onto the cores of the output fibers. If those fibers have a step-index profile that is the same as that of the input fiber, the efficiency of the star coupler is equal to the packing fraction, apart from reflection losses at the walls of the tube and the faces of the fibers. The efficiency is yet lower for graded-index fibers, because not all the light that falls onto the core falls within the local numerical aperture.

A bundle of single-mode fibers displays a far lower packing fraction than a bundle of multimode fibers. A star coupler made of a simple tube is therefore extremely inefficient. A better star coupler for single-mode fibers is an integrated-optical device similar to a branch (Sect. 12.1.3), but where the input waveguide expands to many times its own width before being divided into a number of output waveguides. Unfortunately, the fibers must be connected very precisely to such a coupler, so the simplicity of the tubular star coupler is lost.

Problems

10.1 Uniform illumination with high numerical aperture is shone onto the edge of a slab waveguide that has constant index of refraction. The power coupled into the waveguide is proportional to the parameter $2aNA$, where $2a$ is the thickness of the slab. Find the relationship between this parameter and the power coupled into the waveguide when the index profile is parabolic. (In two dimensions the square of $2aNA$ is the appropriate parameter.)

10.2 A point source is imaged onto the end of an optical fiber. The axis of the beam is accurately parallel to that of the fiber. The beam is to be scanned across the core of the fiber. To what accuracy must the end of the fiber be cleaved perpendicular to the axis? Express your result as a function of the numerical aperture of the focusing optics and evaluate the function for typical microscope objectives.

10.3 For chopping or modulating a beam, we image the output of one fiber onto the input of another using two microscope objectives.

(a) Assume that the two fibers are identical, with 50-μm core and a numerical aperture of 0.2. Explain why coupling efficiency is best at unit magnification, that is, equal focal lengths.

(b) Suppose that the input fiber has a core diameter of 100 μm and a numerical aperture of 0.15; the second fiber remains as in (a). Show that coupling efficiency can be improved by using a value of $m < 1$ and, further, that the value of m is not critical. (Note, though, that the second fiber will always be overfilled at launch.)

10.4   A uniform point source is located in contact with a glass plate 1 mm thick. A step-index fiber is butted against the plate, its axis intersecting the point source. The numerical aperture of the fiber is 0.15, its core diameter is 50 μm, and the index of refraction of the plate is 1.5.

(a) What fraction of the light emitted by the source is guided by the fiber?

(b) What if the plate is 250 μm thick? 125 μm?

(c) In the 250-μm case, suppose that the source is a plane, Lambertian source 50 μm in diameter. What is its effective radius (consider meridional rays only)?

10.5   (a) Find the spatial period L of the ray trajectory in a fiber with a parabolic index profile. Evaluate L for the case that $a = 50$ μm and $\Delta n = 0.01$.

(b) Relate the slope of the trajectory at $r = 0$ to the numerical aperture of the fiber. Using the ray model, explain qualitatively why local numerical aperture decreases uniformly to 0 at the core-cladding interface.

10.6   Assume that attenuation per unit path length is independent of the trajectory in a parabolic-profile fiber. Compare the attenuation of an extreme ray ($r_m = a$) with that of the axial ray. [Hint: Use the binomial expansion to show explicitly that the rays' attenuations differ in second order of a/L.] What happens in the case of a single-mode fiber?

10.7   A plane wave (with infinite extent) is shone onto the end of a planar, multimode waveguide at angle ϕ to the axis of the waveguide. This wave is truncated when it enters the waveguide aperture.

(a) Call the angle that corresponds to the mth waveguide mode θ_m, as in (10.7). Find a relationship between the far-field diffraction angle $\Delta\phi$ of the truncated plane wave and the angle $\Delta\theta$ between adjacent waveguide modes.

(b) Suppose that the wave is incident at precisely the angle $\phi = \theta_m$ that corresponds to an even-numbered waveguide mode. Explain why only one or a few modes are likely to be excited.

(c) The fraction of the power actually coupled into any mode is related to the overlap integral

$$\int E_i(x)E_m(x)dx$$

between the incident field $E_i(x)$ and the mode field $E_m(x)$, where x is the dimension across the waveguide [cf. (10.52) and (10.53)]. Show why even-order modes are excited more efficiently than odd, and that orders $m + 1$ and $m - 1$ in (b) are not excited at all. [Note: If the angle of incidence is not precisely θ_m, or if the incident beam is not uniform, then other modes may be excited.]

10.8 Consider a star coupler with one input fiber and 19 output fibers. The output fibers are closely packed in a hexagonal array. All fibers have 125 μm outside (cladding) diameters, 62.5 μm core diameters, and a numerical aperture of 0.2. Assume further that they have step-index profiles, so their radiance is constant as a function of angle, then falls abruptly to 0. (a) Sketch the array of output fibers. Estimate the shortest star coupler that will excite the output fibers with roughly equal efficiency. Estimate the overall efficiency of the coupler; that is, estimate what fraction of the light from the input fiber is coupled into the cores of the output fibers. (b) Suppose that the fibers are graded-index fibers with the same parameters. Explain qualitatively why the coupler will have to be much longer than your answer to (a). Why will the efficiency also be lower, apart from increased reflection losses?

10.9 A He-Ne laser beam has a beam width w of 1 mm. It is focused onto the core of a single-mode fiber by a diffraction-limited lens that has a focal length of 20 mm. The mode-field radius of the fiber is 5 μm. Calculate the fraction of the total power that is coupled into the fiber. [Hint: How is this problem similar to coupling two dissimilar fibers?] The result, incidentally, may be misleading because the diode-laser beams usually used in optical communications are elliptical and therefore difficult to couple efficiently into a fiber.

10.10 Verify (10.51).

10.11 Show that, in one dimension, the result analogous to (10.51) is

$$A = (2/\pi)^{1/4}/w^{1/2}. \tag{10.58}$$

10.12 Two identical single-mode fibers are offset by a distance δ. Show that the transmittance of the joint is $e^{-\delta^2/w^2}$, where w is the mode-field radius of the fiber. Hint: Let δ be offset in the x direction and perform the calculation in rectangular coordinates. $\int_0^\infty e^{-u^2}\, du = \sqrt{\pi}/2$.

11. Optical-Fiber Measurements

In this chapter, we shall discuss a variety of field and laboratory measurements on telecommunication fibers. These are mostly optical measurements, such as attenuation, the distortion of a pulse due to the properties of the waveguide or the material of which it is made, and the distribution of index of refraction in the core. Except for the effect of the fiber on the electrical bandwidth of a signal, we do not consider electronic measurements here. Likewise, source and detector problems have been discussed in Chap. 4 and are omitted from this chapter.

11.1 Launching Conditions

Certain measurements on multimode fibers, particularly attenuation and bandwidth, often yield different results for different launching conditions; that is, the outcome of a measurement depends on the manner in which the core of the fiber is illuminated. For example, rays that propagate nearly parallel to the axis of a step fiber travel a shorter path than rays that propagate the critical angle, or near cutoff. In the mode picture, this means that low-order modes suffer less loss than high-order modes. Thus, for example, if a step fiber is excited by a beam that has a low numerical aperture, primarily low-order modes are excited, and a measurement of attenuation will yield an artificially low result.

There has been no little effort, therefore, to determine a set of launching conditions that will give substantially the same results among laboratories as well as give representative results when compared to field measurements. Because the fibers used for telecommunications are comparatively long, say 1 to 5 km, the industry has largely agreed to make certain measurements using either the 70-70 condition or some sort of *equilibrium mode simulator*.

The need for an equilibrium mode simulator is illustrated in Fig. 11.1. Here $f(L)$ represents some quantity, such as attenuation per unit length, that we wish to measure. If we launch a mode distribution that has more power in the high-order modes than the equilibrium distribution, we say that the fiber is *overfilled*; if we launch lower power in those modes, the fiber is *underfilled*. In the example, an overfilled condition results in higher values of $f(L)$ than does an equilibrium distribution. As the length of the fiber is increased, the mode distribution approaches the equilibrium distribution. When $f(L)$ approaches the equilibrium value, we say that L is equal to the equilibrium length. Clearly,

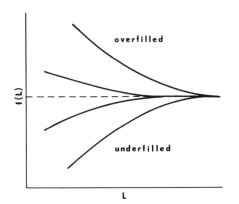

Fig. 11.1. Approach to equilibrium mode distribution

equilibrium length is a function of launch conditions, as is shown by the several curves, which represent different degrees of under- or overfilling. The term "overfilled" can also be used to mean that the fiber is excited by a beam whose numerical aperture exceeds that of the fiber, whose diameter exceeds the diameter of the core, or both. The two uses are, practically, equivalent.

Precisely the same arguments hold for an underfilled launch, except that here this condition results in measurements that are less than the equilibrium values. The object of many fiber measurements is to determine an approximation to the equilibrium value of a quantity, even for short lengths of fiber.

11.1.1 Beam-Optics Launch

The 70-70 condition is usually met by projecting an image of a diffuse source S onto the end of the fiber with an optical system like that shown in Fig. 11.2. This is often called a *beam-optics launch*. The source may be a tungsten lamp, preferably one that has a flat, *ribbon filament* and gives more-uniform illumination than an ordinary, coiled filament. Its image is projected into the plane of an aperture with a pair of condensing lenses. The lenses need not have high quality. The beam is roughly collimated between the lenses so that interference filters F may be used to select the wavelength of the measurement.

A high-quality lens such as a microscope objective projects the image of the *source aperture SA* onto the face of the fiber. The magnification and the diameter of the source aperture are chosen so that the image fills approximately 70% of the fiber core. Similarly, the aperture stop AS is adjusted so that the beam's numerical aperture is approximately 70% of the fiber's numerical aperture.

11.1.2 Equilibrium Mode Simulator

The equilibrium mode distribution arises because of microbending and scattering. It can be approximated in a short length of fiber by wrapping the fiber several times around a post or *mandrel* a centimeter or so in diameter, or by

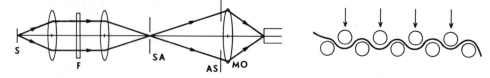

Fig. 11.2. Beam-optics launch, showing source, filter, source aperture and aperture stop

Fig. 11.3. Equilibrium mode simulator by serpentine bend

using a series of posts to create a *serpentine bend* 10 cm or so long (Fig. 11.3). These sharp bends rapidly discriminate against high-order modes (the upper-left corner of the triangle in Fig. 10.10) although they do little to filter out modes with large values of *r*. Nevertheless, these *equilibrium mode simulators* usually give results that are consistent with both the beam-optics launch and the equilibrium mode distribution.

The equilibrium mode simulator is attached to the first few centimeters of the test fiber. After the measurement is made, the fiber is often cut off just beyond the simulator, and the measurement is repeated. The quantity of interest is then taken to be the ratio of the two values. The measurement is done in this way to eliminate the effects of the simulator in cases where they might be significant. Thus, methods that use equilibrium mode simulators may be destructive, whereas the beam-optics method may not be. See Sect. 11.2.1.

Naturally, not any mandrel wrap or series of serpentine bends will suffice. An equilibrium mode simulator must be *qualified* by comparing the output light distribution of a short length of fiber with that of the length of test fiber. The *output numerical aperture* or *radiation angle* is measured for both fibers (Sect. 11.6); the mode simulator is adjusted until the two measurements agree within a certain precision, at which point the filter is said to be qualified. This procedure ensures that the distribution of power among the modes along the length of the fiber is at least roughly constant; it has been shown to yield measurement results that are repeatable from one laboratory to another.

The serpentine-bend mode filter consists of two series of posts; one is moveable, as indicated in the figure. The filter is adjusted by changing the pressure of the moveable posts on the fiber. A mandrel-wrap mode filter may be adjusted by increasing or decreasing the number of turns around the mandrel.

11.1.3 Cladding-Mode Stripper

Light may get into the cladding of a fiber, either because of leaky or refracted rays or, in the case of an overfilled launch, directly from the source. If the cladding is surrounded by air or by a protective coating with a low index of refraction, the light in the cladding will be trapped and propagate down the fiber. Light so guided has mode structure, just as does light guided in the core. Hence, we call light trapped in the cladding *cladding modes*.

If measurements are to be meaningful, cladding modes must usually be eliminated. This is accomplished with a *cladding-mode stripper*. This device is generally just a felt pad saturated with an oil that has a higher index of refraction than the cladding. A few centimeters of fiber are stripped of any protective coating or *buffer* and pressed against the felt pad. Because the index of refraction of the oil is higher than that of the cladding, light incident on the cladding boundary is refracted, rather than reflected. All the light is removed from the cladding in a few centimeters.

As light propagates down the fiber, leaky modes and scattering will cause light to couple gradually into the cladding. The coating of the fiber may absorb cladding modes, or it may have a higher index of refraction than the cladding and act as a cladding-mode stripper. Still, after a long distance, there may again be significant power in the cladding. Therefore, a cladding-mode stripper may be needed at the far end of the fiber as at the near end.

11.2 Attenuation

The diminishing of average power as light propagates along a waveguide is called *attenuation*. Several factors cause attenuation in optical fibers. The fundamental limit, below which attenuation cannot be reduced, is caused by *Rayleigh scattering*. This is scattering that is intrinsic to the material and is caused by microscopic index fluctuations that were frozen into the material when it solidified. These fluctuations are caused by thermodynamic effects and cannot be reduced beyond a certain limit, except perhaps by discovering materials, such as zinc selenide, that harden at a lower temperature than vitreous silica.

Scattering by an index fluctuation is caused by diffraction. To describe such scattering, we therefore return to (7.21), with $g(x)$ a complex number. We take the index fluctuation to have dimension $b \ll \lambda$; the complex-exponential function $\exp(-ikx\sin\theta)$ in (7.21) is therefore equal to 1 over the range of integration of interest.

We next assume that the index of refraction is everywhere n, except at the location of the fluctuation, where it is $n + \Delta n$. Figure 11.4 depicts the situation in a highly schematic way in which the fluctuation is depicted for ease of drawing as a bump on a surface; in fact, it is a density fluctuation inside a solid. The

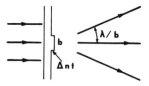

Fig. 11.4. Scattering by an index fluctuation

function $g(x)$ in (7.21) may be replaced by

$$g(x) = e^{-ik\Delta nt} \tag{11.1}$$

where t is the thickness of the fluctuation.

Since the fluctuation is small, we may assume that $\Delta nt \ll \lambda$. We therefore approximate $g(x)$ by

$$g(x) = 1 - ik\Delta nt , \tag{11.2}$$

which is the first two terms of the`Taylor-series expansion of the exponential function. The leading term, 1, describes the light that is transmitted without diffraction and does not interest us. The second term is the scattering term, and we apply it to (7.21) to find that

$$E_s \propto (1/\lambda) \int_{-b/2}^{b/2} ik\Delta nt \, dx \tag{11.3}$$

where E_s is the scattered electric-field strength. We have assumed that the integrand is a constant, so we find immediately that the scattered intensity is

$$I_S \propto \Delta nt/\lambda^4 , \tag{11.4}$$

independent of angle. Therefore, total scattering into 4π sr is proportional to I_s. The result is completely general and describes scattering by any small index fluctuation; it was first derived by Lord Rayleigh to explain the blueness of the sky.

Scattering loss by an ensemble of small fluctuations is therefore proportional to λ^{-4}. Experimentally, we find that scattering loss in a fiber is numerically about equal to $1 \cdot \lambda^{-4}$ when wavelength is measured in micrometers and scattering loss is measured in decibels per kilometer.

The major source of attenuation in the earliest fibers was absorption by impurities, especially water. When the water content is reduced to a few parts per million, the absorption loss may be reduced below scattering loss at certain wavelengths. Other sources of loss are microbending loss, which we have already discussed, and scattering by imperfections. Microbending loss seems to contribute more to mode coupling than to attenuation, and imperfections have largely been removed from high-quality communication fibers. Therefore, we may now obtain fibers whose attenuation loss is very nearly that determined by Rayleigh scattering.

11.2.1 Attenuation Measurements

These are made using either an equilibrium mode simulator or a beam-optics launch with the 70-70 condition. For high-loss fibers, a simple *insertion-loss measurement* suffices. Light is launched into the fiber and the output power measured. Then the fiber is removed from the apparatus and the input power is measured. The transmittance of the fiber is assumed to be the ratio of the two measurements.

The insertion-loss method must be used for fibers that are already attached to connectors. However, for isolated fibers, it ignores possible errors that may result from coupling to the fiber; that is, it assumes that *coupling loss* is far less than attenuation. This is not necessarily the case with the low-loss fibers intended for long-distance communication. For measurements on such fibers, the *cutback method* must be used: The fiber is inserted into the measuring system and the transmitted power is measured. Then the fiber is cut off a few tens of centimeters from the input end and the transmitted power is measured again. The transmittance of the fiber is the ratio of these two measurements. The input coupling loss is the same for both measurements, so its possible effects are eliminated. (Output coupling loss is usually negligible.)

Some researchers, for convenience, use a modified cutback method; they place side by side the fiber being tested and a short piece of the same fiber. Attenuation is assumed to be the ratio of the powers transmitted by the two fibers. This technique is attractive because the measurement can be repeated many times to ensure better precision. However, it is not as accurate as the true cutback method because the two fiber ends may not exhibit exactly the same input coupling loss.

Attenuation is conventionally measured in decibels. If the transmittance of the fiber is T, the absorption coefficient α is given implicitly by

$$T = e^{-\alpha L} \tag{11.5}$$

(Sect. 8.1). Because $e \cong 10^{-0.45}$, we find that the optical density of the fiber is $0.45\,\alpha L$. We assume that the absorption coefficient does not depend on the length of the fiber; this is, after all, the purpose of using the restricted launch conditions.

Optical density is the same as the loss expressed in *bels* or logarithmic units. However, the use of decibels, or tenths of bels, is more common than bels. The attenuation coefficient in decibels may be expressed as

$$\alpha' = (10 \log_{10} T)/L \,, \tag{10.6}$$

where T is the transmittance of the fiber.

Figure 11.5 is a sketch of the attenuation of typical fibers as a function of wavelength. The upper curve describes a high-loss fiber, whereas the lower curve describes a telecommunication fiber designed for relatively long-distance transmission. The peaks are the result of absorption by water impurities. The dashed line shows the theoretical limit imposed by Rayleigh scattering.

The wavelengths of greatest interest for communications are 850 nm, 1.3 μm, and 1.55 μm because they fall between the absorption bands, where attenuation is least. These wavelength regions are known as the *first, second* and *third windows*; the first window was developed first because of the ready availability of near-ir sources and silicon detectors. The second and third windows have

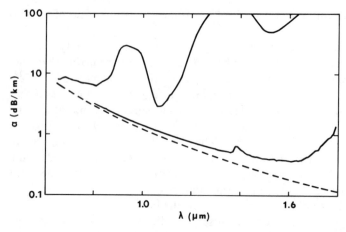

Fig. 11.5. Optical fiber loss coefficient as a function of wavelength. [Courtesy of J. M. Dick, Corning Glass Works]

ultimately greater interest for long-distance transmission such as undersea cables.

11.3 Fiber Bandwidth

Suppose that a pulse or an intensity-modulated signal is transmitted by an optical fiber. If the waveform that emerges from the fiber differs from what was transmitted (except for attenuation), we say that the signal has been *distorted*. Strictly speaking, distortion must be distinguished from *dispersion*, which is the variation of some physical quantity with wavelength. Dispersion is only one of the causes of distortion. Dispersion is sometimes redundantly called *chromatic dispersion*, perhaps to distinguish it from *distortion*.

11.3.1 Distortion

When rays travel different optical paths and therefore take different times to travel the length of the fiber, we speak of *intermodal distortion*. This is the form of distortion we talked about earlier, in connection with (10.4); it is analogous to multipath distortion in television. Intermodal distortion is the factor that limits step-index fibers to either low bandwidths or short distances. As we noted earlier, it may be reduced by using graded-index fibers or be eliminated entirely with single-mode fibers.

Even if intermodal distortion were eliminated entirely, *intramodal distortion* would still exist. Intramodal distortion is distortion that occurs within modes, rather than between, and is important in single-mode fibers and in graded-index fibers in which the source linewidth is large, as with LED sources.

11.3.2 Material Dispersion

One of the causes of intramodal distortion is *material dispersion*; this refers to the variation of the index of refraction with wavelength and is usually called simply dispersion in other branches of optics. Material dispersion should be about the same for all fibers that have roughly the same glass composition, that is, for nearly all fibers that have a vitreous-silica core.

Suppose we use a source whose line width is $\Delta\lambda$ to propagate a narrow pulse down a fiber. We assume that intramodal distortion dominates intermodal distortion in this fiber and consequently ignore differences of group velocity among modes. This is the same as assuming that $\cos\theta \simeq 1$ in the case of a step fiber. Thus, we assume that, for all modes, the pulse propagates with velocity c/n_g, where n_g is the group index of refraction given by (5.33). The pulse traverses a fiber whose length is L in a time

$$\tau(\lambda) = L n_g / c = (L/c)\left(n - \lambda \frac{dn}{d\lambda}\right), \tag{11.7}$$

where $\tau(\lambda)$ is the transit time or *group delay* and varies with λ if n varies with λ.

For signal distortion, we are not interested in the group delay but in the variation of the group delay with wavelength. If the line width of the source is $\Delta\lambda$, the variation $\Delta\tau$ of the group delay as a result of dispersion is

$$\Delta\tau = (d\tau/d\lambda)\Delta\lambda, \tag{11.8}$$

$$\frac{\Delta\tau}{\tau} = -\frac{\lambda}{n}\frac{d^2 n}{d\lambda^2}\Delta\lambda \tag{11.9}$$

where the derivative is to be evaluated at the central wavelength of the source. The electrical *bandwidth* Δf of the fiber is approximately

$$\Delta f = 1/\Delta\tau. \tag{11.10}$$

The second derivative in (11.9) is 0 at $\sim 1.25\ \mu m$; distortion that results from material dispersion can be largely corrected by operating near this wavelength.

11.3.3 Waveguide Dispersion

Another form of intramodal distortion results from *waveguide dispersion*. This comes about because the angle θ is a function of wavelength; therefore, as the wavelength changes slightly, the group delay of a given mode changes slightly, whether or not there is significant material dispersion.

The effective velocity with which a given ray propagates along the fiber is the component of the group velocity parallel to the fiber axis. That is,

$$v_{\parallel} = v_g \cos\theta. \tag{11.11}$$

Because λ and θ are related [through (10.7)], $v_{\parallel}$ depends on wavelength.

In the absence of material dispersion, the group velocity is the same as the phase velocity, so

$$v_{\parallel} = (c/n_1)\cos\theta \tag{11.12}$$

in a step fiber. The group delay is therefore

$$\tau(\lambda) = Ln_1/(c\cos\theta) . \tag{11.13}$$

As before, we use (11.8) to calculate the variation of the group delay over the wavelength interval $\Delta\lambda$; we find that the variation of the group delay with wavelength is

$$\Delta\tau = -\frac{Ln_1}{c}\sin\theta\,\frac{d\theta}{d\lambda}\,\Delta\lambda , \tag{11.14}$$

where we use the approximation $\cos\theta = 1$ after calculating the derivative. To evaluate $d\theta/d\lambda$, we use (10.5) with $\cos i = \sin\theta$ to find that

$$m\lambda = 2d\sin\theta \tag{11.15}$$

and finally that

$$\Delta\tau/\tau \cong -\sin^2\theta(\Delta\lambda/\lambda) . \tag{11.16}$$

The maximum value of $\Delta\tau$ occurs when θ has its maximum value θ_m; this is approximately the total variation of group delay for the fiber as a whole.

Analysis of graded-index fibers that have arbitrary values of the profile parameter g shows that waveguide dispersion is very nearly 0 when

$$g = 2 - 2\Delta - 2\frac{n_1}{n_{g1}}\frac{\lambda}{\Delta}\frac{d\Delta}{d\lambda}\,(1 - \Delta/2), \tag{11.17}$$

where n_{g1} is the group index in the core. The optimal value of g can differ significantly from 2 and depends on the wavelength and the doping of the core of the fiber.

A single-mode fiber suffers only from intramodal distortion; the total intramodal distortion is approximately the sum of (11.9) and (11.16). The second derivative in (11.9) changes sign near 1.25 μm; at some greater wavelength, total distortion can be made very nearly 0. A single-mode fiber can therefore be combined with a narrow-line source such as a laser diode to provide a communication channel with a very large bandwidth. This is one of the reasons for the interest in using such fibers with long-wavelength sources.

11.3.4 Bandwidth Measurements

These also need to be made with some kind of standard launch conditions, most commonly, both the numerical aperture and the entrance face are overfilled with a uniform intensity distribution. A cladding-mode stripper is therefore also required. Measurements may be made in either the *time domain* or the *frequency domain*.

In frequency-domain measurements, the source intensity is modulated at a given frequency. The signal is detected at the source and at the far end of the fiber. The ratio of the two measurements is taken; such ratios taken at a large number of frequencies, when plotted against frequency, make up the *frequency-response* curve of the fiber. Generally, the curve is normalized so that the low-frequency values are set equal to 1. The bandwidth of the fiber is generally defined as the lowest frequency at which the frequency response falls to one-half the low-frequency value or, in electrical-engineering parlance, at which it falls by 3 dB.

Time-domain measurements are performed by transmitting a pulse along the fiber. The frequency response may be calculated by using Fourier-transform methods. Approximate measurements may be made by measuring the *duration* of the input and output pulses; this is usually defined as the time the pulse takes to rise from half its maximum value, pass through the maximum, and fall to half the maximum value.

If the input and output pulses are roughly Gaussian functions of time, the broadening $\Delta\tau$ due to propagation along the fiber may be estimated from the equation

$$\Delta\tau_o^2 \cong \Delta\tau_i^2 + \Delta\tau^2 , \tag{11.18}$$

where $\Delta\tau_i$ is the duration of the input pulse and $\Delta\tau_o$ is the duration of the output pulse. The bandwidth of the fiber is roughly $1/\Delta\tau$.

11.3.5 Coherence Length of the Source

The detectors in optical-communication systems are power detectors; they are sensitive to intensity, rather than electric-field strength. As a result, we are, in effect, modulating and detecting intensity rather than electric field. Thus, for measurements to be meaningful, we require the fibers to be linear with respect to intensity, rather than electric field. That way, when the intensity of the source is modulated in a given fashion, the output of the detector will be proportional to the output of the source (except for distortion and decreased bandwidth).

What are the conditions that make the fiber a system that is linear in intensity? The coherence length of the source must be sufficiently short that we are at all times adding intensities and not fields. Suppose that the bandwidth of the source is Δv. Then the coherence length is approximately $c/(n_1\Delta v)$, where c/n_1 is the velocity of light in the fiber core.

Suppose that the source is modulated at frequency f_m. This is the highest electrical frequency in which we will be interested. The shortest time interval over which we will, in effect, make a measurement is $1/f_m$. Since the signal travels with the velocity c/n_1, this time corresponds to a distance $c/(n_1 f_m)$. If this distance is greater than the coherence length of the source, the measurement is incoherent. This leads immediately to the relation that

$$f_m < \Delta v ; \tag{11.19}$$

the modulation frequency of the signal must always be less than the bandwidth of the source. (If the signal consists of a series of pulses, then f_m is approximately the reciprocal of the pulse duration.) If this condition is not satisfied, interference among modes will become important, and the detector output will not be a faithful representation of the source output. The fiber cannot, in this case, be regarded as a linear system.

11.4 Optical Time-Domain Reflectometry

This technique, also known as *backscatter*, consists of launching a short pulse into a long fiber and detecting whatever power is scattered in the reverse direction, trapped by the fiber, and returned to the vicinity of the source. The backscattered power may be detected in any of several ways; Fig. 11.6 shows a particularly effective system, which uses a polarizing beam splitter to discriminate against the reflection from the entrance face of the fiber.

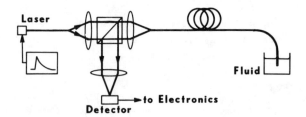

Fig. 11.6. Optical time-domain reflectometer

In this system, a laser diode is excited with a short current pulse. The emission from the laser is passed through a short piece of step-index fiber, in part to ensure a uniform, unpolarized source. The emerging beam is collimated, passed through a beam splitter, and focused into the fiber.

A mandrel-wrap mode filter is used to simulate the equilibrium mode distribution. It is especially important to mode-filter the backscattered light. This is so because the backscattered light overfills the numerical aperture of the fiber. If there is mode coupling, light scattered far down the fiber approaches the equilibrium mode distribution as it returns toward the source. This effect can cause spurious results.

Finally, if the beam splitter is the type that reflects one polarization and transmits the other, it will help to discriminate against the comparatively large Fresnel reflection by the input face of the fiber. (The index-matching oil at the other end reduces reflection there to a small value.) The scattered light becomes unpolarized after traversing a short length of fiber, so the beam splitter directs half this light to the detector.

Optical time-domain reflectometry, or OTDR, is especially useful when only one end of a fiber is accessible, as may be the case when diagnostics are necessary on an installed system.

Suppose that a pulse is injected into a fiber. At the distance z along the fiber axis, the power is reduced to $\exp(-\alpha z)$, where α is the attenuation coefficient of the fiber. Some of the power scattered by the fiber in the region around z is trapped by the fiber and returned to the source. That light has traveled a total distance of $2z$ and is therefore attenuated by $\exp(-2\alpha z)$ if the mode distributions in both directions are nearly the same.

If we measure the power returned from the distance z and from a point near the source, we may calculate the attenuation coefficient from the relation that

$$P(z)/P(0) = e^{-2\alpha z} , \tag{11.20}$$

where $P(z)$ is the energy of the pulse scattered from the region around z and $P(0)$ is the energy of the pulse scattered from the region near 0. If we take logarithms, we find that

$$\log_e[P(z)/P(0)] = -2\alpha z , \tag{11.21}$$

or, in engineering units,

$$\alpha' z = -5\log_{10}[P(z)/P(0)] , \tag{11.22}$$

where $\alpha' z$ is the loss of the fiber in decibels. The factor 5 appears (instead of 10) because the measurement involves propagation in both directions.

Figure 11.7 shows a semilogarithmic plot of backscattered power as a function of time; time here means time after the emission of the pulse by the laser. The upper curve was taken without the mandrel-wrap mode filter; near the

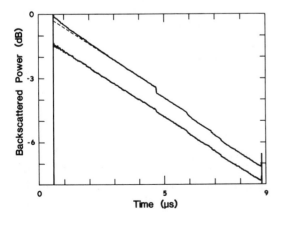

Fig. 11.7. Backscattered power as a function of time for two different launching conditions. [After B. L. Danielson, NBS Technical Note 1065]

source the curve is not linear but approaches linearity after about 3 μs. The lower curve was taken with a mode filter and shows good linearity.

The glitch near $t = 5$ μs is the result of a fusion splice; its effect seems to be exaggerated in the upper curve. The other glitches are the result of unknown defects in the fiber. This illustrates the usefulness of OTDR: It may be used to locate faults or splices as well as to determine the approximate attenuation coefficient of the fiber. If we measure the round-trip transit time Δt of the signal returned from an imperfection, we may determine the location of the imperfection as

$$z = c\Delta t/2n_1 , \tag{11.23}$$

where c/n_1 is approximately the group velocity in the fiber. Such distance measurements are accurate to about c/n_1 multiplied by the pulse duration, which usually comes to several tens of centimeters.

11.5 Index Profile

Sometimes, it is necessary to measure the *index profile* of a fiber, not just the value of Δn. Generally, the measurement takes the form of a plot of index of refraction as a function of position along a diameter of the fiber. At least a half-dozen methods have been developed for carrying out this measurement; most of these require automatic data acquisition and sophisticated computer analysis, and will only be mentioned here.

Techniques for measuring the index profile may conveniently be classified into two groups: transverse methods, in which the fiber is illuminated in a direction perpendicular to its axis, and longitudinal methods, in which the fiber is examined at a cleaved or otherwise prepared end.

Transverse methods have several advantages over longitudinal methods. Because we do not need access to an end of the fiber, transverse methods are inherently nondestructive. They may be performed very nearly in real time, for example, to monitor the fiber's properties as it is being manufactured. Longitudinal methods, on the other hand, require the fiber to be cut and an end to be prepared. They are therefore destructive and somewhat time consuming. However, certain longitudinal methods are precise and easy to implement and are therefore often the methods of choice when absolute measurements are necessary.

11.5.1 Transverse Methods

There are primarily two kinds of transverse methods: interferometry and diffraction or scattering. In the first case, the fiber is placed in a cell filled with a fluid whose index of refraction is very nearly equal to that of the cladding of the fiber. The windows of the cell are plane and parallel. The cell is located in one arm of an interferometer or interference microscope, and the intensity of the resulting fringe pattern is measured.

Figure 11.8 shows the path of a ray through a graded-index fiber. The path is curved because the index of refraction is not constant throughout the core. Even if the curvature can be ignored, the phase shift ϕ of the ray must be calculated from the expression

$$\phi = k \int n(s)ds \, , \qquad\qquad (11.24)$$

where s is the dimension shown in the figure. Inferring ϕ from the fringe pattern and inverting this equation to calculate the index profile $n(r)$ requires a complicated procedure that will not be discussed here.

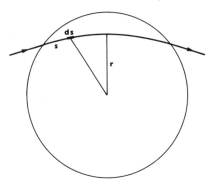

Fig. 11.8. Ray path through a transverse cross section of a graded-index fiber

Other transverse methods use examination of the scattering or diffraction pattern created when the fiber (also index matched) is illuminated with a laser beam perpendicular to its axis. When measuring the far-field pattern, we refer to *transverse scattering*. A related method is called *transverse backscattering* because the beam that is reflected backward, toward the source, is examined. Another related method, called the *focusing method*, measures the near-field diffraction pattern and is most useful for evaluating *preforms*, the rods from which the fibers are manufactured.

All the transverse methods require a scan of a diffraction or interference pattern, and all require inverting an integral to evaluate the index profile. As a practical matter, they therefore require automatic data acquisition and computer processing. In addition, most of the calculations assume that the fiber core has radial symmetry; this assumption may or may not be appropriate. The overall accuracy of these methods is perhaps 10%.

11.5.2 Longitudinal Methods

Longitudinal methods illuminate the fiber or a specially prepared sample in the direction parallel to the axis. One of the earliest of these methods was *slab interference microscopy.* Here, a thin slice of fiber is placed in an interferometer or *interference microscope.* The core-cladding index difference may be determined by a direct examination of the interferogram. The optical-path difference between the ray that passes through the core center and those that pass through the cladding is simply Δnt, where t is the thickness of the sample. If the interferometer is the type where the ray passes through the sample one time, each interference fringe corresponds to one wavelength of optical path difference. As long as t is known accurately enough, Δn may be calculated directly. The index profile may be calculated by hand or with an automatic data-acquisition scheme and computer processing.

One of the difficulties with the transverse-interference method is that the sample must be thin enough that the rays travel through it without appreciable bending owing to the graded-index profile. This usually means that the sample must be no more than a few tens of micrometers thick. In addition, it is difficult to determine what errors arise from polishing the sample: Polishing may change the surface composition of the material, and the surface may deviate from flatness as a result of variation of polishing rate with glass composition (which, of course, varies as the index of refraction varies).

Another early method was a direct measurement of the reflectance of the end face of the fiber. This method is direct and precise and has high spatial resolution; it is somewhat difficult because the changes of reflectance are small and highly stable sources and detectors are required. In addition, it is possibly susceptible to error due to contamination of the surface by a layer with different index of refraction from the bulk.

11.5.3 Near-Field Scanning

This is one of two important methods that depend on the local numerical aperture, given by (10.34). Because the numerical aperture of the fiber varies with radius, so does the light-gathering capability. According to (10.3) and the discussion immediately following, the collection efficiency of a step-index fiber is proportional to the square of the numerical aperture when the source is Lambertian. By precisely similar reasoning, we conclude that the local collection efficiency of a graded-index fiber is given by the square of the local numerical aperture; this observation provides us with a method for measuring the index profile.

The input face of a fiber is fully illuminated by a Lambertian source with condensing optics whose numerical aperture exceeds the maximum numerical aperture of the fiber. If the fiber is held straight and is less than a meter or so long, mode coupling will be negligible. The radiation pattern at the output end of the fiber will exactly match the acceptance pattern determined by the local numerical aperture. Therefore, the output power as a function of position is precisely proportional to the square of the local numerical aperture, or to $[n^2(r) - n_2^2]$. Because $n(r)$ is very close to n_2, we may approximate this square with the expression,

$$NA^2 = 2n_2 \Delta n(r) ; \tag{11.25}$$

therefore, the output power distribution very closely approximates the index profile of the fiber.

The index profile is measured by magnifying the output face of the fiber with a microscope objective that has a higher numerical aperture than the fiber. A detector is scanned across the magnified image and the photocurrent acquired with a computer or a pen recorder.

The data obtained in this way would be a precise representation of the index profile of the fiber, but for the presence of leaky rays. The fiber will accept leaky rays, which will cause the relative intensity at any radius other than 0 to be greater than the expected value. In measurements on step-index fibers, this is a serious problem that can sometimes be corrected analytically. The leaky-ray problem is not so severe with graded-index fibers and, in addition, most researchers believe that leaky rays are eliminated as a result of slight but inevitable core ellipticity. Consequently, near-field scanning has proven to be a useful tool for measuring index profile. Unfortunately, this method yields only a relative index profile; the system cannot readily be calibrated to yield absolute values of the index of refraction.

One disadvantage of near-field scanning is that it cannot be used to determine the profile of a single-mode fiber. This is so because the fiber has a very low numerical aperture. In a sense, when we perform a near-field scan on a single-mode fiber, we are viewing the entrance face through the fiber itself, and the numerical aperture is so low that we cannot resolve the structure of the entrance face. The radiation pattern at the output face is characteristic of the mode of the fiber, not of the local numerical aperture. A near-field scan may be a useful measurement for determining the distribution of light in the fiber, but it will not give information about the index profile.

11.5.4 Refracted-Ray Method

With this method, we illuminate the fiber with a focused beam whose numerical aperture is two or three times that of the fiber; we examine not the rays that are trapped by the fiber but, rather, those that are not. The refracted-ray method yields absolute index-of-refraction measurements and resolution equal to that of ordinary microscopy. In addition, it usually eliminates entirely the effects of leaky rays.

The principle of the method is shown in Fig. 11.9, which shows a bare fiber in an oil-filled cell. A ray enters the cell at angle θ and impinges on the face of the fiber at angle θ_1, which greatly exceeds the acceptance angle of the fiber. That ray eventually emerges from the cell at angle θ' to the rear face of the cell. We wish to trace the ray through the fiber and determine θ' as a function of θ, the index of refraction n of the fiber, and the index of refraction n_L of the oil.

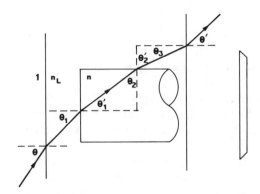

Fig. 11.9. Refracted-ray method of measuring fiber index profile

We begin by applying Snell's law at the point where the ray hits the fiber end,

$$n_L \sin \theta_1 = n \sin \theta_1' \; . \tag{11.26}$$

If the fiber end is accurately perpendicular to the axis, θ_2 is complementary to θ_1'; squaring the equation we find that

$$n_L^2 \sin^2 \theta_1 = n^2 (1 - \sin^2 \theta_2) \; . \tag{11.27}$$

Applying Snell's law to the angles θ_2 and θ_2' and assuming that the fiber is perpendicular to the cell's faces, we find that

$$n^2 - n_L^2 = n_L^2 (\sin^2 \theta_1 - \sin^2 \theta_3) \; , \tag{11.28}$$

or, in terms of the angles of incidence and refraction from the air to the cell,

$$n^2 - n_L^2 = \sin^2 \theta - \sin^2 \theta' \; . \tag{11.29}$$

If the index of refraction of the liquid closely enough matches that of the fiber, we write

$$2n_L(n - n_L) = \sin^2 \theta - \sin^2 \theta' \; . \tag{11.30}$$

If θ is the angle of incidence of the most-oblique ray that enters the cell, $n - n_L$ is a linear function of $\sin^2 \theta'$, where θ' is the angle of emergence of that ray. We must now show that the power transmitted by the fiber is also a linear function of $\sin^2 \theta'$ because, if that is so, $n - n_L$ will also be a linear function of power.

We place an opaque stop behind the cell. The stop is centered about the axis of the fiber and subtends angle θ_s as seen from the face of the fiber; θ_s greatly exceeds the acceptance angle of the fiber, so most leaky rays will be intercepted by the stop. If the source is Lambertian, the power radiated into a cone is, according to (4.17), proportional to the square of the sine of the cone's vertex angle. Therefore, the relative power transmitted around the opaque stop is proportional to $\sin^2 \theta' - \sin^2 \theta_s$. In short, when the stop is in place, the power transmitted beyond the stop is indeed a linear function of $n - n_L$.

This result may be generalized to a graded-index fiber by applying Snell's law in its general form (10.23). Therefore, the index profile of a fiber may be measured by measuring the transmitted power as a function of radius. The experiment is unfortunately not as easy as near-field scanning, but it can yield some results that cannot be determined from near-field scanning. In addition, the opaque stop may be made large enough to eliminate almost completely the effects of leaky rays in graded-index fibers.

Index profiles determined by refracted-ray scans may be made absolute by calibrating the system in one of several ways. Probably the simplest is to use a vitreous-silica fiber, whose index of refraction is known very precisely. If the index of refraction of the oil is known, measurements of the corresponding transmitted powers may be used to relate power to index of refraction. The system may also be shown to be linear by using several known oils instead of just one.

The spatial limit of resolution of a refracted-ray scanner is about equal to that of the microscope objective used to focus the laser beam onto the fiber end, provided that the opaque stop is not too large. The same is not the case with near-field scans. In that case, as we noted above, we are in effect viewing the end of the fiber through the fiber itself. At any radius, the limit of resolution is determined by the local numerical aperture at that radius and becomes very large as the measurement approaches the core-cladding boundary.

The refracted-ray method is one of the only techniques that can be used to measure the profiles of single-mode fibers, because it has a sufficiently high spatial limit of resolution and high enough precision to measure the small index differences characteristic of single-mode fibers.

Figure 11.10 shows a typical refracted-ray scan of a graded-index fiber. The dip at the center of the core is common to many such fibers and is an artifact of their manufacture. This fiber also exhibits a nonuniform cladding and a narrow low-index layer surrounding the core.

11.5.5 Core-Diameter Measurements

Closely related to index-profile measurements, core-diameter measurements are probably most important to manufacturers of fiber-to-fiber connectors and other connecting devices. Core diameter must also be known, at least approximately, for measurements that require a beam-optics launch.

Core diameter is usually measured by performing some kind of index-profile measurement, especially near-field scan and refracted-ray method. The problem is how to define core diameter when the measurement might suffer from the presence of leaky rays or other defects. In particular, near-field scans of graded-index fibers almost always exhibit a substantial loss of resolution near the core-cladding boundary; this happens because the local NA approaches 0 as r approaches a.

Figure 11.11 is a near-field scan of the fiber of Fig. 11.10; the horizontal scales are about the same. The measured index profile is badly rounded near the

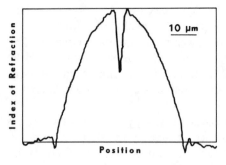

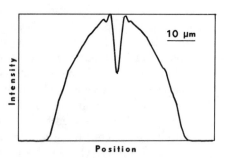

Fig. 11.10. Index profile of a graded-index fiber by the refracted-ray method

Fig. 11.11. Index profile of the fiber of Fig. 11.10 by near-field scanning. [Courtesy of E. M. Kim]

boundary, where the true profile is somewhat sharp. To eliminate this effect, the core diameter is estimated either by extrapolating the better-resolved portions of the curve to the base line, or by defining the core diameter as the width of the scan between the points at which the index of refraction is 2.5% of the peak value Δn. These approaches usually give good agreement as well as results that are accurate enough to be practical.

The problem is somewhat different with single-mode fibers, where the radiation pattern greatly differs from the index profile. In that case, the engineer must distinguish between the core diameter and the spot size of the radiation pattern. For connector problems, spot size is the important parameter, and core diameter is largely irrelevant. Therefore, a near-field scan will give meaningful results, whereas a refracted-ray scan will not.

11.6 Numerical-Aperture Measurements

Numerical-aperture measurements are needed for several reasons. The collection efficiency of a fiber connected to a source such as a light-emitting diode depends on the numerical aperture (Sect. 10.1). Coupling between two fibers usually depends on their having the same numerical apertures; sensitivity of a connection to angular misalignment also depends on the numerical apertures of the fibers. In addition, the numerical aperture must be known for certain measurements, such as attenuation, the require standardized launch conditions. Whether the launch uses the beam-optics method or some kind of mode filter, the numerical aperture of the fiber must be measured precisely.

A measurement of the numerical aperture of a fiber is in reality a measurement of the far-field radiation pattern of the fiber. A detector is scanned in an arc at whose center is located the exit face of the fiber. To ensure that the size of the exit face of the fiber has no role in the measurement, the detector must be located at least several times the distance a^2/λ from the fiber end; that is, the detector must be located in the far field (Sect. 5.5.4) of the fiber's exit face. Otherwise, the measurement may depend on the radius of the arc. This may be tested by making several scans with different radii and comparing the shapes (but not the intensities) of the resulting curves.

The size of the detector is important; it must be small enough to permit undistorted recording of the radiation pattern, but not so small that the photocurrent is unmanageably small. We may derive a rule of thumb by noting that the finest detail in the radiation pattern is determined by the largest dimension of the fiber, that is, the radius. The radiation pattern can have no angular detail finer than about λ/a; the detector must subtend a somewhat smaller angle than this value to ensure sufficient resolution. This condition may also be tested by making several scans with different detector apertures.

For attenuation measurements, a fiber's numerical aperture must be measured with the beam-optics launch or with an equilibrium mode simulator that has been properly qualified. Measurements related to connectors may require one of these standard launch conditions, depending on the application. For example, connecting a short-haul fiber to an LED may suggest a measurement of the radiation numerical aperture with overfilled launch conditions if that appears to simulate the use of the fiber better.

Problems

11.1 Sketch an optical system that will image the filament of a tungsten lamp onto a fiber. The fiber must be illuminated uniformly with a sharp, circular spot with a diameter of 35 μm, and the numerical aperture of the optical system should be 0.14. The tungsten lamp's filament is flat and has dimensions 1.5 cm by 1.5 mm; its temperature is uniform over 90% of its width. The lamp has a 2-cm diameter envelope.

Lay out the system with thin lenses but do not forget to specify the general class of lens used to image the beam onto the fiber. Assume that the tolerances on numerical aperture and spot size are $\pm 5\%$.

11.2 (a) The radiant emittance M_λ of the lamp in Problem 11.1 is about $10 \text{ W} \cdot \text{cm}^{-2} \cdot \mu\text{m}^{-1}$ when $\lambda = 0.85 \ \mu\text{m}$. The light is filtered with a bandpass filter with $\Delta\lambda = 10$ nm. How much power is coupled into the fiber? [Answer: approximately 0.2 μW.]

(b) The light is mechanically chopped with frequency $f = 100$ Hz. A detector claims NEP $= 10^{-11}$ W when the bandwidth is 1 Hz. An electronic circuit with a bandwidth of 10 Hz amplifies the output of the detector. (i) What is the NEP of the system? (ii) The poorest fiber to be tested has a loss of 20 dB, or a transmittance of 0.01. Will the detector be adequate?

11.3 A cladding mode is excited in an optical fiber whose cladding diameter is 125 μm. The fiber is immersed in a cladding-mode stripper. Assume as a wild guess that the reflectance at the surface of the cladding is 1%. After what distance will the intensity of the cladding modes be reduced by a factor of 10^{-6}? What if the reflectance is 10% (remember that the ray hits the surface at nearly glancing incidence)?

11.4   A certain fiber has an attenuation of 1 dB/km at 0.85 μm. A 1-mW laser is coupled to it with an efficiency of 30%. The fiber is 5 km long and is cemented with low loss to a detector for which $NEP/\sqrt{\Delta f} = 10^{-10}$ W$/\sqrt{Hz}$. If the electrical bandwidth of the signal is 10 MHz, is the detector adequate?

11.5   The index of refraction of vitreous silica can be described by a *dispersion equation*

$$n(\lambda) = 1.45084 - 0.00334\lambda^2 + 0.00292/\lambda^2 . \tag{11.31}$$

Assume that the dispersion of a single-mode fiber is adequately described by (11.31), even though the core is doped with impurities. Consider a single-mode fiber that is 3 km long and is coupled to a light-emitting diode whose wavelength is 1.3 μm and whose linewidth is 100 nm. Calculate the variation of group delay that results from (a) material dispersion and (b) waveguide dispersion.

11.6   The cladding of a certain graded-index fiber may be described by (11.31). Suppose that the core is doped with impurities so that

$$\Delta n_0(\lambda) = 0.01202 + 0.00003\lambda^2 + 0.00028/\lambda^2 , \tag{11.32}$$

where the subscript 0 means the value of Δn at the center of the core. Calculate the optimum value of g at 0.85, 1.3, and 1.55 μm.

11.7   A typical light-emitting diode that operates at 1.3 μm may have a linewidth of 100 nm. Calculate the electrical bandwidth that corresponds to this linewidth. How many video channels could we transmit with this diode? Ignore the properties of the detector and assume that the channels are 10 MHz apart. How many audio channels separated by 10 kHz could we transmit?

11.8   At the wavelength of 0.85 μm, the scattering loss of a certain single-mode fiber is 1 dB/km, and the absorption loss is the same. Assume, for convenience, that the scattered radiation is Lambertian. (a) If the numerical aperture of the fiber is 0.1, roughly what fraction of the light scattered from any point is guided back toward the source? (b) An optical time-domain reflectometer uses a polarizing beam splitter, as in Fig. 11.6. What fraction of the power incident on the fiber returns to the detector from a point 5 km distant? [Hint: Don't forget the absorption loss.] (c) The duration of the pulses is 10 ns, and the peak power is 1 mW. What is the power scattered back to the detector from a point 5 km from the detector? What value of $NEP/\sqrt{\Delta f}$ will be required of the detector? Assume that the electrical bandwidth is the reciprocal of the pulse duration (and that we wish to resolve times of the order of 10 ns).

11.9 We want to determine the index profile of a waveguide by measuring the difference of Fresnel reflection between the core and the cladding. If the waveguide has a numerical aperture of 0.1 and its cladding index is 1.457, what is the ratio of the maximum reflectance to that from the cladding? What if we immerse the fiber in a fluid whose index of refraction is 1.48? Describe what might happen if the immersion fluid displayed an index of refraction between those of the core and the cladding.

11.10 Suppose that we put a screen 10 cm behind the cell of Fig. 11.9. The numerical aperture of the incident beam is 0.5, and the index of refraction of the fluid in the cell is 1.464. (a) What is the radius of the illuminated spot on the screen when the beam passes through the fluid but not the fiber? Ignore the thickness of the cell.

The index of refraction of the cladding of the fiber is 1.457, and that of the core is 1.470. What is the radius of the spot when the beam is incident on (b) the cladding, (c) the core?

The optimum opaque stop subtends an angle given by the numerical aperture of the microscope objective divided by $\sqrt{2}$. (d) Assuming that the stop is 10 cm from the cell, calculate its radius. Is it smaller than the radius calculated in (c)?

11.11 We want to measure the numerical aperture of a multimode fiber that has a 62.5 μm (diameter) core. The fiber is designed for 0.85 μm. We have available an instrument that scans a detector in an arc 5 cm in radius. The detector is 5 mm in diameter. Will the instrument suffice? If not, what modifications are necessary?

12. Integrated Optics

In the semiconductor industry, an electronic integrated circuit is one that is manufactured entirely on a wafer or chip of silicon or some other semiconductor. An optical device that is manufactured on a flat substrate and performs functions similar to electronic circuits is called by analogy an *optical integrated circuit*. Devices such as Fourier-transform optical processors can also be manufactured on flat substrates. To distinguish such devices from optical integrated circuits, we shall call them *planar optical devices*. Both planar optical devices and optical integrated circuits fall into the general classification of *integrated optics*.

12.1 Optical Integrated Circuits

These are generally expected to play a major role in optical communications. The ideal is to build complete transmitters, repeaters, and receivers on single substrates and connect them directly to single-mode optical fibers. In a repeater, for example, such integration will eliminate the need to convert from an optical signal to an electronic signal and then back to optical.

Devices that are manufactured on a single chip or substrate are known as *monolithic* devices. Monolithic optical circuits are usually made on a substrate of gallium arsenide; the devices manufactured on the substrate are various concoctions of gallium, aluminum, arsenic, phosphorus, and other dopants, depending on the wavelength at which the device operates. For example, lasers that operate at 0.85 μm are made of an alloy of gallium arsenide and aluminum arsenide, or GaAlAs. Pure gallium arsenide has a band gap (Sect. 4.2.4) that corresponds to the wavelength of 0.91 μm, and aluminum arsenide, to 0.65 μm. If the concentrations of gallium arsenide and aluminum arsenide are chosen properly, the alloy can in principle be made to emit light at any wavelength in between.

Gallium and arsenic are located in the third and fifth columns of the periodic table; compounds made of elements found in these columns are called *III–V compounds*. Many such compounds can be combined and deposited on a substrate of gallium arsenide. For example, a laser made of gallium indium arsenide phosphide, or GaInAsP, can be tailored to the wavelength of 1.3 μm, which is of considerable importance in optical communications.

Unlike silicon, gallium arsenide and its relatives can be made to emit light or exhibit optical gain. Devices made of these compounds are therefore called *active devices*. The III–V compounds are the leading candidates for the manufacture of monolithic devices because both lasers and detectors, as well as other devices, can in principle be manufactured on a single substrate.

Circuits that are not monolithic are called *hybrid*. These are usually manufactured on substrates of glass, silicon, lithium niobate, and, sometimes, polymers. Lithium niobate may be used as a substrate because of its high electro-optic coefficient (Sect. 9.4), and silicon because of its usefulness as a detector, for example. Glass and acrylic plastic (polymethyl methacrylate or PMMA) are cheap and plentiful, and lasers can be manufactured in certain glasses such as glass doped with neodymium. In general, however, monolithic devices cannot be manufactured on these materials, so, for example, a GaAlAs laser might have to be epoxied to the substrate rather than integrated onto the surface.

Optical integrated circuits have several advantages over circuits manufactured from discrete electronic circuits. They display very fast operation, or, equivalently, very high electrical bandwidth. This is so not only because of the very high frequency of the light wave, but also because the components are smaller than electronic components and therefore display lower capacitance. Such low capacitance allows fast switching or high modulation frequency. In addition, also because of the size of the components, fairly low voltages can be used to create the high electric fields required to induce the electro-optic effect and other means to control the light in the waveguide. Similarly, a waveguide can confine light tightly over a considerable length, so nonlinear effects can be induced with comparatively low optical power. Because of the integration of a number of functions onto a single substrate, optical integrated circuits are stable and, once connected, free from alignment problems. Finally, integration offers the potential for mass production and low prices.

Figure 12.1 is a sketch of a simplified transmitter and receiver integrated onto two separate substrates and connected by single-mode optical fibers. The

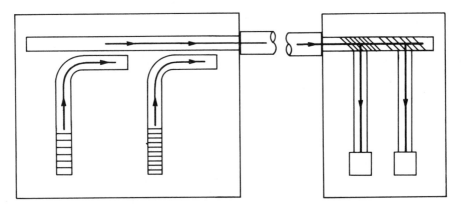

Fig. 12.1. Integrated-optical transmitter and receiver

transmitter consists, in this case, of two lasers that operate at two different wavelengths. They are connected by means of *directional couplers* to a rectangular-strip waveguide, which is, in turn, connected to a single-mode fiber. The receiver is also an optical integrated circuit. It consists, in this case, of another strip waveguide connected directly to the end of the fiber. Bragg reflectors are etched into this waveguide; each reflector selectively reflects one wavelength into a short waveguide connected to a detector. The current from the detector may then be used directly or reprocessed and retransmitted into another single mode fiber. Ideally, the reprocessing would be carried out on the same substrate as the detection.

We will not discuss the system, manufacturing, or electronic aspects of optical integrated circuits but will concentrate on the unique optical parts that are now being developed for integrated optics. These include lasers, couplers, detectors, mirrors, and modulators.

12.1.1 Channel or Strip Waveguides

In Sect. 10.2, we tacitly assumed that the waves in the waveguide were unconfined in the directions parallel to the substrate; that is, we treated first the case of a *slab waveguide*. Optical integrated circuits, however, use *channel* or *strip* waveguides such as that shown in Fig. 12.2. Such waveguides confine the light to a narrow strip a few micrometers across and perhaps 1 or 2 μm deep.

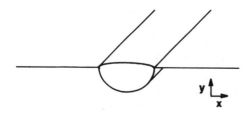

Fig. 12.2. Channel waveguide diffused or ion-exchanged into a substrate

The electric-field distribution in a channel waveguide is a function of both x and y, as opposed to a slab waveguide, where the modes are a function of one variable only. Often, however, we can assume that the modes are separable, that is, that the electric field may be expressed as a product of two functions, one purely a function of x and the other of y. When this is so, we have to consider pairs of equations like (10.5) and (10.6), and we have to use two mode numbers m and m' to describe each mode (see also Sect. 10.2.5). The mode of the waveguide is therefore approximately the product of the modes of two slab waveguides, one parallel to the substrate and one perpendicular. In most of what follows, we will assume that both m and m' are 0, so the waveguide can support only a single, lowest-order mode.

The waveguide shown in Fig. 12.2 is typical of a *diffused* waveguide such as might be manufactured by diffusing a narrow stripe of titanium into a lithium niobate substrate or a stripe of silver into a glass substrate at a high temperature. Similar waveguides may be made chemically by *ion exchange* in a solution or by *ion implantation* in a high vacuum. For example, the sodium ions in soda-lime glass (the stuff of which microscope slides are made) may be chemically exchanged for potassium ions to form a high-index layer or a channel waveguide. Sometimes a waveguide is covered with another layer, or, during the ion exchange, it may be drawn downward, into the substrate, by an electric field. Such a waveguide is called a *buried waveguide*.

12.1.2 Ridge Waveguide

Figure 12.3 shows a *ridge waveguide*. This is manufactured by depositing a narrow strip onto an existing slab waveguide.

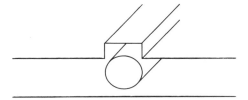

Fig. 12.3. Ridge waveguide, with a cylinder representing the mode in the waveguide

In Sect. 10.2 we found that the effective index of refraction of a ray inside a waveguide is $n_1 \cos \theta$, where θ is the angle between the ray and the axis of the waveguide and n_1 is the index of refraction of the waveguide material. In reality, the ray represents a mode of the waveguide; therefore, the angle θ is determined by the conditions for constructive interference inside the waveguide. That angle changes if the thickness of the waveguide changes. To calculate the change of angle, we assume that any mode is far from cutoff, so $\Phi \cong \pi/2$, and rewrite (10.6) in the form

$$(m + 1)\lambda = 2n_1 d \sin \theta , \qquad (12.1)$$

where m is the mode number of a particular mode and θ is again the complement of i. Let us assume that the waveguide thickness changes slightly by an amount Δd. We differentiate (12.1) to find that

$$\Delta m\lambda = 2n_1 \Delta d \sin \theta + 2n_1 d \cos \theta \Delta\theta . \qquad (12.2)$$

We assume that the mode number m is unchanged by the change in d; this is plausible only if Δd is small and the waveguide is not too near the cutoff of the mode. In this case, $\Delta m = 0$, and we find that

$$\Delta d/d = -\Delta\theta/\tan\theta \cong -\Delta\theta/\theta . \tag{12.3}$$

The corresponding change in n_e may be found by differentiating (10.10),

$$\Delta n_e = -n_1 \sin\theta\Delta\theta . \tag{12.4}$$

Using (12.3) to relate $\Delta\theta$ to Δd, we learn immediately that

$$\Delta n_e \cong [(m+1)\lambda^2/4n_1 d^3]\Delta d ; \tag{12.5}$$

that is, an increase of thickness increases the effective index of a waveguide.

Because it increases the effective index of the underlying waveguide, a ridge such as that shown in Fig. 12.3 therefore behaves as a channel waveguide and confines a mode just as would an increase of the actual index of refraction of the material. If the ridge is not too thick, only a single mode will be excited, and it will be confined mostly to the underlying slab waveguide, not the ridge itself, as indicated by the ellipse in Fig. 12.3. The material of the ridge may be the same as that of the slab waveguide, as tacitly assumed in the derivation of (12.5), or it may be another material. The area on top of the ridge or adjacent to it may be filled with a lower-index material if it is necessary to integrate other devices on top of the ridge waveguide.

Ridge waveguides are used mostly with III–V compounds or polymers. Substrates or layers of III–V compounds cannot easily be altered by diffusion or ion exchange, but ridge waveguides can be readily deposited onto appropriate substrates.

12.1.3 Branches

Figure 12.4 shows a *branch* in a waveguide. It consists of a single-mode waveguide, a short tapered section, and two output waveguides. Even if the waveguides are identical and the angle is very small, the loss at such a branch is significant. In Sect. 10.4.2, we calculated the loss between single-mode waveguides that differed slightly in thickness, for example. The transmittance of such

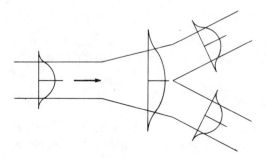

Fig. 12.4. Electric-field distributions at a branch

a joint depended on the overlap integral (10.53) of the two modes electric field distributions. In Figure 12.4, because of the tapered section between the input and output waveguides, the mode that impinges on the output waveguides expands relatively losslessly until its width w has roughly doubled. Further, the mode in either of the two output waveguides is offset by roughly a distance w from the mode of the input waveguide. For these reasons, well under 50% of the input power is coupled into each of the output waveguides, and loss can amount to several decibels even for angles of $2°$ or less. At larger angles, coupling loss is increased still farther because of the angular misalignment (10.57). More-efficient splitters can be made with *directional couplers* (Sect. 12.1.5), but branches are still important for dividing a beam into many channels for telephony, for example.

A star coupler (Sect. 10.4.3) with 2^n output fibers can be made simply by cascading n branches in sequence. Alternatively, the tapered section in Fig. 12.4 can be allowed to expand until its width is much larger than the width of a single waveguide before it is made to branch into many waveguides. Such couplers are useful for connecting the output of one fiber into many fibers or detectors, for example.

12.1.4 Distributed-Feedback Lasers

Ordinary semiconductor lasers are made by cleaving the laser crystal to create two parallel faces that serve as the laser mirrors. The wavelength range of the laser can be adjusted by adjusting the relative amounts of gallium and aluminum or arsenic and phosphorus. In general, however, these lasers oscillate in many spectral modes and are impossible to incorporate into a monolithic device. For communication, we wish to devise a method for manufacturing a laser that oscillates in a single mode, requires no cleaving, can be manufactured directly on the surface of a substrate, and connects directly to a waveguide on the surface of the substrate.

One solution to this problem is the *distributed-feedback laser*. This is a waveguide laser in that the active material has the form of a waveguide on the surface of the substrate. Instead of locating a mirror at each end of a short waveguide, we etch a grating into the upper surface of the waveguide, as shown in Fig. 12.1. The grating spacing d is chosen so that the Bragg condition is satisfied only for reflection in the reverse direction. If the effective index of refraction of the waveguide is n_e, this means that $2n_e d = \lambda$. The condition holds for rays incident from either direction. Thus, there is feedback in both directions, but it is distributed along the entire length of the laser. Such a laser can be manufactured directly on the surface of the optical integrated circuit, with no need for cleaving the faces and connecting the laser to the chip.

The index of refraction of gallium arsenide is 3.6; n_e is approximately the same. Therefore, the grating spacing d must be about 120 nm for the wavelength of 850 nm. Because of the high index of refraction of the material, the grating must be etched into, rather than deposited onto, the upper surface of the waveguide. This is so because most of the materials that may be deposited onto the surface have substantially lower index of refraction than the gallium arsenide itself and do not cause sufficient perturbation of the waveguide mode.

12.1.5 Couplers

Couplers serve a variety of purposes. They may connect a laser beam to a waveguide, a waveguide to a detector, or a waveguide to another waveguide. The device shown in Fig. 12.1 uses several kinds of couplers.

End couplers or *butt couplers* are those that require a polished or cleaved waveguide edge to couple power into or out of the waveguide. For these couplers to operate efficiently, the mode of the waveguide must be matched as closely as possible to the mode of a fiber or diode laser or to the spot size of a focused Gaussian beam. The latter problem is similar to mode matching of optical cavities (Sect. 8.3.3). Unfortunately, lasers and optical fibers are rotationally symmetric, whereas integrated-optical waveguides are rectangular. There is nearly always a mismatch between the modes of the fiber and any single-mode devices to which it is coupled. In addition, butt coupling is not ideal because it requires a carefully prepared waveguide edge. Nevertheless, there are situations where butt coupling is required, and the effort must be made to optimize the mode matching.

We have discussed prism and grating couplers in Chap. 10. These are generally, but not necessarily, used with slab waveguides, in which the light is unconfined in the direction parallel to the surface.

Another type of coupler is the *tapered output coupler*. This is shown in Fig. 12.5. It is most easily explained on the basis of ray optics. Because of the taper, rays that are reflected from the upper surface of the waveguide hit the lower surface with successively smaller angle of incidence. Eventually, the angle of incidence no longer exceeds the critical angle, and the ray escapes from the waveguide. The wave that escapes does not retain the mode structure that is characteristic of the waveguide, so the light cannot be coupled efficiently to another guiding structure. Nevertheless, the tapered output coupler is sometimes useful, because of its simplicity, for coupling to a detector.

Another type of coupler, but one that is not an input or output coupler, is the *directional coupler*. The directional coupler shown in the transmitter in Fig. 12.1

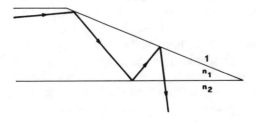

Fig. 12.5 Tapered output coupler

works in precisely the same way as the prism couplers discussed in Chap. 10 in that coupling is effected through the penetration of the evanescent wave into the adjacent waveguide. If the two waveguides are identical, mode matching is automatic. The main design problem then is to adjust the coupling length for the desired power transfer. The wave that is excited in the second waveguide propagates in a particular direction, to the right in the example shown. These directional couplers are very similar to those used in microwave electronics. Figure 12.6 shows two examples of directional couplers used as branches.

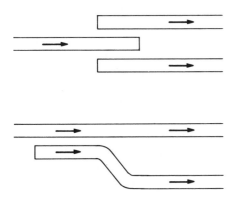

Fig. 12.6. Directional couplers used as branches

The receiver section of Fig. 12.1 shows another kind of directional coupler, one based on Bragg reflection. These may be tuned for any wavelength and therefore used to discriminate among signal channels that have different wavelengths. They are designed by applying the Bragg condition with $\theta = 45°$; this shows immediately that the spacing d' of the coupling grating must be equal to $\sqrt{2}d$, where d is the grating spacing of the corresponding distributed-feedback laser.

The reflectance of a Bragg reflector is maximum at the design wavelength. The spectral width of the reflectance is determined by the number of reflecting beams, which is the number of grooves in the grating. Because the waveguide is apt to be no more than 10 μm wide, the number of grooves must be less than, say, 50. Therefore, the finesse of the reflectance curve will be of the order of 50. This factor is defined precisely as the finesse of a Fabry-Perot interferometer and restricts the number of independent wavelengths that may be transmitted simultaneously through the fiber.

12.1.6 Modulators and Switches

At present, nearly all optical-communication systems function by modulating the intensity of the light beam that carries the signal. Although there are methods for modulating phase, polarization, or even optical frequency, these will not likely find widespread use in the very near future. Therefore, we restrict the discussion to intensity modulation. Certain of the modulation devices can also operate as switches, so we discuss them together.

The most straightforward way to modulate the intensity of a light wave is to modulate the source – in this case, by modulating the current to the laser. However, this may not always be the best way, and researchers are examining methods that may require less power consumption or allow higher modulation frequencies (or, in digital-electronics terminology, higher bit rates).

When the modulator is outside the laser, it is called an *external modulator*. External modulators of intensity fall into two classes: electro-optic and acousto-optic. Bulk modulators have been discussed in Sect. 9.4; here we briefly describe their optical-waveguide counterparts.

Acousto-optic modulation may be accomplished in either a rectangular or a slab waveguide by exciting a *surface acoustic wave* (SAW wave) and allowing some or all of the light inside the waveguide to be diffracted by the acoustic wave.

The acoustic wave is generated with the *interdigitating transducer* shown in Fig. 12.7. An electrical signal with the right frequency is applied to the conducting fingers of the transducer. Electrostriction causes the surface of the material to be alternately compressed and relaxed, thereby launching an acoustic wave in the direction perpendicular to the fingers. The wavelength of the acoustic wave is equal to the distance between adjacent fingers, and the frequency is determined by the velocity of sound in the medium.

This device is a modulator in the sense that the power of the transmitted beam is lowered in proportion to the net diffracted power. However, it may also be set up as a beam deflector by making use of the diffracted beam instead of, or in addition to, the transmitted beam. Usually, when the device is operated as a beam deflector, the ultrasonic wave will be set up as a thick grating, in the sense that certain holograms are thick (Sects. 7.1, 9.4). Such a beam deflector may be used to switch nearly 100% of a beam from one waveguide to another.

The electro-optic effect may be used instead of the acousto-optic effect to modulate or deflect the beam within a waveguide. One method is to induce a grating into a waveguide by using a set of interdigitating electrodes to change

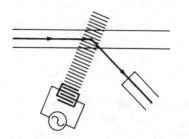

Fig. 12.7 Acousto-optic output coupler

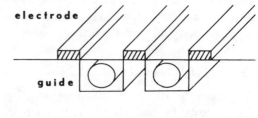

Fig. 12.8 Electro-optic switch between two waveguides

the index of refraction within the waveguide. Since the index of refraction depends on electric-field strength, these interdigitating electrodes induce a stationary phase grating in the material. This grating may be used for modulation or deflection, as in the case of the acousto-optic modulator.

Another electro-optic modulator or switch may be constructed by bringing two waveguides into near contact, as in the directional coupler. Metal electrodes are located between and on both sides of the waveguides, as shown in Fig. 12.8. Suppose that the waveguides are not identical, that is, that the phase-matching condition does not apply. Then virtually no power will be coupled from one waveguide to the other. To switch the beam, we apply voltages to the electrodes. The index of refraction is altered so that the phase-matching condition now applies, and the light is switched to the second waveguide. Such a system may be used as a switch or as a modulator.

Figure 12.9 shows two Mach-Zehnder interferometers (Sect. 6.2.2) that use integrated optics, rather than bulk optics. In the upper drawing, the two branches take the place of the mirrors in the bulk interferometer, and the incident beam is split equally between the two arms. The waveguides are manufactured on a material, such as lithium niobate, that has a high electro-optic coefficient. The lower drawing shows the same interferometer, but with directional couplers instead of branches.

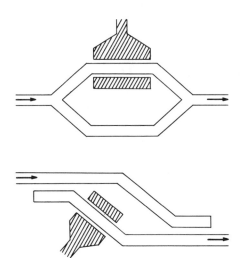

Fig. 12.9. Integrated Mach-Zehnder inter-ferometers. Shaded areas are electrodes

If the paths of the two arms are equal, all the incident light is transmitted through the interferometer. It is possible, however, to change the path in one of the arms by applying a voltage across the waveguide in that arm; the electrodes for doing so are shown shaded. The resulting electric field changes the index of refraction in one arm of the interferometer only. If the optical path in that arm is changed by exactly $\lambda/2$, the two beams that reach the output waveguide will be exactly out of phase and cancel each other. The transmittance of the interferometer is then 0. In fact, the transmittance can be set to any value between 0 and 1 by adjusting the voltage on the electrode. The Mach-Zehnder interferometer can therefore be used as either a switch or a modulator.

The electrical bandwidth of such a modulator is determined, in part, by the transit time of the light past the electrode; the rise or fall time of the switch can be no less than that transit time. To decrease the rise or fall time, the electrode may be made part of a microwave transmission line. Then, the electrical signal travels along the optical waveguide at roughly the speed of the light in the waveguide, and the speed of the modulator is no longer limited by the transit time. Such a modulator is called a *traveling-wave modulator*.

Thin-film modulators will probably find their greatest use in optical-communication systems, with bulk modulators usually reserved for other purposes; still, there will probably continue to be instances of bulk modulators also used in optical communications, and other instances of waveguide modulators used in other applications where low electrical power is required.

Most of the devices so far constructed have been built on substrates of glass, vitreous silica, and lithium niobate or similar substances. If they are to be used in monolithic circuits, they will have to be perfected on gallium arsenide substrates as well.

12.2 Planar Optical Devices

Although the integrated-optical circuits we have just discussed are planar devices, we shall reserve the term *planar optical devices* for optical systems other than communication systems that are constructed of thin-film waveguides on planar substrates. Often, these are planar-waveguide versions of ordinary optical devices such as interferometers and optical processors.

Figure 12.10 sketches a planar-optical spectrum analyzer that will serve as the archetype for our discussion of such devices. The device consists of a substrate on whose entire surface lies a uniform optical waveguide. In this case, a diode laser is butt-coupled to one edge of the waveguide; in other examples,

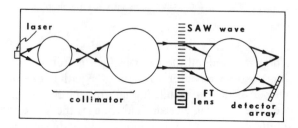

Fig. 12.10. Planar-optical processor

a He-Ne laser may be coupled to the waveguide with a lens. Two *planar lenses* collimate the beam from the laser; they are shown as circles.

The function of this device is to generate the Fourier transform or frequency spectrum of an electronic signal. Here the term "Fourier transform" means with respect to electrical frequency in hertz, rather than spatial frequency in lines per millimeter. The electronic signal is used to modulate the amplitude of a surface acoustic wave, which propagates through the collimated optical beam. We need not go into detail to realize that the electrical signal is thereby changed from a function of time into a function of length. Thus, measuring the spatial Fourier transform with the optical device is equivalent, with suitable scaling, to measuring the temporal Fourier transform.

The spectrum analyzer contains a third lens for determining the Fourier transform. An array of detectors is located in the secondary focal plane of this lens and is probably connected to a computer for further data analysis and processing. The Fourier transform is calculated optically instead of by digital computer because of its nearly instantaneous action at many frequencies simultaneously. This is known as *parallel processing.*

As before, we will not discuss the system or electronic aspects of planar optical devices, but will concentrate on the optical components, in this case, the lenses. There are several ways of making *waveguide lenses*, and each has its advantages and disadvantages. We shall conclude this chapter with a discussion of four kinds of waveguide lens.

12.2.1 Mode-Index Lenses

We learned in Sect. 12.1.2 that the effective index in a waveguide increases with increasing thickness according to (12.5). The change of effective index that results from a change of waveguide thickness depends on mode number m. A lens that makes use of the change of n_e with waveguide thickness will therefore suffer from a sort of chromatic aberration unless the waveguide propagates only one mode. We shall consequently assume a single-mode guide hereafter. If the index difference between the substrate and the waveguide is 1%, the thickness of a single-mode waveguide will be about five wavelengths. If the wavelength is about 1 μm, the change of effective index is about 0.002 per micrometer of thickness change. Suppose that we wish to make a "len" in the waveguide by increasing the thickness of the waveguide in the manner shown in Fig. 12.11. According to the "len" equation (2.17) the required radius of curvature R is

$$R = f'(\Delta n/n) , \tag{12.6}$$

where f' is the desired focal length. Again for the wavelength of 1 μm, we find that the radius is very roughly equal to $f'\Delta d/1000$, when Δd is measured in micrometers. This relationship shows that it is perfectly plausible to design a "len" or a lens that has a focal length in the millimeter range by making a waveguide-thickness change in the micrometer range.

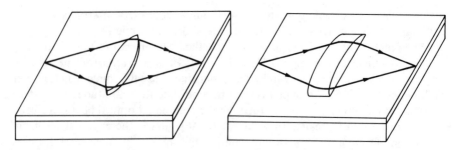

Fig. 12.11. Mode-index lens   Fig. 12.12. Graded-index lens

Lenses manufactured by depositing one or more index steps in an arc are called *mode-index lenses*. They are possibly the simplest type of waveguide lens to manufacture and analyze but have at least two disadvantages. The first we have already noted: They will have aberrations in multimode guides. In addition, such lenses are somewhat lossy because the edge of the waveguide is unpolished and therefore somewhat rough on the micrometer scale. This roughness causes scattering and results in attenuation and possibly loss of contrast as well. Nevertheless, because they are simple to make, these lenses find use in both planar optical devices and optical integrated circuits.

We noted in connection with distributed-feedback lasers (Sect. 12.1.1) that it is often preferable to etch into a planar waveguide than to deposit a layer on top of it. This is so because of the high index of refraction of many waveguide materials. Etching the waveguide, however, lowers the effective index of refraction. A positive lens made by etching is therefore concave, rather than convex. Such lenses may be more practical than convex lenses made by deposition because a larger index change may be effected by etching the waveguide than by depositing a layer on top of it.

12.2.2 Luneburg Lenses

Graded-index lenses are also possible. Figure 12.12 shows a simple example of such a lens. A tapered strip is deposited on top of the waveguide. The index of refraction of the strip must be greater than or equal to that of the waveguide, so there will be no total reflection at the boundary between the original waveguide and the strip. The upper surface of the strip therefore becomes the upper surface of the waveguide. As before, the effective index of refraction inside the waveguide increases with increasing optical thickness. The waveguide therefore displays a higher index of refraction at the center of the strip than elsewhere; the index decreases gradually away from the center of the strip. The strip therefore can focus light in much the same manner as a graded-index waveguide. If the index profile of the strip is chosen properly, the strip becomes a *graded-index lens*. It is a kind of mode-index lens; such lenses can be made by sputtering or evaporating material onto the substrate and using a mask to control the amount of material that is deposited at each point.

The lens may also be deposited with radial symmetry. Such a lens is known as a *Luneburg lens*. Its operation is similar to the one-dimensional lens, except that ray paths must be analyzed in two dimensions instead of one. Fermat's principle is used to calculate the precise index profile for different conjugates; the calculation is too complicated to discuss here, but we can still draw one important conclusion about these lenses.

Figure 12.13 shows a top view of a Luneburg lens. Suppose that the lens images one point A without aberration onto another point B. The lens has radial symmetry; therefore there is no preferred optical axis. Any other point A' that lies on the circle that passes through A and whose center is C is indistinguishable from A. Its image B' is free of aberrations and indistinguishable from B. Just as A and A' lie on a circle, B and B' also lie on a circle. Therefore, we conclude that the Luneburg lens images concentric circles onto one another. If the lens is designed to be without aberration for conjugates A and B, it will be without aberration for all equivalent points A' and B' that lie on the two circles.

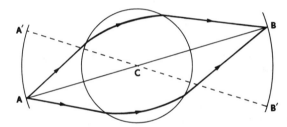

Fig. 12.13. Luneburg lens

The Luneburg lens is an example of a *perfect imaging system*. Except that the object and image lie on circles and not on lines, the Luneburg lens is completely free from aberration when used with the proper conjugates.

12.2.3 Geodesic Lenses

Luneburg lenses are special cases of mode-index lenses. They therefore operate best in single-mode guides or, at least, in waveguides in which only one mode propagates. Another kind of lens is known as a *geodesic lens* because the rays follow a *geodesic*, that is, the shortest distance between two points on a surface. Figure 12.14 shows top and side views of a geodesic lens in a waveguide. The lens consists of a spherical depression in the substrate. The waveguide follows the depression but unlike the previous cases has uniform thickness throughout. The ray that follows a diameter of the depression travels a greater optical path than the ray that grazes the lens; other rays travel intermediate distances. This is precisely the case with an ordinary positive lens: The lens is optically thickest in the center; as we saw in Sect. 5.5, the focusing action is equivalent to the rays'

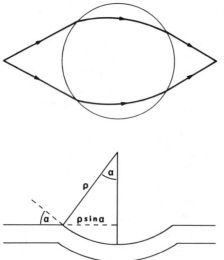

Fig. 12.14. Geodesic lens

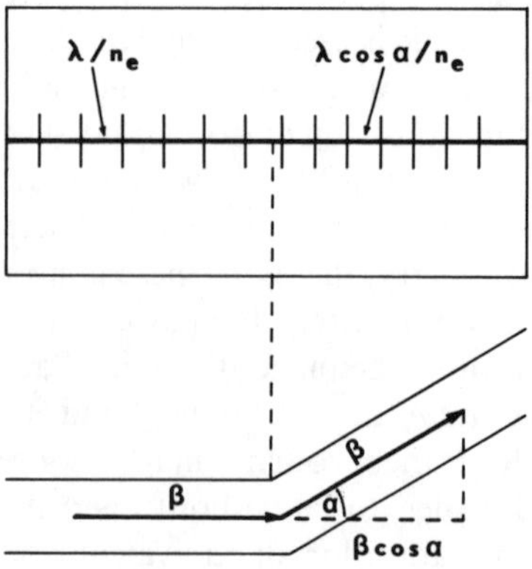

Fig. 12.15. Propagation at a bend in the waveguide

following equal optical paths. Because the lens is optically thickest in the center, the peripheral rays must travel greater geometrical (as opposed to optical) paths.

The same principle is applied to the design of geodesic lenses. Ray trajectories are calculated by a procedure that is equivalent to minimizing the optical-path length between conjugate points. Like Luneburg lenses, geodesic lenses may be designed so that they image concentric circles without aberration. The shape of the depression must in this case be aspheric.

We can derive some of the properties of a geodesic lens by looking at it from another point of view. Consider first an optical waveguide that has a sharp bend, as shown in Fig. 12.15. The angle of the bend is α. A wave propagates in the horizontal section of the waveguide in the direction perpendicular to the bend. When the wave reaches the bend, it will continue to propagate in the sloped section of the waveguide, except possibly for some bending loss. If the thickness and index of refraction of the waveguide remain constant beyond the bend, the wave vector will remain the same, except for a change of direction. Thus, the wavelength λ/n_e will be the same in both sections of waveguide.

For planar-optical applications, we are interested in the projection β_x of the wave vector β in the horizontal plane. In the horizontal section, the projection is equal to the wave vector itself; in the sloped section, the projection is given by

$$\beta_x = \beta \cos \alpha . \tag{12.7}$$

Thus, when viewed from the top, the wave appears to slow by the factor $\cos \alpha$. That is, for small angles of incidence, the waveguide beyond the fold appears to have an effective index of refraction

$$n_e' = n_e/\cos\alpha .$$ (12.8)

(This result may also be derived by considering the wavefronts in the two sections. If λ/n_e is the effective wavelength in the horizontal section, then the wavelength in the sloped section appears from above to be reduced to $(\lambda/n_e)\cos\alpha$.)

We may derive the paraxial focal length of a spherical geodesic lens in the following way. The radius of curvature of the depression is ρ. The radius of the depression, as seen from the top, is $\rho\sin\alpha$, where α is the angle shown in Fig. 12.14. A ray that intersects the geodesic lens along a diameter experiences a bend whose angle is also α. Thus, in paraxial approximation, the geodesic lens may be regarded as a thick, biconvex lens whose radius of curvature is $\rho\sin\alpha$, whose surfaces are separated by twice that length, and whose relative index of refraction is $n_e'/n_e = 1/\cos\alpha$.

We may calculate the focal length of such a lens by using (2.32) for a thick lens with one element whose index of refraction is n. For our case, both radii of curvature are equal to R, and the separation between the surfaces is equal to $2R$; therefore,

$$f' = R/2[1 - (1/n)] .$$ (12.9)

This result is, incidentally, the same as that for a spherical lens with index of refraction n. For our case, n is replaced by $1/\cos\alpha$ and R is equal to $\rho\sin\alpha$. The focal length of a spherical geodesic lens is therefore

$$f' = \rho\sin\alpha/2(1 - \cos\alpha) .$$ (12.10)

This is the result derived by more-conventional methods.

In practice, the lens will suffer from spherical aberration if it is used at a high aperture. In addition, if α is large, there is significant loss at the edge of the lens. Real geodesic lenses must be tapered to reduce this loss at the edges; the profile of the depression is therefore aspheric.

Geodesic lenses may be made by depositing a bump on the surface of the substrate and depositing the waveguide over the bump; because the optical path along a diameter of the bump is greater than that near the edge, the bump also gives rise to a positive lens. A more easily controllable way to manufacture a geodesic lens, however, is to machine a depression using a computer-controlled lathe with a diamond-point cutting tool. This process is sometimes called *diamond turning* and may be used to machine the depression to precisely any shape.

Another advantage of geodesic lenses is that their focal length does not depend on mode number because n_e'/n_e does not depend on mode number. Therefore, they may be used with multimode waveguides with no aberration resulting from the range of mode numbers.

12.2.4 Gratings

Gratings may be etched or deposited onto a planar waveguide for any of several purposes. A *diffraction lens* is equivalent to a one-dimensional Fresnel zone plate superimposed onto or etched into the surface of the waveguide. Such a lens focuses light as described by the formalism we derived in Sect. 5.5.3. The half-period zones, however, are not alternately clear and opaque, as in a conventional zone plate. Rather, alternate zones cause a phase shift of π of the incident wave. This phase shift exactly compensates the path difference between rays from alternate zones, so all the zones contribute to the intensity of the image. In the conventional zone plate, only half the zones contribute; as a result, a *phase zone plate* in principle yields an image that is 4 times as intense as a conventional zone plate.

Figure 12.16a shows a thin diffraction lens in a highly schematic top view. The zones may be created, for example, by etching the waveguide layer to lower the effective index within each zone. The wavelength inside the waveguide is equal to the vacuum wavelength divided by the effective index, so the effective index in the waveguide must be known before the parameters of the lens can be calculated accurately.

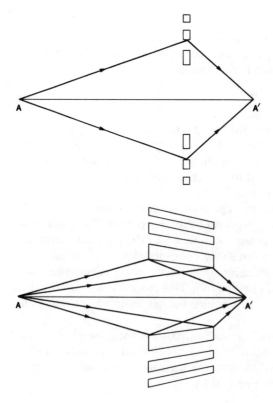

Fig. 12.16a, b. Diffraction lenses. (a) Thin lens. (b) Thick lens

For higher efficiency, a diffraction lens may be made thick, like a thick hologram. Such a lens is sketched in Fig. 12.16b. That figure shows two rays that originate from one conjugate point, reflect off one groove, and pass through the other conjugate point. Because those rays (and all others) must have the same optical path, we conclude that the grooves are segments of ellipses, though often line segments suffice. The efficiency of a diffraction lens will approach 100% provided that the parameter

$$Q = 2\pi\lambda t/nd^2 \tag{12.11}$$

exceeds 10 (Problem 7.3).

Waveguide lenses other than diffraction lenses depend on changing some property of the waveguide such as the effective index of refraction. Geodesic and Luneburg lenses are comparatively hard to manufacture and do not use conventional lithographic techniques such as those borrowed from the integrated-circuit industry. Mode-index lenses are comparatively lossy owing to scattering from the edges of the lens. For these reasons, diffraction lenses may prove to be the most useful in the long run.

Gratings may also be used to couple light out of the plane of the waveguide; if the grooves are suitably curved, the light may be made to focus in the air above the waveguide, for example. Such gratings have application in compact-disk reading heads and in communication from one electronic integrated circuit to another, as in a computer.

Consider the waveguide grating of Fig. 12.17. The grating could be etched or superimposed onto the upper surface of the waveguide, or it could be a volume grating that is created throughout the waveguide layer by suitable periodic doping of the material. In either case, we begin with (10.12), which we derived in connection with the grating coupler. We look first at a *substrate mode*, or a wave diffracted into the substrate. Equation (10.12) becomes

$$m\lambda = n_s d \sin i_s - n_g d \sin i_g , \tag{12.12}$$

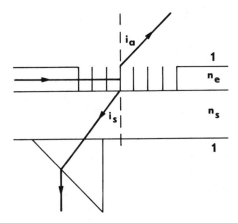

Fig. 12.17. Grating in a waveguide, showing air mode, substrate mode and coupling prism

where the subscript s means substrate and g means the waveguide layer itself. The symbol m here means the order of diffraction; we assume that the waveguide supports only a single mode. If we use (10.11) for the effective index of refraction, we find that

$$m\lambda = n_s d \sin i_s - n_e d \ . \tag{12.13}$$

Comparison of (12.12) and (12.13) shows that the wave inside the waveguide may be described by either of two equivalent pictures: as a wave that propagates at angle i_g to the surface of the waveguide and in a material whose index of refraction is n_g, or as a wave that propagates parallel to the surface of the waveguide but in a material whose index of refraction is equal to the effective index n_e.

Since n_e in (12.13) is always less than n_s, m must always be less than 0; that is, there can be only negative diffraction orders. Therefore, for the lowest order of diffraction ($m = -1$),

$$n_s \sin i_s = n_e - \lambda/d \ . \tag{12.14}$$

Similarly, in the air,

$$\sin i_a = n_e - \lambda/d \ . \tag{12.15}$$

where the subscript a stands for air and $n_a = 1$. If, now, we solve (12.15) for n_e and substitute into (12.14), we find that

$$\sin i_s = (\sin i_a)/n_s \ . \tag{12.16}$$

Because n_s is always greater than 1, $\sin i_s$ is less than 1 whenever $\sin i_a$ is less than 1, though the reverse is not necessarily true. Therefore, we conclude that whenever a wave is diffracted into the air, a wave is also diffracted into the substrate.

For efficiency, we may prefer to limit the radiation pattern to one mode only. We could, for example, suppress the wave in the air, or the air mode, by requiring that $\sin i_a$ in (12.15) be less than -1. According to (12.16), then, $i_s < 0$; the diffracted wave travels in the reverse direction, that is, downward and to the left in Fig. 12.17. Depending on the angle of diffraction, that wave may be trapped in the substrate by total internal reflection, or it may emerge from the substrate nearly parallel to the surface; this is especially possible in a high-index material such as gallium arsenide. It may therefore be necessary to contact a prism to the underside of the substrate in order to couple the light out.

For this and other reasons, we might prefer an air mode to a substrate mode. A grating that diffracts primarily an air mode can be manufactured either by blazing the grating (Sect. 6.1.1), or by depositing a thick grating that acts as a Bragg reflector. Blazing a grating is not easy but can be done by manufacturing a grating and *ion-milling* it with an ion beam that is directed onto the surface at an angle and erodes the grating selectively at an angle. In either case, a single order of diffraction can be selected in much the same manner as a blazed grating or a thick hologram selects a single order.

Waveguide gratings are, however, very different from bulk gratings. As Fig. 12.17 shows, the radiation is incident on the grating at glancing incidence. Consequently, the amplitude of the electric field decreases exponentially with distance from the left edge of the grating; the proof is the same as the proof that led to (8.6). If we try to focus the light diffracted by the grating to a point, we will have to deal not with a uniform beam but rather with a beam that decays exponentially across the entrance pupil of the lens.

Consider a one-dimensional grating that extends in the x direction from 0 to ∞. The diffracted amplitude varies with x as $\exp(-x/w)$, where w is the characteristic decay length and depends in a complicated way on the depth of the grooves. Imaging the diffracted wave with a lens is equivalent to imaging a collimated beam with the same amplitude dependence, so we use (7.31) with $g(x) = \exp(-x/w)$ when $x > 0$, and $g(x) = 0$ otherwise:

$$E(f_x) = \int_0^\infty e^{-x/w} e^{-2\pi i f_x x} \, dx \ . \tag{12.17}$$

Straightforward integration shows that the intensity is described by the Lorentzian function

$$I(f_x) = w^2/[1 + (2\pi w \sin \theta/\lambda)^2 \ . \tag{12.18}$$

A rectangular aperture whose width is $3w$ gives about the same half-width as the Lorentzian function. Problem 12.12, however, shows that a Lorentzian function with the same peak intensity displays substantially higher intensity a few resolution limits from the center of the image. This is one reason that it is hard to manufacture a waveguide grating with diffraction-limited resolution.

12.2.5 Surface-Emitting Lasers

Ordinary junction lasers (Sect. 8.4.8) are sometimes termed *edge emitters* because they emit light from the edge of a facet; that is, the active medium is a layer of semiconductor, the oscillation takes place parallel to the plane of the layer, and light is emitted from the edge of the layer. Distributed-feedback lasers (Sect. 12.1.1) are also, in a sense, edge emitters.

A *surface-emitting laser* may be fabricated by combining the methods of integrated optics and multilayer mirrors (Sects. 6.4.2 and 6.4.3). That is, a surface-emitting laser is similar in concept to an MDM interference filter, except that the dielectric layer is replaced by an active medium, most commonly gallium arsenide or a related alloy such as GaInAs, GaAlAs, and GaInP, depending on the required wavelength.

To make the laser, a quarter-wave stack is grown onto a substrate, not by vacuum evaporation but by chemical-vapor deposition in a high vacuum. The substrate may be GaAs, and the layers in the stack are alloys of III–V compounds. The active medium, a 1–2 μm layer, is grown on top of the quarter-wave stack and followed by another stack. Often, there is also an inactive layer, or buffer layer, between the active layer and each stack. The stacks serve as the laser mirrors, and the laser oscillates in the direction perpendicular to the plane of the layers.

To attain stable emission or single-mode operation, the laser has to be restricted in the direction transverse to the direction of propagation. To this end, the material is damaged by proton bombardment everywhere except in the regions where laser emission is required. A conducting electrode may then be deposited over each of these regions, which are a few micrometers in diameter. It is possible in this way to fabricate an array of many thousands of lasers per square centimeter. With external optics, these lasers can be operated in phase with one another, so they form the elements in a single, more-powerful laser. Alternatively, the individual lasers on the substrate can be left isolated from one another and excited separately in order to excite individual fibers in a bundle, for use in a telephone or community television system, or as a matrix element in an optical computer.

Problems

12.1   Gallium arsenide has an index of refraction of about 3.6. Suppose that we make a channel waveguide by increasing the index of refraction of a narrow strip by about 1%. The waveguide thus formed is 2 μm deep. If the index of refraction inside the waveguide is constant, that is, if the waveguide displays an index step, what is the greatest width that will allow single-mode operation at 1.3 μm?

12.2   On top of a slab waveguide 2 μm thick is a ridge waveguide an additional 2 μm thick; that is, the ridge waveguide is 4 μm thick in all. The index of refraction of the substrate is 1.5, and that of the slab is 1.503. If the ridge has the same index of refraction as the slab, can it support only one mode, or more, at the wavelength of 0.85 nm?

12.3   A certain slab waveguide is 4 μm thick and has an index of refraction of 1.5 and a numerical aperture of 0.1. This waveguide contains a step that is 4 μm high and lies perpendicular to the direction of propagation of the lowest-order mode in the waveguide. Use the effective index of refraction to estimate the reflectance of the step if the wavelength of the light is 0.85 μm.

12.4   Consider a waveguide branch in one dimension. Assume for convenience that the mode-field width of the tapered section is w, that in either output waveguide is $w/2$, and that the two modes are displaced from one another by w (Fig. 12.4). Ignore the angle between the input and output waveguides. Set up an equation similar to (10.53), and show that the fraction of the input power coupled into either output waveguide is $(4/5)e^{-8/5}$. What is the overall efficiency of the branch? The missing power is partly reflected back into the input waveguide and partly scattered out of the waveguides entirely. [Hints: Use the result of Problem 10.11. Change the variables of integration to $u = x/w$; then manipulate. $\int_0^\infty e^{-u^2}\, du = \sqrt{\pi}/2$.]

12.5 (a) A slab waveguide with a numerical aperture of 0.2 is 6λ thick and ends in a gradual taper. At what thickness does the lowest-order bound mode become radiative? Ignore phase change on reflection.

12.6 A waveguide designed for 0.85 μm is 8 μm wide. It contains a Bragg reflector oriented at $45°$ to the waveguide axis; this reflector couples light into an identical waveguide oriented at right angles to the original waveguide (Fig. 12.1). (a) Express the Bragg condition (9.32) in terms of the period d of the grating and calculate d. Assume that the wavelength inside the waveguide is close to λ/n, where $n = 1.5$. (b) Roughly how many lines or grooves actually contribute to the light that is coupled into the second grating? (c) Assume that the answer to (b) is equal to the finesse, or the effective number of interfering beams, of the coupler. Estimate the spectral width of the reflected wave, assuming that the light in the waveguide is spectrally broad. (d) If the light has a spectral width of 100 nm, how many distinct channels can be separated by a set of such gratings? (e) How could you increase the number of channels? You may change any waveguide parameters, including the substrate, but must use only single-mode waveguides.

12.7 A mode-index "len" is made on a substrate whose index of refraction at 0.85 μm is 3.6. A slab waveguide 1 μm thick is formed on that substrate and then selectively etched to form a "len". The thickness of the waveguide beyond the "len" is 0.5 μm. The index of refraction of the film is 0.025 higher than that of the substrate, so only a single mode propagates in the waveguide. (a) Calculate the radius of curvature required to give the "len" a focal length of 5 mm. (b) Perform the same calculation for a polymer waveguide whose index of refraction is 1.5. Assume that a slab waveguide 8 μm thick is reduced to 2 μm to form the "len".

12.8 Show that the effective F number of a Luneburg lens may be as great as 0.5.

12.9 Consider a diffraction lens that has a total of m Fresnel zones. Use (5.68) along with the binomial expansion $(1 - 1/m)^{1/2} = 1 - (1/2m)$ to show that the width of the outermost Fresnel zone is $\Delta s_m = s_1/2\sqrt{m}$.

12.10 A diffraction lens is used at unit magnification with $l' = -l = 5$ mm at the wavelength of 1 μm. The effective index of refraction in the waveguide is 1.5. The lens has approximately 25 000 Fresnel zones in all and is 1 mm thick. Show that the lens is nearly 100% efficient. [Hint: Use the result of Problem 12.9. For what portion of the zone plate is the parameter Q less than 10?]

12.11   Consider a slab waveguide made of a material into which a volume hologram can be exposed. The developed fringes of the hologram lie at 45° to the surface of the waveguide and extend through the slab to the substrate. The Bragg condition is satisfied at whatever angle of diffraction corresponds to specular reflection from one of the fringes or Bragg planes. (a) Calculate the angle of diffraction of the air mode and show that it is not equal to 0. [Hint: The ray that is bound to the waveguide behaves as if it travels parallel to the surface of the waveguide and experiences an index of refraction of n_e. After reflection from a Bragg plane, the ray does not undergo multiple reflections and therefore experiences an index of refraction of n_1, so first write an equation analogous to (2.29). This problem involves only geometry, Snell's law, and your equation.] (b) Find the grating period d that corresponds to this value of i_a. If we wanted any other period or angle of diffraction, we would have to change the angle of the Bragg planes accordingly.

12.12   Consider a waveguide grating that displays a decay length w (Sect. 12.2.4). After a distance $3w$, the amplitude inside the waveguide decays to about 5% of the incident amplitude. Compare the far-field diffraction pattern of this grating to that of a slit whose width is $3w$. That is, make a rough sketch of the two diffraction patterns, and show that the grating displays higher intensity a few resolution limits from the center of the diffraction pattern. Assume that both patterns have the same intensity at the origin. [Note: It is necessary to plot only the first few zeros and secondary maxima of the diffraction pattern of the slit.]

Suggested Reading Material

Chapter 1

Sobel, M. I.: *Light* (University of Chicago Press, Chicago 1987)

Chapter 2

Hecht, E.: *Optics*, 2nd edn. (Addison-Wesley, Reading, Mass. 1987) Chaps. 4–6

Iizuka, K.: *Engineering Optics*, 2nd edn., Springer Ser. Opt. Sci., Vol. 35 (Springer, Berlin, Heidelberg 1987)

Jenkins, F. A., White, H. E.: *Fundamentals of Optics*, 4th edn. (McGraw-Hill, New York 1976) Chaps. 1–9

Longhurst, R. S.: *Geometrical and Physical Optics*, 3rd edn. (Longmans, London 1973) Chaps. 1, 2

Martin, L. C.: *Technical Optics*, Vol. 1 (Pitman, London 1960) Chaps. 1–4, 8

Smith, W. J.: *Modern Optical Engineering* (McGraw-Hill, New York 1966)

Stavroudis, O. N.: *Modular Optical Design*, Springer Ser. Opt. Sci., Vol. 28 (Springer, Berlin, Heidelberg 1982)

Chapter 3

Iizuka, K.: *Engineering Optics*, 2nd edn., Springer Ser. Opt. Sci., Vol. 35 (Springer, Berlin, Heidelberg 1987)

Kingslake, R. (ed.): *Applied Optics and Optical Engineering* (Academic, New York 1965) Vols. 1–5

Kingslake, R.: *Optical System Design* (Academic, Orlando, Fla. 1983)

Longhurst, R. S.: *Geometrical and Physical Optics*, 3rd edn. (Longmans, London 1973) Chaps. 3, 4, 14–16

Malacara, D. (ed.): *Physical Optics and Light Measurement*, Methods of Exp. Physics, Vol. 26 (Academic, Boston 1988)

Martin, L. C.: *Technical Optics* (Pitman, London 1966) Vol. 1, Chap. 5; Vol. 2

O'Shea, D. C.: *Elements of Modern Optical Design* (Wiley-Interscience, New York 1985)

Pawley, J. B. (ed.): *Handbook of Biological Confocal Microscopy*, rev. edn. (Plenum, New York 1990)

Shannon, R. R., Wyant, J. C. (eds.): *Applied Optics and Optical Engineering*, Vol. 7 (Academic, New York 1979)

Smith, W. J.: *Modern Optical Engineering* (McGraw-Hill, New York 1966)

Wilson, T. (ed.): *Canfocal Microscopy* (Academic, London 1990)

Wilson, T., Sheppard, C.: *Theory and Practice of Scanning Optical Microscopy* (Academic, Boston 1984)

Chapter 4

Dereniak, E. L., Crowe, D. G.: *Optical Radiation Detectors* (Wiley, New York 1984)

Electro-Optics Handbook (RCA Corporation, Harrison, N.J. 1968)

Garbuny, M.: *Optical Physics* (Academic, New York 1965)

Keyes, R. J. (ed.): *Optical and Infrared Detectors*, 2nd edn., Topics Appl. Phys., Vol. 19 (Springer, Berlin, Heidelberg 1980)

Kingston, R. H.: *Detection of Optical and Infrared Radiation*, Springer Ser. Opt. Sci., Vol. 10 (Springer, Berlin, Heidelberg 1978)

Kressel, H. (ed.): *Semiconductor Devices*, 2nd edn., Topics Appl. Phys., Vol. 39 (Springer, Berlin, Heidelberg 1982) Chaps. 2, 3, 10, 11

Kruse, P. W., McGlauchlin, L. D., McQuistan, R. B.: *Elements of Infrared Technology* (Wiley, New York 1962) Chaps. 2, 6–10

Malacara, D. (ed.): *Geometrical and Instrumental Optics*, Methods of Exp. Physics. Vol. 25 (Academic, Boston 1988)

Mauro, J. A. (ed.): *Optical Engineering Handbook* (General Electric Company, Scranton, Pa. 1963)

Pankove, J. I. (ed.): *Display Devices*, Topics Appl. Phys., Vol. 40 (Springer, Berlin, Heidelberg 1980) Chap. 2

Walsh, J. W. T.: *Photometry*, 3rd edn. (Dover, New York 1958)

Wolfe, W. L.: Radiometry, in *Applied Optics and Optical Engineering*, Vol. 8, ed. by Shannon, R. R., Wyant, J. C. (Academic, New York 1980)

Wolfe, W. L., Zissis, G. J.: *The Infrared Handbook* (Environmental Research Institute of Michigan, Ann Arbor, Mich. 1978)

Chapter 5

Babić, V. M., Kirpičnikova, N. Y.: *The Boundary Layer Method in Diffraction Problems*, Springer Ser. Electrophys., Vol. 3 (Springer, Berlin, Heidelberg 1979)

Dainty, J. C. (ed.): *Laser Speckle and Related Phenomena*, 2nd edn., Topics Appl. Phys., Vol. 9 (Springer, Berlin, Heidelberg 1984)

Ditchburn, R. W.: *Light*, 2nd edn. (Wiley-Interscience, New York 1983) Chaps. 1–6

Frieden, B. R.: *Probability, Statistical Optics, and Data Testing*, Springer Ser. Inf. Sci., Vol. 10 (Springer, Berlin, Heidelberg 1983)

Goodman, J. W.: *Statistical Optics* (Wiley-Interscience, New York 1985)

Hecht, E.: *Optics*, 2nd edn. (Addison-Wesley, Reading, Mass. 1987) Chaps. 2, 3, 7, 9, 10

Jenkins, F. A., White, H. E.: *Fundamentals of Optics*, 4th edn. (McGraw-Hill, New York 1976) Chaps. 11–18

Klein, M. V.: *Optics* (Wiley, New York 1970) Chaps. 7–11

Möller, K. D.: *Optics* (University Science Books, Mill Valley, Calif. 1988)

Reynolds, G. O., DeVelis, J. B., Parrent, G. B., Jr., Thompson, B. J.: *The New Physical Optics Notebook* (SPIE Optical Engineering Press, Bellingham, Wash. 1989)

Chapter 6

Born, M., Wolf, E.: *Principles of Optics*, 6th edn. (Pergamon, New York 1980) Chaps. 7, 8

Candler, C.: *Modern Interferometers* (Hilger and Watts, Glasgow 1951)

Ditchburn, R. W.: *Light*, 2nd edn. (Wiley-Interscience, New York 1966) Chaps. 5, 6, 9

Françon, M.: *Optical Interferometry* (Academic, New York 1966)

Petit, R. (ed.): *Electromagnetic Theory of Gratings*, Topics Curr. Phys., Vol. 22 (Springer, Berlin, Heidelberg 1980)

Sawyer, R. A.: *Experimental Spectroscopy*, 3rd edn. (Dover, New York 1963)

Tolansky, S.: *Introduction to Interferometry*, 2nd edn. (Wiley, New York 1973)

Chapter 7

Ballard, D. H., Brown, C. M.: *Computer Vision* (Prentice-Hall, Englewood Cliffs, N.J. 1982)

Bjelkhagen, H. I.: *Silver-Halide Recording Materials for Holography*, Springer Ser. Opt. Sci., Vol. 66 (Springer, Berlin, Heidelberg 1993)

Casasent, D. (ed.): *Optical Data Processing, Applications*, Topics Appl. Phys., Vol. 23 (Springer, Berlin, Heidelberg 1978)

Castleman, K. R.: *Digital Image Processing* (Prentice-Hall, Englewood Cliffs, N.J. 1979)

Cathey, W. T.: *Optical Information Processing and Holography* (Wiley, New York 1974)

Caulfield, H. J., Lu, S.: *The Applications of Holography* (Wiley-Interscience, New York 1970)

Das, P. K.: *Optical Signal Processing: Fundamentals* (Springer, Berlin, Heidelberg 1990)

DeVelis, J.B., Reynolds, G. O.: *Theory and Applications of Holography* (Addison-Wesley, Reading, Mass. 1967)

Duffieux, P. M.: *The Fourier Transform and Its Applications to Optics*, 2nd edn. (Wiley, New York 1983)

Ekstrom, M. P.: *Digital Image Processing* (Academic, Boston 1984)

Gaskill, J. D.: *Linear Systems, Fourier Transforms, and Optics* (Wiley, New York 1978)

Gonzalez, R. C.: *Digital Image Processing*, 2nd edn. (Addison-Wesley, Reading, Mass. 1987)

Goodman, J. W.: *Introduction to Fourier Optics* (McGraw-Hill, New York 1968)

Horner, J. L.: *Optical Signal Processing* (Academic, Boston 1987)

Huang, T. S. (ed.): *Picture Processing and Digital Filtering*, 2nd edn., Topics Appl. Phys., Vol. 6 (Springer, Berlin, Heidelberg 1979)

Inoue, S.: *Video Microscopy* (Plenum, New York 1986)

Lee, S. H. (ed.): *Optical Information Processing*, Topics Appl. Phys. Vol. 48 (Springer, Berlin, Heidelberg 1981)

Mechels, S. E., Young, M.: Video microscopy with submicrometer resolution; Appl. Opt. **30**, 2202–2211 (1991)

Nussbaumer, H. J.: *Fast Fourier Transform and Convolution Algorithms*, 2nd edn., Springer Ser. Inf. Sci., Vol. 2 (Springer, Berlin, Heidelberg 1982)

Ostrovsky, Yu. I., Butusov, M. M., Ostrovskaya, G. V.: *Interferometry by Holography*, Springer Ser. Opt. Phys., Vol. 20 (Springer, Berlin, Heidelberg 1980)

Rosenfeld, A., Kac, A. C.: *Digital Picture Processing*, 2nd edn., Vols. 1, 2 (Academic, New York 1982)

Smith, H. M. (ed.): *Holographic Recording Materials*, Topics Appl. Phys., Vol. 20 (Springer, Berlin, Heidelberg 1977)

Smith, H. M.: *Principles of Holography*, 2nd edn. (Wiley-Interscience, New York 1975)

Stark, H. (ed.): *Applications of Optical Fourier Transforms* (Academic, New York 1982)

Steward, E. G.: *Fourier Optics: An Introduction* (Ellis Horwood, Chichester, U.K. 1983)

Thompson, B. J.: Principles and Applications of Holography, in *Applied Optics and Optical Engineering*, Vol. 6, ed. by R. Kingslake, B. J. Thompson (Academic, New York 1980)

Wilson, T. and Sheppard, C.: *Theory and Practice of Scanning Optical Microscopy* (Academic, New York 1984)

Yaroslavsky, L. P.: *Digital Picture Processing. An Introduction*, Springer Ser. Inf. Sci., Vol. 9 (Springer, Berlin, Heidelberg 1985)

Yu, F. T. S.: *Optical Information Processing* (Wiley, New York 1983)

Chapter 8

Arecchi, F. T., Schulz-Dubois, E. O.: *Lasers Handbook* (North-Holland, Amsterdam, and American Elsevier, New York 1972)

Charschan, S. S. (ed.): *Lasers in Industry* (Van Nostrand Reinhold, New York 1972)

Das, P. K.: *Lasers and Optical Engineering* (Springer, Berlin, Heidelberg 1991)

Kaiser, W.: *Ultrashort Laser Tubes and Applications*, 2nd edn., Topics Appl. Phys., Vol. 60 (Springer, Berlin, Heidelberg 1993)

Kingslake, R., Thompson, B. J. (eds.): *Applied Optics and Optical Engineering*, Vol. 6 (Academic, New York 1980) Chaps. 1–3

Koechner, W.: *Solid-State Laser Engineering*, 3rd edn., Springer Ser. Opt. Sci., Vol. 1 (Springer, Berlin, Heidelberg 1991)

Kogelnik, H.: Modes in Optical Resonators, in *Lasers*, Vol. 1, ed. by A. K. Levine (Dekker, New York 1966)

Lengyel, B. A.: *Lasers*, 2nd edn. (Wiley-Interscience, New York 1971)

Mollenauer, L. F., White, J. C., Pollock, C. R.: *Tunable Lasers*, 2nd edn., Topics Appl. Phys., Vol. 59 (Springer, Berlin, Heidelberg 1992)

Rhodes, Ch. K. (ed.): *Excimer Lasers*, 2nd edn., Topics Appl. Phys., Vol. 30 (Springer, Berlin, Heidelberg 1984)

Schäfer, F. P. (ed.): *Dye Lasers*, 3rd edn., Topics Appl. Phys., Vol. 1 (Springer, Berlin, Heidelberg 1990)

Shapiro, S. L. (ed.): *Ultrashort Light Pulses*, 2nd edn., Topics Appl. Phys., Vol. 18 (Springer, Berlin, Heidelberg 1984)

Shimoda, K.: *Introduction to Laser Physics*, 2nd edn., Springer Ser. Opt. Sci., Vol. 44 (Springer, Berlin, Heidelberg 1990)

Siegman, A. E.: *Lasers* (University Science Books, Mill Valley, Calif. 1986)

Sinclair, D. C., Bell, W. E.: *Gas Laser Technology* (Holt, Rinehart and Winston, New York 1969) Chaps. 4–7

Sliney, D., Wohlbarsht, M.: *Safety with Lasers and Other Optical Sources* (Plenum, New York 1980)

See articles on specific laser systems in *Lasers*, ed. by A. K. Levine, Vol. 1 (1966), Vol. 2 (1968), Vol. 3 (1971), ed. by A. K. Levine, and A. DeMaria (Dekker, New York)

Chapter 9

Ditchburn, R. W.: *Light*, 2nd edn. (Wiley-Interscience, New York 1963) Chaps. 12–16

Jenkins, F. A., White, H. E.: *Fundamentals of Optics*, 4th edn. (McGraw-Hill, New York 1976) Chaps. 20, 24–28, 32

Lotsch, H. K. V.: Beam Displacement of Total Reflection: The Goos-Hänchen Effect, Optik **32**, 116–137, 189–204, 299–319, and 553–569 (1970, 1971)

Meltzer, R. J.: Polarization, in *Applied Optics and Optical Engineering*, Vol. 1 (Academic, New York 1965)

Mills, D. L.: *Nonlinear Optics, Basic Concepts* (Springer, Berlin, Heidelberg 1991)

Shen, Y.-R. (ed.): *Nonlinear Infrared Generation*, Topics Appl. Phys., Vol. 16 (Springer, Berlin, Heidelberg 1977)

Terhune, R. W., Maker, P. D.: Nonlinear Optics, in *Lasers*, Vol. 2, ed. by A. K. Levine (Dekker, New York 1968)

Yariv, A.: *Introduction to Optical Electronics* (Holt, Rinehart and Winston, New York 1971) Chaps. 8, 9, 12

Chapter 10

Allard, F. C. (ed.): *Fiber Optics Handbook for Engineers and Scientists* (McGraw-Hill, New York 1990)

Cherin, A. H.: *Introduction to Optical Fibers* (McGraw-Hill, New York 1983)

Enoch, J. M., Tobey, F. L., Jr. (eds.): *Vertebrate Photoreceptor Optics*, Springer Ser. Opt. Sci., Vol. 23 (Springer, Berlin, Heidelberg 1981) Chap. 5

Geckeler, S., *Optical Fiber Transmission Systems* (Artech House, Norwood, Mass. 1987)

Ghatak, A., Thyagarajan, K.: *Optical Electronics* (Cambridge University Press, Cambridge 1989)

Jeunhomme, L. B.: *Single-Mode Fiber Optics, Principles and Applications* (Dekker, New York 1990)

Kressel, H. (ed.): *Semiconductor Devices*, 2nd edn., Topics Appl. Phys., Vol. 39 (Springer, Berlin, Heidelberg 1982)

Marcuse, D.: Loss analysis of single-mode fiber splices; Bell Syst. Tech. J. **56**, 703–718 (1977)

Midwinter, J.: *Optical Fibers for Transmission* (Wiley, New York 1979)

Miller, S. E., Chynoweth, A. G.: *Optical Fiber Telecommunications* (Academic, New York 1979)

Senior, J. M.: *Optical Fiber Communications* (Prentice-Hall, Englewood Cliffs, N.J. 1985)

Sharma, A. B., Halme, S. J., Butusov, M. M.: *Optical Fiber Systems and Their Components*, Springer Ser. Opt. Phys., Vol. 24 (Springer, Berlin, Heidelberg 1981)

Suematsu, Y., Iga, K.: *Introduction to Optical Fiber Communications* (Wiley, New York 1982)
Technical Staff of CSELT: *Optical Fibre Communication* (McGraw-Hill, New York 1980)

Chapter 11

Danielson, B. L., Day, G. W., Franzen, D. L., Kim, E. M., Young, M.: *Optical Fiber Characterization*, NBS Special Publ. 637 (Public Information Office, National Bureau of Standards, Boulder, Colo. 1982) Vol. 1
Chamberlain, G. E., Day, G. W., Franzen, D. L., Gallawa, R. L., Kim, E. M., Young, M.: *Optical Fiber Characterization*, NBS Special Publ. 637 (Public Information Office, National Bureau of Standards, Boulder, Colo. 1983) Vol. 2
Jeunhomme, L. B.: *Single-Mode Fiber Optics, Principles and Applications* (Dekker, New York 1990)
Marcuse, D.: *Principles of Optical Fiber Measurements* (Academic, New York 1981)

Chapter 12

Barnoski, M. K. (ed.): *Introduction to Integrated Optics* (Plenum, New York 1974)
Hunsperger, R. G.: *Integrated Optics: Theory and Technology*, 2nd edn., Springer Ser. Opt. Sci., Vol. 33 (Springer, Berlin, Heidelberg 1984)
Hutcheson, L. D. (ed.): *Integrated Optical Circuits and Applications* (Dekker, New York 1987)
Iga, K.: *Fundamentals of Micro-optics* (Academic, New York 1984)
Nishihara, H., Haruna, M., Suhara, T.: *Optical Integrated Circuits* (McGraw-Hill, New York 1989)
Nolting, H. P., Ulrich, R. (eds.): *Integrated Optics* (Springer, Berlin, Heidelberg 1985)
Okoshi, T.: *Planar Circuits for Microwaves and Lightwaves*, Springer Ser. Electrophys., Vol. 18 (Springer, Berlin, Heidelberg 1985)
Tamir, T.: *Integrated Optics*, 2nd edn., Topics Appl. Phys., Vol. 7 (Springer, Berlin, Heidelberg 1979)
Tamir, T. (ed.): *Guided-Wave Optoelectronics*, 2nd edn., Springer Ser. Electronics Photonics, Vol. 26 (Springer, Berlin, Heidelberg 1990)

Subject Index

Entries are listed alphabetically by section. More-important entries or definitions are indicated by **boldface**.